Schematic Capture with MicroSim™ PSpice®

FOURTH EDITION
FOR WINDOWS VERSION 3.1
Includes PC Board Layout Using PADS-PERFORM

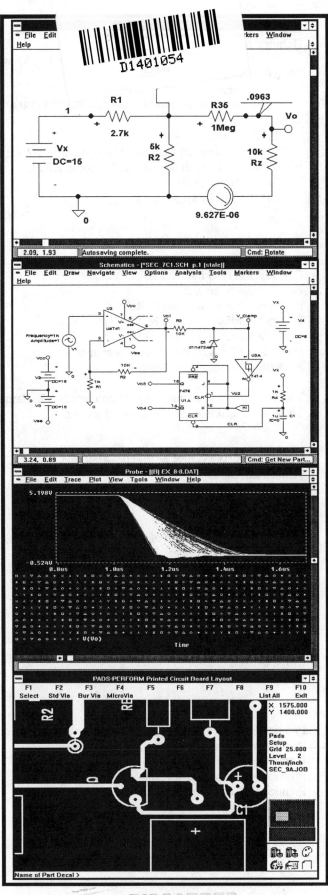

Marc E. Herniter

Associate Professor
Electrical Engineering Department
Northern Arizona University

Chief Engineer
Fire, Wind, and Rain Technologies LLC

Prentice Hall

Upper Saddle River, New Jersey
Columbus, Ohio

Library of Congress Cataloging-in-Publication Data
Herniter, Marc E.
      Schematic capture with MicroSim PSpice : for windows 3.1
Marc E. Herniter. – 4th ed.
      p.   cm.
   "Includes PC board layout using PADS-PERFORM."
   Includes index.
   ISBN 0-13-021261-X
   1.   Electric circuit analysis—Data processing. 2.
Electronics—Charts, diagrams, etc. 3. PSpice. I. Title
   TK454.H445 2000
   621.3815'0285'5369—dc21

                                        98-51765
                                           CIP

Cover photo: FPG International
Editor: Scott Sambucci
Production Editor: Stephen C. Robb
Design Coordinator: Karrie M. Converse-Jones
Cover Designer: Mark Shumaker
Production Manager: Patricia A. Tonneman
Marketing Manager: Ben Leonard

This book was set in Times New Roman by Marc E. Herniter and was printed and bound by Victor Graphics, Inc.
The cover was printed by Victor Graphics, Inc.

Printed in the United States of America

10 9 8 7 6 5 4 3 2

ISBN: 0-13-021261-X

Prentice-Hall International (UK) Limited, *London*
Prentice-Hall of Australia Pty. Limited, *Sydney*
Prentice-Hall Canada, Inc., *Toronto*
Prentice-Hall Hispanoamericana, S. A., *Mexico*
Prentice-Hall of India Private Limited, *New Dehli*
Prentice-Hall of Japan, Inc., *Tokyo*
Prentice-Hall (Singapore) Pte. Ltd., *Singapore*
Editora Prentice-Hall do Brasil, Ltda., *Rio de Janeiro*

Warning and Disclaimer

This book is designed to provide tutorial information about the MicroSim™ and PADS-PERFORM set of computer programs. Every effort has been made to make this book as complete and accurate as possible, but no warranty or fitness is implied.

The information is provided on an "as is" basis. The author and Prentice Hall shall have neither liability nor responsibility to any person or entity with respect to any loss or damages arising from the information contained in this book or from the use of the disks and programs that may accompany it.

Trademarks

"Adobe" and "Acrobat" are registered trademarks of Adobe Systems Incorporated.

"MicroSim™ PSpice®" is a registered trademark and "Designlab" is a trademark of MicroSim Corporation.

"Probe," "Schematics," "StmEd," "Stimulus Editor," "Parts," and "Monte Carlo" are trademarks of MicroSim Corporation.

"Microsoft," "MS," and "MS-DOS" are registered trademarks and "Windows" and "Win32s" are trademarks of Microsoft Corporation.

"IBM" is a registered trademark of International Business Machines Corporation.

"Sun" is a registered trademark of Sun Microsystems Incorporated.

"Open Windows" is a trademark of Sun Microsystems Incorporated.

"PADS" and "PADS-PERFORM" are registered trademarks of PADS-SOFTWARE Incorporated.

To
my wife, Corena,
my daughters, Katarina Alexis Sierra and Laina Calysta Dine'h
my parents, Mom and Dad,
my cats, Zak, Chester, Tipper, and Cajsa,
my dogs, Samantha, Costa, and Sage,
my saltwater fish, Flake, Bloat, and Tomato.

Preface

This manual is designed to show students how to use the PSpice circuit simulation program from MicroSim with the schematic capture front end, Schematics. It is a collection of examples that show students how to create a circuit, how to run the different analyses, and how to obtain the results from those analyses. This manual does not attempt to teach students circuit theory or electronics; that task is left for the main text. Instead, the manual takes the approach of showing students how to simulate many circuits found throughout the engineering curriculum. An example could be the DC circuit shown below.

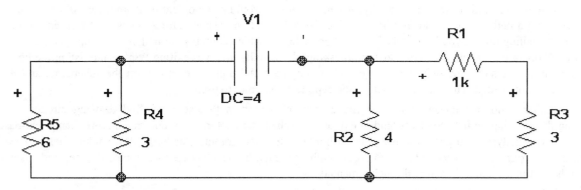

It is assumed that the student has been given enough information to completely analyze the circuit. This manual assumes that the student wishes to check his or her answers (or intuition) with this program. The student would construct the circuit as shown in Part 1 and then run either the node voltage analysis in Part 3 or the DC Sweep in Part 4. This circuit is different from the circuits in Parts 3 and 4, but the procedure given in those parts can be applied to the circuit.

This manual was designed to be used by students for their entire educational career and beyond. Since the parts are arranged by analysis type, they contain a range of examples from circuits covered in first semester circuit theory courses to senior level amplifier and switching circuits. Sections that are too advanced for beginning students may be skipped without loss of continuity. All parts contain both simple circuits and advanced circuits to illustrate the analysis types. Sections do not have to be covered sequentially. Individual examples can be identified that apply to specific courses. However, the following sequence is suggested for first-time users. All beginning users should follow Part 1 completely to learn how to draw, print, and save schematics. All students should follow some of the examples in Parts 2 and 3 that are relevant to the course and also cover a few of the examples that may apply to earlier courses (if any). The early examples in these parts have the most step-by-step detail of how to use the software.

This manual contains examples that apply to courses throughout the engineering curriculum. Introductory circuits classes usually cover DC circuits, AC circuits with phasors, and transient circuits with a single capacitor or inductor and a switch. Examples are given to cover these types of problems. After reviewing the examples in this manual, a student should be able to simulate similar problems. A typical first electronics course may cover transistor biasing, amplifier gain, and amplifier frequency response. Examples of these analyses are also given. Higher level electronics courses would cover Monte Carlo analysis, worst case analysis, and distortion. Examples of these types of analyses are included.

Exercises are given at the end of each section. These exercises specify a circuit and give the simulation results. The students are encouraged to work these problems to see if they can obtain the same simulation results. The exercises are intended to give students practice in using the software, not to teach them circuits. Since this software covers such a wide variety of courses, problems are not given. These problems are best left to the instructor or main text. Using PSpice on problems specific to the class material is far more instructional than using it on problems designed to teach PSpice. My philosophy is that PSpice should be used only to verify one's own calculations or intuition. In my classes I assign problems that are worked by hand calculation, simulated with PSpice, and then tested in the lab. The students then compare the measured results to the hand calculations and PSpice simulations. Without hand calculation, it is impossible to know if the PSpice simulations are correct.

The book is written as if the instructor were giving a class demonstration on how to use the software. Intermediate windows are shown, and all mouse selections are specified. When I first started teaching the schematic capture version, I brought the students into the computer lab. I gave a lecture using an LCD projection screen and an overhead projector. The students could see the screens projected by the overhead and could follow along using their own computers. This required too much lecture time to cover the wide scope of the Schematics software, so I wrote this manual using the philosophy that the screen captures presented in this manual would be the same as if I were presenting the software in a lecture.

The main advantage of the schematic capture front end of PSpice is its ease of use. At first this may not be apparent. When I first started to use Schematics, I tested it with a simple three- or four-node circuit. Since I was familiar with writing netlists, it was far easier to write a simple four-line netlist than to search through the many menus of Schematics and create a schematic. As I became more proficient at using the program and remembering the standard parts, I could create a schematic faster than I could type a netlist. The schematic version becomes much easier when you use parts with which you are not familiar, such as an exponential or pulsed voltage source. How many of us can remember the order of the parameters in a pulsed voltage source? Usually you have to look them up in the manual. With Schematics a manual is not necessary. You get a part called **Vpulse**. The parameters of the source are listed in the part's attributes, and the order is not needed since Schematics takes care of the order automatically. Another example would be an operational amplifier. If you were describing the circuit using a netlist, you would first have to find the order and number of calling nodes of the op-amp subcircuit. To figure out the calling nodes, you have to look at the library listing which contains the netlist of the op-amp subcircuit. Since the MS-DOS operating system does not allow multitasking, this usually involves exiting PSpice and listing the library. In the schematic capture version, you only need to get the op-amp part. All nodes are shown on the schematic, and the correct calling order is not needed. This makes the schematic capture far simpler to use.

Another major advantage of the schematic capture version is that students find drawing circuits much more interesting than writing netlists. Students tend to explore the schematic capture front end much more than a text-based shell. Since all of the analyses and parts are available on-line as windows, graphics, or help files, students tend to explore the abilities of the program and they don't have to dig through a manual. In the text-based version of PSpice, students would first come to the instructor rather than look through a manual.

There are many other advantages to using the schematic capture version. The enormous popularity of the Microsoft Windows operating system should attest to the ease of use of a graphics-based interface compared to a text-based interface. One such advantage is automatic documentation. When you simulate a circuit, you automatically have a circuit schematic. This schematic can be incorporated into lab notebooks and reports. In a corporate environment, documentation of this type is extremely important. With the Windows operating system, the schematics can be incorporated into written documents using screen captures. This manual is an example of what can be accomplished.

At Northern Arizona University, both the schematic capture and text-based versions of PSpice are available. Originally, the text-based version was taught, and the students had a hard time accepting the schematic capture version. The main reason was that they were used to the MS-DOS operating system and did not have any experience using Windows. After students realized the simplicity of the schematic capture version, they rarely used the text-based version. Now, the only time the text-based version is used is the night before a large PSpice homework assignment is due, when all of the Windows-based PCs are being used.

The version of Schematics described in this manual is Version 7.1. Currently, the schematic capture version of PSpice is available for Windows-based PCs and Sun work stations running Open Windows. The parts libraries described in this manual have been changed slightly from MicroSim's distribution libraries. The libraries have been consolidated, and new parts have been added. The new parts were added to make the program simpler for beginning students to use. Please note that the libraries contained in this manual are slightly different from the factory distribution libraries.

Software Included with the Manual

The software is provided on a CD-ROM. The following items are included:

- MicroSim Evaluation Version 7.1 – This is the version of Schematics and PSpice that was used to compose this manual. **This is the last version of Schematics and PSpice that will run under Windows 3.x.**

- MicroSim Evaluation Version 6.1 – An older version of Schematics and PSpice that was used in the second edition of this text.

- Win32s – Required for running versions 6.1 and 7.1 of Schematics and PSpice under Windows 3.x.

- MicroSim Evaluation Version 5.2 – An older version of Schematics and PSpice that has fewer hardware requirements. This version does not require a coprocessor.

- Adobe Acrobat Reader Version 2.1.

- Chapter 10 of this manual in Adobe Acrobat Reader format.

- PADS-PERFORM – PC board layout software from PADS-SOFTWARE Inc.

- All schematic files for examples and exercises used in this manual. These are provided so that if the manual does not have enough detail, students can open the actual schematic file used to create the example and examine the setup. Most of the screen captures indicate the name of the file at the top of the screen capture.

- Floppy disk copy files. If the user does not have a CD-ROM, the installation files are broken into small portions that can be easily copied to high-density 3.5″ floppy disks. Thus, as long as there is a local facility that has a CD-ROM, the files can be copied to floppy disks for distribution.

Software Updates

Periodic updates of the software can be obtained from OrCAD's web site at www.orcad.com. Updates to the libraries for this manual can be obtained from the author's web site at www.cse.nau.edu/~meh.

Comments and Suggestions

The author would appreciate any comments or suggestions on this manual. Comments and suggestions from students are especially welcome. Please feel free to contact the author using any of the methods listed below:

- **E-mail:** *Marc.Herniter@nau.edu.*
- **Phone:** (520) 523-4440
- **FAX:** (520) 523-2300
- **Mail:** Northern Arizona University, Electrical Engineering Department, Box 15600, Flagstaff, Arizona, 86011-1560.

Acknowledgments

I would like to thank my students at Northern Arizona University for giving me continued inspiration to improve this manual. Without their constant curiosity, this book would not be necessary. I am grateful to my colleagues at Northern Arizona University for their support in developing CAD at Northern Arizona University. Many thanks to Professors Jerry M. Hatfield and George W. Hoyle. I am extremely grateful to OrCAD Corporation for allowing us to distribute the evaluation version of Schematics and PSpice with this manual. I would like to thank Mike Bosworth of OrCAD corporation for helping the academic community by continuing to distribute evaluation versions of PSpice, Schematics, and future OrCAD products to students. I would like to thank Phil Kilcoin of OrCAD Corporation for his help with this text and on future projects. Finally, I would like to express my deepest appreciation to my wife, Corena, who no longer lets me sit in front of my computer twenty-four hours a day.

Before You Begin

System Requirements

- A typical installation of the MicroSim Designlab version 7.1 software requires approximately 29 MB of hard disk space. A minimum of 8 MB of RAM is required, but 16 MB is recommended. A math coprocessor is required.
- PADS-PERFORM requires 7 MB of hard disk space, 16 MB of RAM, and a math coprocessor. A VGA color monitor is required to distinguish different levels of a board design.

Limitations of the Evaluation Version of Schematics and PSpice Version 7.1

- Schematic capture is limited to one schematic page (A-size or A4)
- A maximum of 50 symbols can be placed on the schematic
- A maximum of 9 symbol libraries can be configured
- A maximum of 20 symbols is allowed in a user-created symbol library
- A maximum of 70 parts can be netlisted for PSpice
- A maximum of 30 components can be netlisted for PCB layout
- A maximum of 50 nets is allowed for PCB layout

General Conventions

- This manual assumes that you have a two- or three-button mouse. The words *LEFT* and *RIGHT* refer to the left and right mouse buttons.
- This manual assumes that you have a color monitor. A color monitor is not necessary for running the software or following the manual. However, the manual will refer to items being highlighted in red when they are "selected." If you do not have a color monitor, you will still be able to see the selected items. The items will not be highlighted in red, however, and may not be as obvious as if they were shown in red.
- All text highlighted in bold refers to menu selections. Examples would be **File** and **Analysis.**
- All text in capital letters refers to keyboard selections. For example, press the ENTER key.
- *All text in this font refers to text you will see on the computer screen. This applies to all text except menu selections*.
- `All text in this font refers to text you will type into the program.`
- The word "select" is interpreted as "click the left mouse button on."

Keyboard Conventions

Throughout the manual many keyboard sequences are given as shortcuts for making menu selections. The explanation of these sequences will be given later. It is important to know the convention used to specify the sequences.

- Many control key sequences will be specified. For example, CTRL-R means hold down the "Ctrl" key and press the "R" key simultaneously. CTRL-A means hold down the "Ctrl" key and press the "A" key simultaneously. Not all keyboards are the same. Some keyboards may have a key labeled "Control" rather than "Ctrl."
- The keyboard sequence ALT-ESC in Microsoft Windows is used to toggle the active window. ALT-ESC means hold down the "Alt" key and press the "Esc" key simultaneously.

Sign Conventions

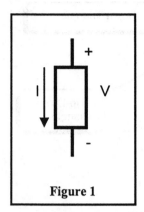

Figure 1

In circuit analysis it is important to know the polarity of voltage and direction of positive current flow. For resistors, capacitors, and inductors, the person doing the analysis usually assigns voltage and current references. The standard convention is that current is always positive entering the positive voltage terminal. This convention is shown in Figure 1. With this convention, if we know which terminal is marked positive for voltage, we know the direction for positive current flow. PSpice follows this convention and always assigns one of the terminals as the positive terminal. Knowing the polarity that PSpice assigns to components is not necessary until you ask for the current through a device or specify an initial condition. If we were interested in the current through a resistor, R1, for example, we would specify this current as I(R1). PSpice will give us the current through R1. The question is, which direction through R1 does PSpice interpret as positive? To solve this problem, a positive sign has been added to the resistor and capacitor graphics, and a dot has been added to the inductor graphic, as shown in Figure 2.

Figure 2

Also note that the dot in the inductor symbol is usually used to indicate polarity for mutual inductance. It is standard not to have a dot in the inductor symbol unless there is mutual coupling between inductors. However, to indicate polarity for a single inductor, we would have had to add a plus sign to the inductor symbol. The plus sign together with a dot could be confusing when talking about mutual inductance. **Thus, the dot is always specified in the inductor symbol. For single inductors, the dot means the same thing as the plus sign in the capacitor and resistor graphics. When mutual inductance is specified, the dot refers to the dot convention in mutual inductance.** The dot does not imply mutual inductance. To specify mutual coupling we must specify two parts, the inductor (L) and the coupling (K).

Nomenclature Used in This Manual

This manual uses many terms associated with Windows. Some of the terms are discussed here:

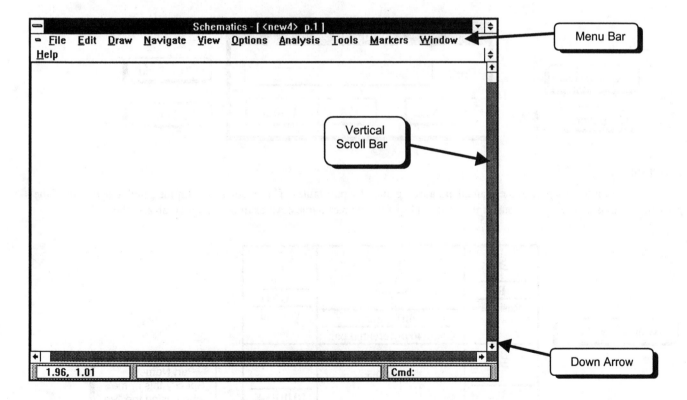

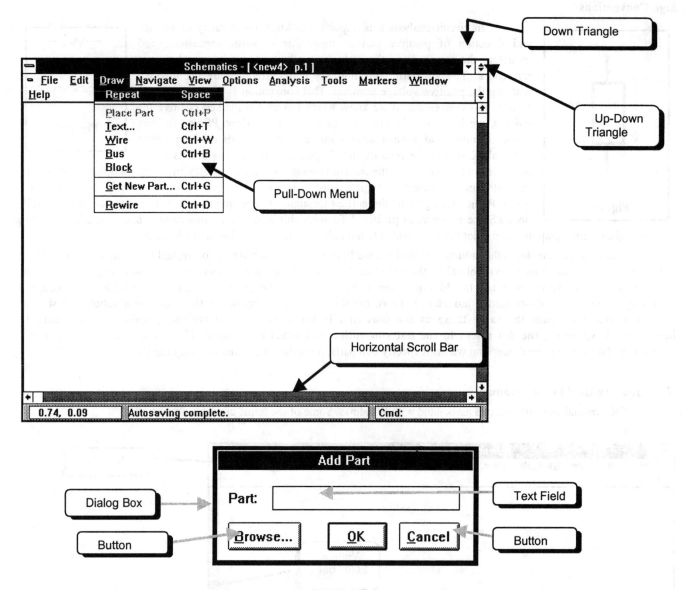

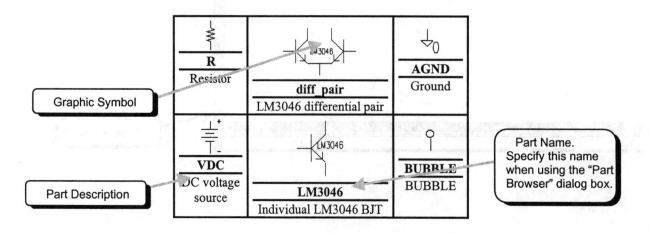

Part Tables

All circuit examples in the manual are accompanied by part tables. These tables contain the graphic symbols of the parts in the circuit, give descriptions of the parts, and specify the part names. An example of a part table is shown below:

Contents

PART 1
Editing a Basic Schematic

This part assumes that the Microsoft Windows 3.x operating system and Schematics are already installed on your computer. If the Schematics software has not been installed, refer to Appendix A for installation instructions. **If you do not follow the installation instructions in Appendix A, the libraries specific to this manual will not be installed correctly.** If Windows has not been installed, refer to the Windows operating system documentation for instructions.

1.A. Starting Schematics

If your computer has Windows installed, Windows may run automatically when you start your PC. If Windows does not run automatically, type "win" at the command prompt. Windows may open in one of many possible forms, and the PSpice set of programs may not be easy to find. Your screen may open in one of the forms described in this part. If the screen shown below does not look like your screen, read further. If your screen looks like this:

double click the *LEFT* mouse button on the Program Manager icon, . Look for the Design Center Eval group icon, , when the Program Manager opens. The location of this icon may not be immediately apparent. You may see a screen as shown below:

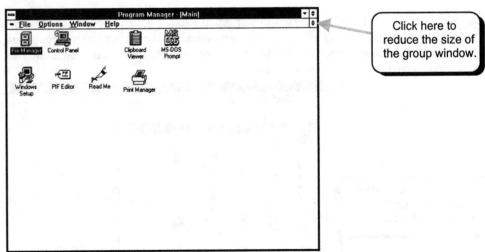

A single group occupies the entire screen of the Program Manager. To see what else is in the Program Manager, click the *LEFT* mouse button on the lower* up-down triangle, , in the upper right corner of the Program Manager. This will reduce the size of the window occupying the entire Program Manager screen. You may now get the following screen:

---

*If you click the upper up-down triangle, the Program Manager itself will change size.

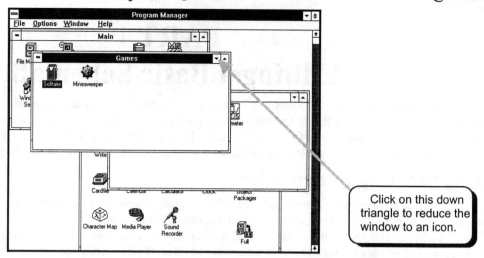

Click on this down triangle to reduce the window to an icon.

The Program Manager now has several windows open, and the Design Center group icon is not easily spotted. To shrink the windows to see the group icons, click the **LEFT** mouse button on the down triangle, ▾, in the upper right corner of all the small windows. This will reduce the open windows to icons:

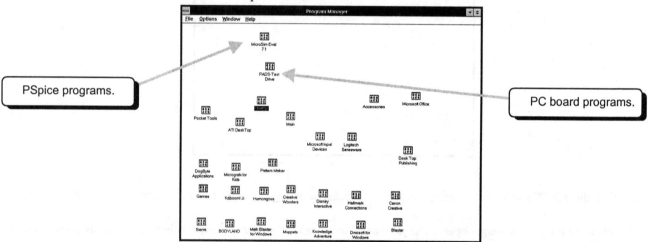

PSpice programs.

PC board programs.

You may now be able to see the Design Center group icon, MicroSim Eval 7.1 . Double click the **LEFT** mouse button on the Design Center icon to open the Design Center group. If you cannot see the Design Center group icon, you may need to click the left, right, up, or down arrows in the Program Manager window. When you double click the **LEFT** mouse button on the Design Center group icon, the screen below should appear:

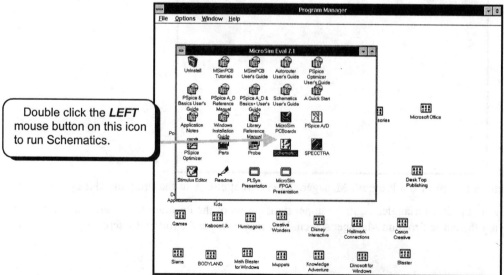

Double click the **LEFT** mouse button on this icon to run Schematics.

This is the Design Center group of programs. We will briefly describe some of the program icons.

- ![icon] - This icon represents the Schematics program. Schematics is used to draw circuits and invoke the PSpice simulation program. Schematics also creates netlists for use by the PSpice program.

- ![icon] - This icon represents the PSpice program. PSpice simulates circuits described by netlists. Netlist file names are of the form "????????.cir," where ???????? is a file name of 8 characters or less.

- ![icon] - This is the Probe program. Probe lets you look at graphical results generated by the PSpice program.

- ![icon] - This is the Parts program. Parts lets you create PSpice models. The evaluation version is limited to creating diode models.

- ![icon] - This is the Stimulus Editor program. The Stimulus Editor helps you generate waveforms for independent voltage and current sources. It is useful for creating pulse, sine wave, exponential, and digital waveforms when you do not know the PSpice commands. The evaluation version allows only sine waves and digital clocks.

Double click the *LEFT* mouse button on the Schematics icon ![icon] to run the schematic capture circuit simulation package. A blank schematic page will appear:

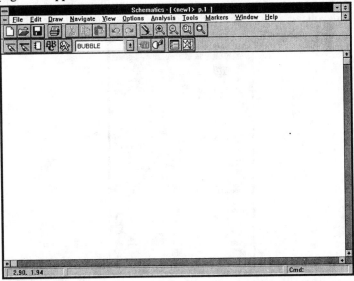

1.B. Placing Parts

We will now create a schematic. The first part we will get is an independent voltage source. All parts are retrieved using the procedure outlined below. First, click the *LEFT* mouse button on the **Draw** menu selection. The **Draw** menu will pull down:

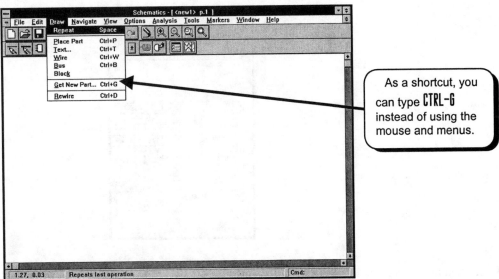

As a shortcut, you can type **CTRL-G** instead of using the mouse and menus.

Click the *LEFT* mouse button on **Get New Part**. The **Part Browser Basic** dialog box will appear:

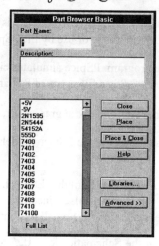

If you know the name of the part you want to place, you can type the part name in the box. This menu shows all the parts available to us. Most independent voltage sources start with the letter v. Type the letter **v**. The list of parts will scroll down to display the parts that begin with the letter v:

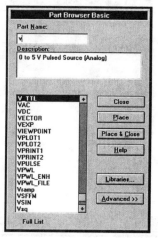

The part presently selected is **V_TTL** and its description is *0 to 5 V Pulsed Source (Analog)*. We see that when we select a part, the dialog box displays a description of the part. To select a part and see its description, click the *LEFT* mouse button on the name of the part. For example, click the *LEFT* mouse button on the text **Vramp**:

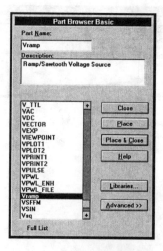

We see that the part named **Vramp** is a *Ramp/Sawtooth Voltage Source*. This was a part created for this textbook to make it easier to generate ideal waveforms. You can look at the descriptions of other parts if you wish.

We wish to select a generic DC voltage supply. Click the *LEFT* mouse button on the text **VDC** to select the part:

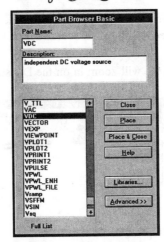

The description of **VDC** is **independent DC voltage source**. To accept the part and place it in your circuit, click the **LEFT** mouse button on the **Place & Close** button. When you click the button, the program will return to the schematic with the graphic for the part attached to the mouse pointer:

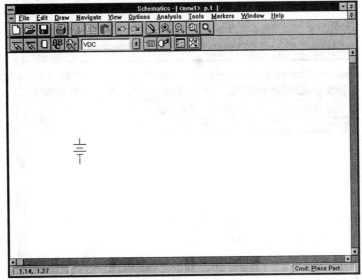

The graphic moves with the mouse. Move the graphic to the place where you want to put the voltage source. Click the **LEFT** mouse button **once** to place the part and then move the mouse pointer. Note on your screen and in the figure below that when you place a part, a second source appears where the mouse pointer should be. This is the auto repeat function for placing parts.

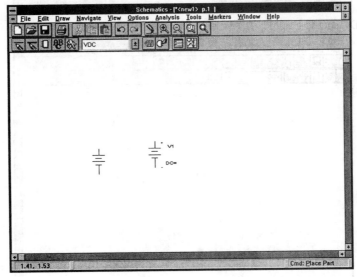

If you want to add another voltage source to your schematic, you can move the mouse to where you want to place the source and click the *LEFT* mouse button. We do not need another DC source at this time, so we click the *RIGHT* mouse button to make the second source disappear.

To make the schematic more readable, we will zoom in on the DC source. Select **View** from the Schematic menu bar:

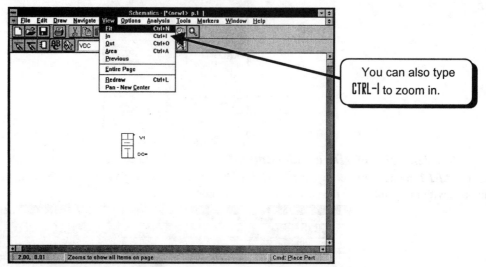

To zoom in, click the *LEFT* mouse button on the **In** menu selection. The mouse pointer will be replaced by a cross hair:

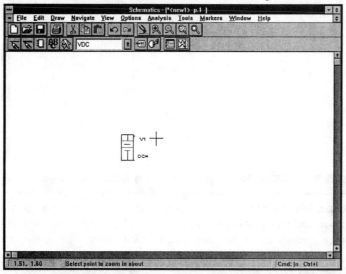

Place the cross hair over the DC source graphic, ⊣⊢, and click the *LEFT* mouse button to zoom in:

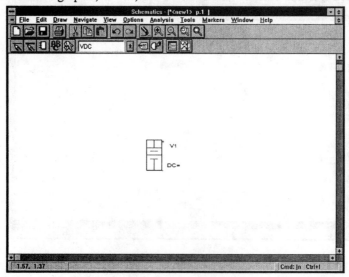

To repeat the last zoom command, press the space bar. The screen will zoom in around the same spot:

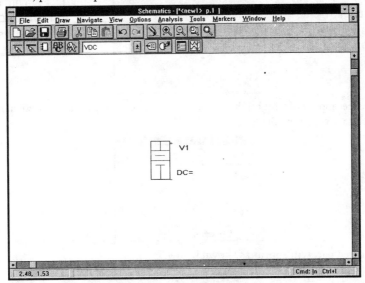

We now must edit the attributes of the source to make it a 12 VDC source. There are two ways to change the attributes of a part. The first way we will look at is editing the individual attributes that are displayed on the screen. We will first change the voltage of the DC source. To do this double click the **LEFT** mouse button on the **DC=** text next to the DC source. The dialog box below will appear:

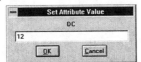

Note that the dialog box says that the attribute we are changing is the **DC** attribute. To change the attribute, type in the desired value. We will create a 12 volt source so type in **12**:

Click the **OK** button to change the attribute. The part will appear in the schematic with the line **DC=12** displayed in the schematic:

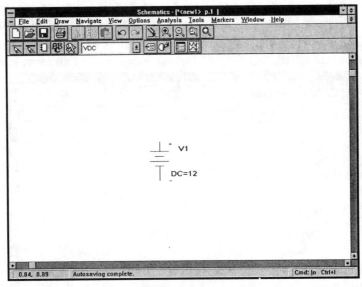

If your text appears garbled, type CTRL-L to redraw the screen. To change the name of the voltage source double click the **LEFT** mouse button on the text **V1**.* The dialog box below will appear:

---

*The name of the source may vary depending on how many sources you have placed in your schematic.

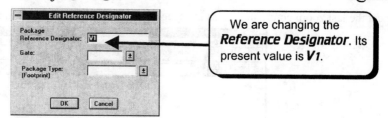

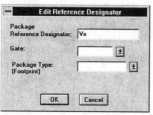

The present name of the source appears in highlighted text. To change it, type the desired name of the source. We will change the name of the source to Vx, so type the text **Vx**:

Click the **OK** button. The updated part will appear as shown below:

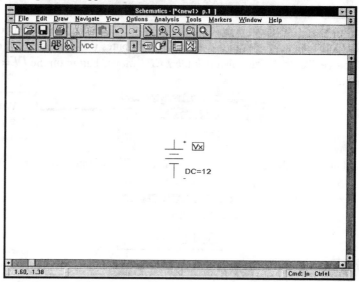

The part now has the desired attributes. To illustrate the second method of changing a part's attributes, we will change the source voltage to 15 volts and the name back to V1. To change the attributes, click the **LEFT** mouse button once on the DC source graphic, ⊣║⊢. The graphic will be highlighted in red when the part is selected. It may take several tries to highlight the desired part. When the part is highlighted, click the **LEFT** mouse button on the **Edit** menu selection:

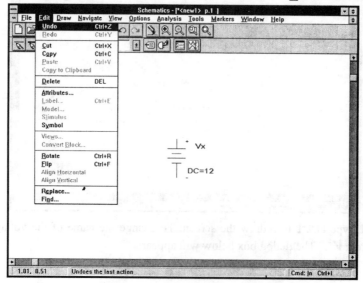

Click the *LEFT* mouse button on the <u>A</u>ttributes menu selection[*]:

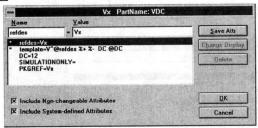

This screen can be used to change many of the attributes of a part. Note that there are more attributes than appear on the schematic. Not all attributes of a part are displayed on the schematic. To change the voltage of the source, click the *LEFT* mouse button on the line *DC=12*. The text *DC* will appear in the *Name* box, and the number *12* will appear in the *Value* box:

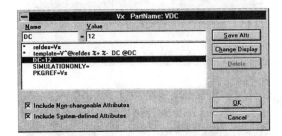

To change the value of the source, move the mouse cursor to the right of the number *12* in the *Value* box and click the *LEFT* mouse button. Press the BACKSPACE key until the number *12* is erased and type in the new value. We will type in the number *15* to make a 15 volt DC source:

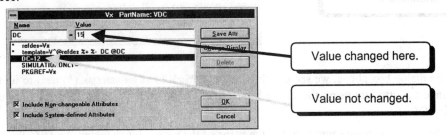

Value changed here.

Value not changed.

Note that line *DC=12* has not yet changed. To change the value of the attribute, click the *LEFT* mouse button on the *Save Attr* button. The value will be changed, and the screen will look like this:

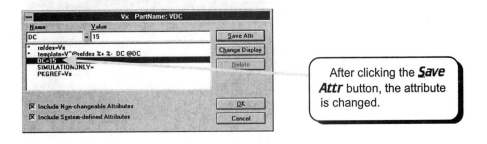

After clicking the *Save Attr* button, the attribute is changed.

---

[*]This dialog box can also be obtained by double clicking the *LEFT* mouse button on the DC voltage source graphic, ─┤╵├─ . Note also that if the text <u>A</u>ttributes appears as Attributes, you have not properly selected the graphic. Click the *LEFT* mouse button on the graphic again until it becomes highlighted in red. When the graphic is highlighted in red, you may edit its attributes.

Next, we will attempt to change the name of the source specified by the **refdes** attribute. To change the attribute, click the *LEFT* mouse button on the line **refdes=Vx**. The text **refdes** will appear in the **Name** box, and the text **Vx** will appear in the **Value** box:

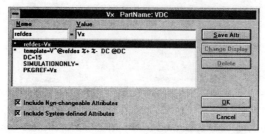

Change the value from **Vx** to **v1** and click the **Save Attr** button. The error warning below will appear:

The error states that we cannot change the name of the source using the above method. Click the **OK** button to acknowledge your mistake:

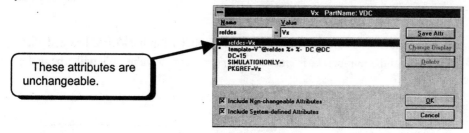

These attributes are unchangeable.

In the dialog box above, we see that the line **refdes=Vx** has an asterisk (*) next to it. The asterisk indicates that we cannot change the attribute using the dialog box above. We can only change attributes that do not have an asterisk, such as the **DC=15** line.

Although we cannot change the name of the source by changing the **refdes** attribute, we can change the name by changing the **PKGREF** attribute. To change the attribute, click the *LEFT* mouse button on the line **PKGREF=Vx**. The text **PKGREF** will appear in the **Name** box, and the text **Vx** will appear in the **Value** box:

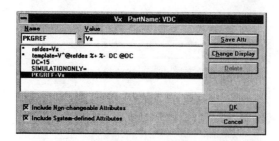

Change the value from **Vx** to **v1** and click the **Save Attr** button. The value of the attribute **PKGREF** will change in the dialog box:

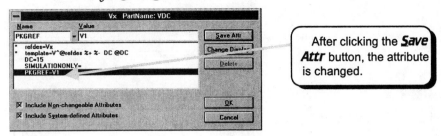

After clicking the **Save Attr** button, the attribute is changed.

Click the **OK** button to return to the schematic. We see that both the line **DC=15** and the name of the source, **V1**, have changed in the schematic:

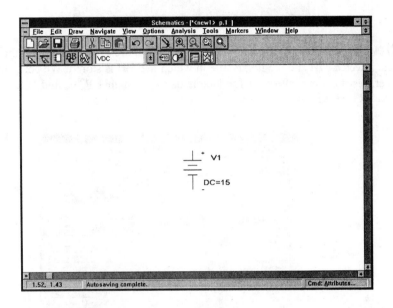

For practice, change the name of the source back to Vx.

      We now need to place some resistors in the circuit. The part name for a resistor is R. We could choose the **Draw** menu selection and then select **Get New Part** to get the resistor. Instead, we will type **CTRL-G**. The **Part Browser Basic** dialog box will appear:

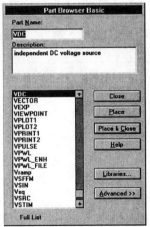

At this point we could just search for a resistor. However, we will first take a look at the advanced browser. Click the **LEFT** mouse button on the **Advanced** button:

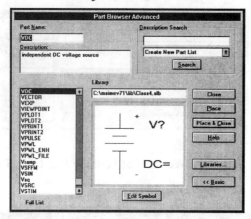

We see that the **Part Browser Advanced** dialog box displays much more information than the basic browser. It displays the graphic symbol for the selected part and the library file in which the part is stored. We could now search through the list to find a resistor. However, resistors, capacitors, and inductors have part names R, C, and L for simplicity. Since we know that a resistor part is called "R" we will just type in **r**:

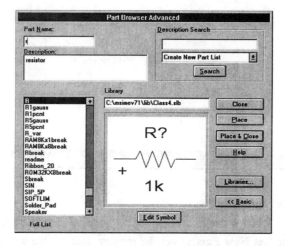

Click the **Place & Close** button with the *LEFT* mouse button to accept the part. The resistor graphic will appear attached to the mouse pointer:

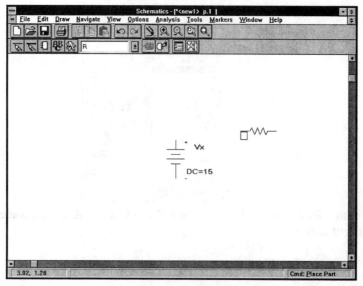

Note that the resistor moves with the mouse. Move the resistor to where you want to place it and click the *LEFT* mouse button. The resistor is placed, and a second resistor appears where the mouse pointer should be:

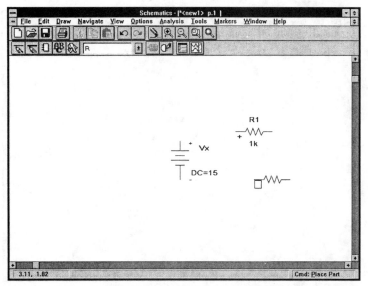

We now wish to add a second resistor to the schematic. The next resistor we want to place will be vertically oriented. Currently the resistor attached to the mouse is horizontal. To rotate the part, type CTRL-R. The part will rotate while attached to the mouse. Orient the part vertically as shown in the screen below using the CTRL-R command.

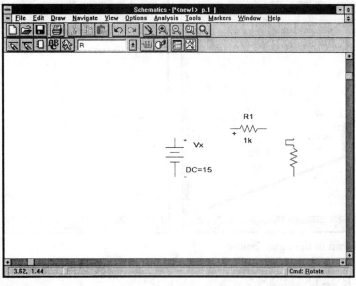

Move the mouse to place the part in the location shown below. To place the part click the *LEFT* mouse button. When you click the *LEFT* mouse button, the part is placed, and a third resistor appears attached to the mouse pointer:

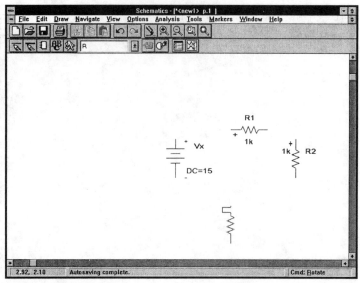

Before we add more resistors, we must scroll the Schematics window to the right. Click the *LEFT* mouse button on the right arrow of the horizontal scroll bar as shown below:

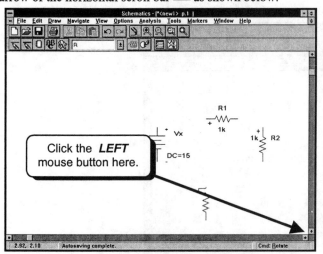 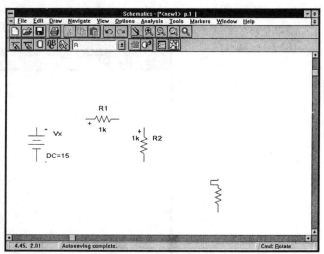

Add two more resistors as shown in the figure below:

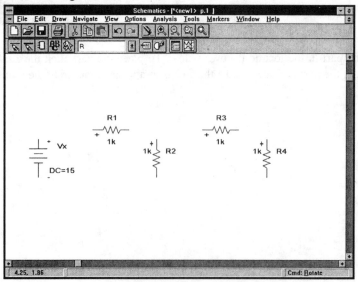

Use the CTRL-R key sequence to rotate the part when necessary and click the *LEFT* mouse button to place the part. When you have placed all four resistors, click the *RIGHT* mouse button to stop placing resistors. When you click the *RIGHT* mouse button, the resistor graphic at the mouse pointer is replaced by the mouse pointer.

Next, we would like to display the circuit a little more clearly. To zoom in on the parts, and make the circuit fill the screen, click the *LEFT* mouse button on the **View** menu selection:

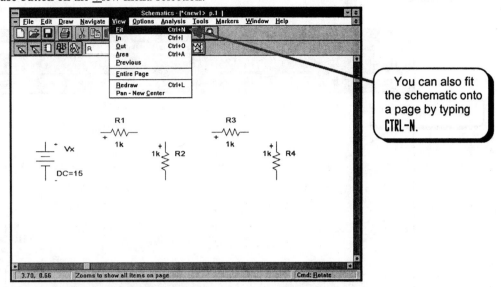

Click the *LEFT* mouse button on **Fit**. The circuit will fill the screen:

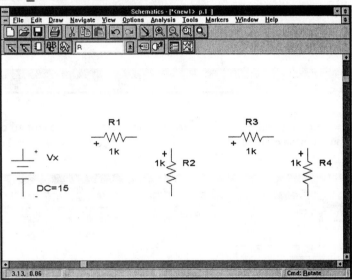

1.C. Correcting Mistakes

At this point you may have some mistakes in your schematic. You may have clicked the *LEFT* mouse button one too many times and placed too many resistors on your schematic, or you may have placed resistors too close to each other. To move parts follow the procedure below. For the moment we will assume that you wish to move a resistor.

1. Click the *LEFT* mouse button on the resistor graphic, ⊣W⊢, you wish to move. When the resistor graphic is highlighted in red, it has been selected. It may take several tries to highlight the resistor.

2. When the graphic is highlighted in red, press and hold the *LEFT* mouse button on the highlighted resistor graphic. Without releasing the button, move the mouse to move the resistor graphic to the desired spot in the schematic. When you find an appropriate spot, release the button.

    If you need to delete a part, follow Step 1 above. When the appropriate part is selected, press the DELETE key.

1.D. Changing Attributes

There are several ways to change the attributes of parts. Some have already been illustrated previously, but we will now go over all of the different methods. The first thing we need to do is to select a part. We will select resistor R1. Place the mouse pointer to the left of and above **R1**. Press and hold the *LEFT* mouse button and move the mouse down and to the right. Notice that a box is drawn. Everything inside this box will be selected. Move the mouse so that the box encloses **R1** and release the mouse button. The resistor graphic, ⊣W⊢, should now appear in red, indicating that it has been selected. This is one method of selecting a part.

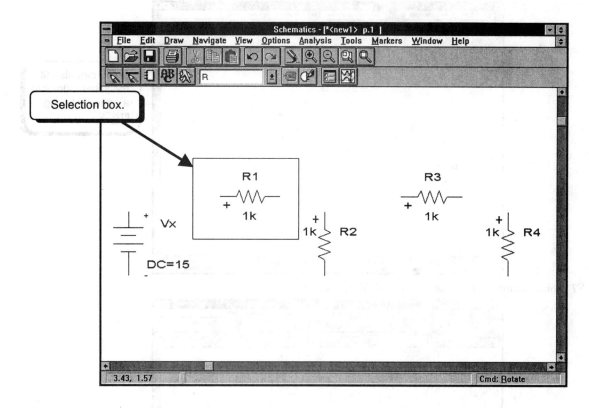

In the screen above, the resistor and some text are enclosed by the solid black box. To edit all of the resistor's attributes at the same time, click the *LEFT* mouse button on the **Edit** menu selection. The **Edit** menu will pull down:

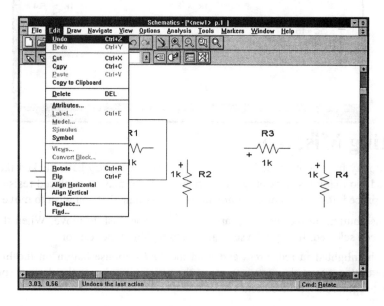

Click the *LEFT* mouse button on **Attributes**.* The dialog box below will appear:

---

*Note that if the text **Attributes** appears grayed out (Attributes), you have not properly selected the graphic.

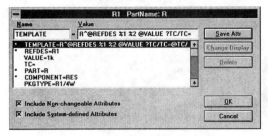

This screen shows all the attributes of R1. Note that not all the attributes are displayed on the screen. The attribute **Template=R^...** has an asterisk (*) next to it, indicating that it cannot be changed. This attribute generates the netlist line for the resistor. The attribute **TC** is a temperature coefficient for the resistor. Its default is zero (no temperature dependence). The attribute **VALUE** is the value of the resistor in ohms. The attribute **REFDES** is the name of the resistor and cannot be changed using this dialog box. The **PART** attribute is the name of the part in the Schematics library. This is the name used to get the part. The remaining attributes are for PC board design.

To change the resistance value, click the *LEFT* mouse button on the line **VALUE=1k**. The text **VALUE** now appears in the **Name** box and the number **1k** appears in the **Value** box:

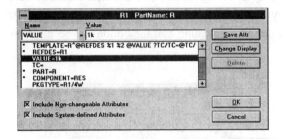

Change the value from **1k** to **2.7k**. **There must not be a space between the number 2.7 and the letter "k."** After you change the value to 2.7k, make sure you click the **Save Attr** button to save the change. The attribute should now have the new value as shown in the next screen. Note that the highlighted line **VALUE=2.7k*** has been changed as well as the number in the **Value** box.

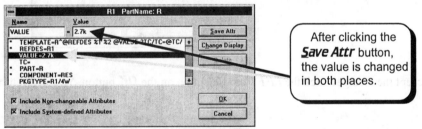

If you remember, the package type of the resistor was not shown on the schematic. This is because the **PKGTYPE** attribute is not being displayed. All attributes can either be displayed or not. We will illustrate how to display the **PKGTYPE** attribute. Click the *LEFT* mouse button on the **PKGTYPE** attribute. The **PKGTYPE** attribute will be highlighted:

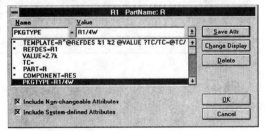

Click the *LEFT* mouse button on the **Change Display** button. The dialog box below will appear:

---

*For a list of multipliers available in PSpice, such as k, u, m, see Appendix B.

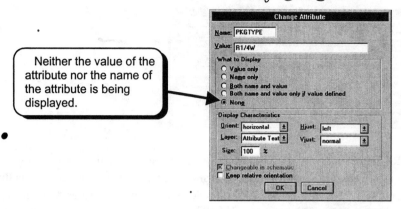

This screen indicates that neither the value **R1/4 W** nor the name **PKGTYPE** will be displayed on the schematic. There are several display options. Selecting the **Value only** option will display the value of the attribute. In this case, the text **R1/4W** will be displayed. Selecting **Name only** will display the name of the attribute. In this case, the text **PKGTYPE** will be displayed. Selecting the option **Both name and value** will display the name of the attribute and its value. In this case the text **PKGTYPE=R1/4W** will be displayed. Presently, **None** is selected, meaning that nothing related to this attribute will be displayed. To select another display option, click the *LEFT* mouse button on the circle ○ next to the option you wish to select. The circle should fill with a dot ◉, indicating that the option has been selected. I will display only the value of the attribute:

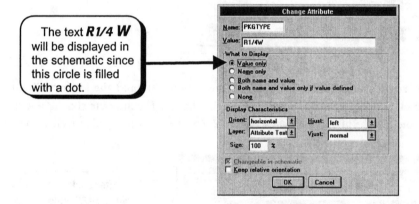

To accept the changes click the *LEFT* mouse button on the **OK** button. You will return to the following screen:

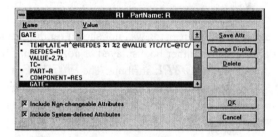

Click the *LEFT* mouse button on the **OK** button to accept all the changes you have just made. You will return to the schematic with all of the attributes updated:

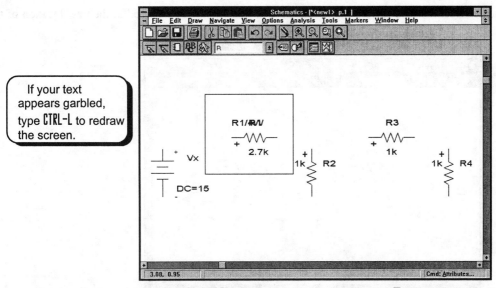

On my display, the text for **R1/4W** was placed on top of the text **R1**. To display both attributes clearly, we must move one of the attributes. Click the *LEFT* mouse button on the text. One of the attributes will become enclosed in a box, indicating that it has been selected:

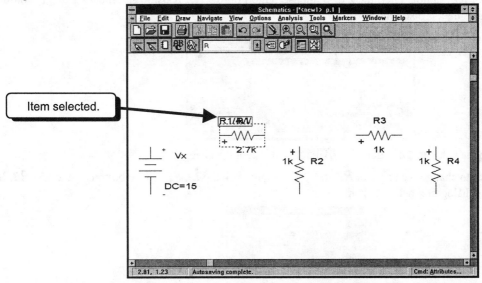

Click and **HOLD** the *LEFT* mouse button on the selected text. While continuing to hold down the button, move the mouse. The box that encircles the selected text will become attached to the mouse pointer and move with the mouse:

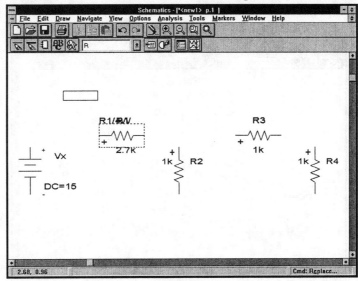

Move the outline to a convenient location and release the mouse button. The text will be placed at the new location of the box:

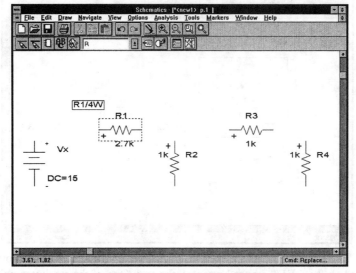

The second way to edit a part's attributes is to double click the *LEFT* mouse button on the graphic symbol for that part. To edit the attributes of *R2,* double click the *LEFT* mouse button on the *R2* resistor graphic, $+\mathord{\sim}\!\!\sim\!\!\sim$. If you double click fast enough, the attributes dialog box will appear:

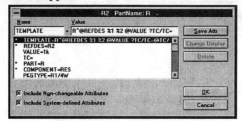

Edit the attributes to change the value to **5k**. Click the *OK* button to accept the changes.

Next, we would like to change the name of *R3* to R35. Double click the *LEFT* mouse button on the text *R3*. If you double click fast enough, the dialog box below will appear:

Change the name to **R35** and click the *OK* button. The name will be changed in the schematic.

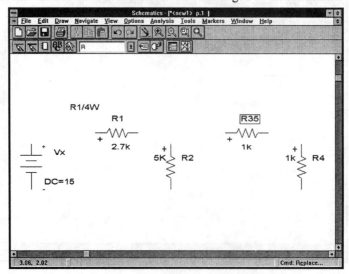

To change the value of **R35**, double click on the text **1k**. If you double click fast enough, a dialog box will appear:

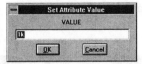

The dialog box indicates that the current value is **1k**. To change the value, type in the new value, **1Meg**, for example:

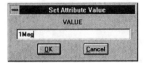

1Meg stands for 1×10^6. There must not be any spaces between the **1** and the **Meg**.* Click the **OK** button to accept the change:

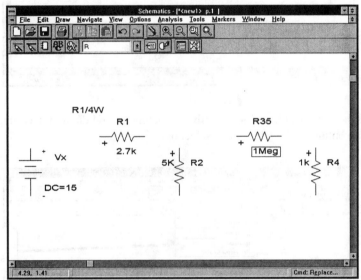

Using one of the methods given above, change the name of **R4** to Rz, and the value to 10k:

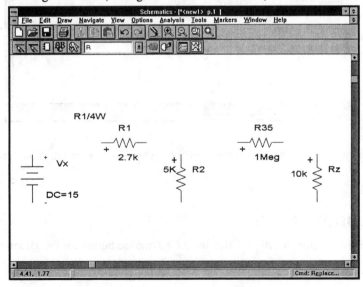

---

*In PSpice the multipliers m and M both refer to "milli." Thus the numbers 1m and 1M both equal 0.001.

We are now finished placing parts and changing attributes. Using the method given on pages 19–20 to move text items, move all the necessary text to make the schematic more readable. When you are done, you should have a schematic that looks something like the next screen:

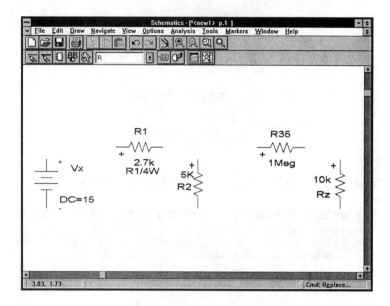

At this point your schematic may have bits of garbage floating around due to the editing. To get a fresh copy of the screen, click the *LEFT* mouse button on the **View** menu selection:

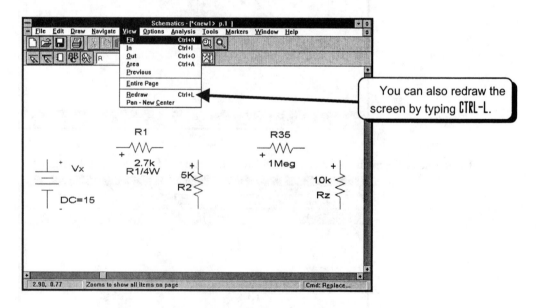

Click the *LEFT* mouse button on **Redraw**. The screen will be cleared and redrawn. You can also redraw the screen by typing CTRL-L.

1.E. Wiring Components

We now must wire the resistors together. Click the *LEFT* mouse button on the **Draw** menu. The **Draw** pull-down menu will appear:

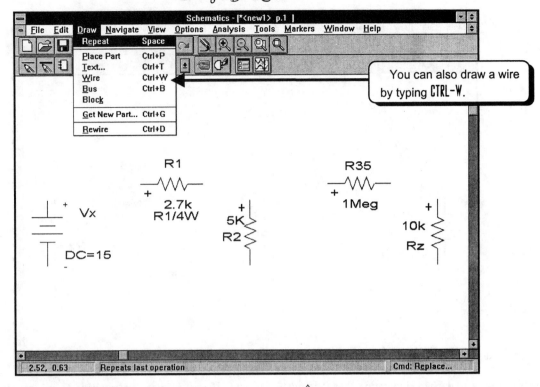

Click the *LEFT* mouse button on the **Wire** menu selection. A pencil, ✐ , will appear where the mouse pointer should be. Move the pencil so that it points toward the top of the positive (+) terminal of the DC voltage source:

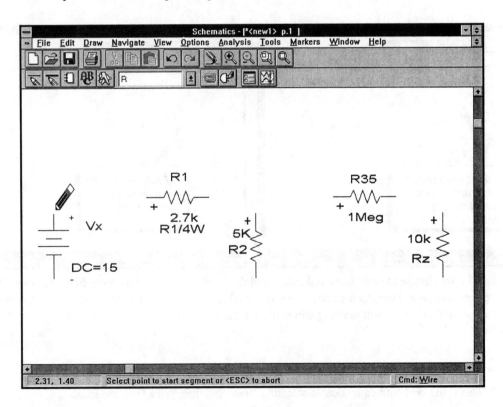

To start drawing a wire, click the *LEFT* mouse button on top of the positive terminal and then move the pencil away:

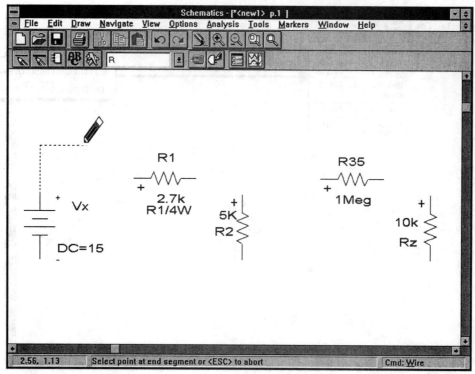

Note that the wire is dashed, indicating that it is not yet a wire. **If you missed the positive terminal, click the *RIGHT* mouse button and start over.** Next, move the pencil to point toward the left terminal of the *2.7k* resistor and click the *LEFT* mouse button.

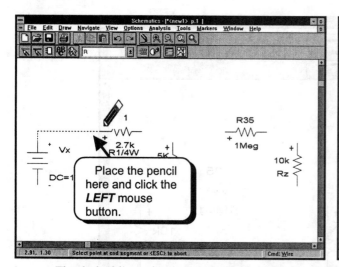

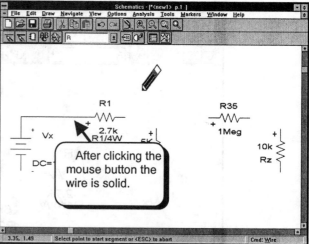

The dashed lines change to solid lines, indicating that the lines are wires. Only solid lines are wires. The schematic shows that the voltage source and the *2.7k* resistor are now wired together. The schematic also shows that the pencil is still on the screen, indicating that we are still drawing wires. If your cursor is not shown as a pencil, you can start drawing wires using three methods:

1. Select **Draw** and then **Wire** from the Schematics menus.

2. Type **CTRL-W**. This is the keyboard equivalent to selecting **Draw** and then **Wire** from the menus.

3. Click the *LEFT* mouse button on the draw wires icon, ▨.

      Move the pencil to the right terminal of the *2.7k* resistor:

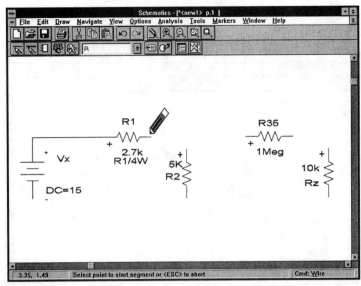

Click the **LEFT** mouse button and then move the mouse. You should have a dashed wire connected to the right terminal of the **2.7k** resistor:

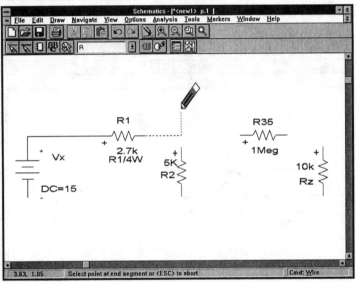

Move the pencil to the left terminal of **R35** and click the **LEFT** mouse button to make the connection. The dashed line should turn solid, indicating that it is a wire:

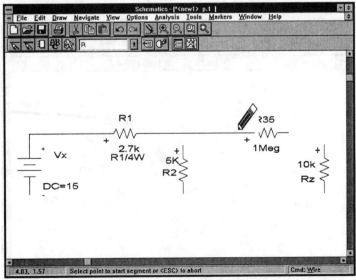

Note that the pencil is still displayed instead of the mouse pointer. This indicates that we can continue drawing wires. Move the pencil to point toward the wire:

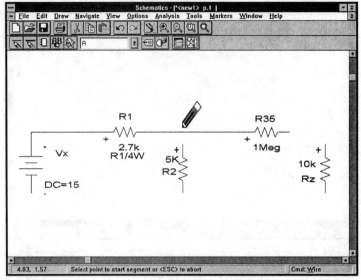

Click the *LEFT* mouse button to start drawing a wire and then move the pencil away from the wire. A dashed line should join with the solid wire:

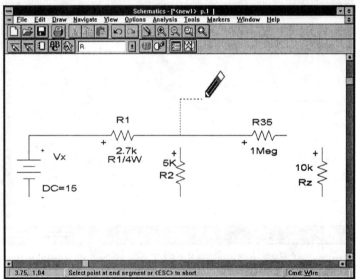

Move the pencil to the top terminal of *R2*:

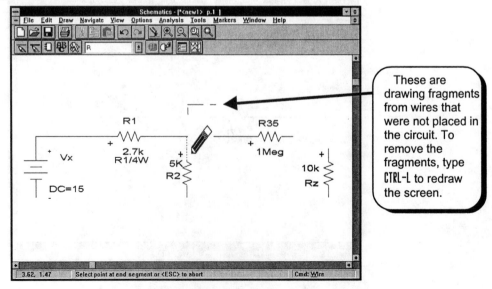

These are drawing fragments from wires that were not placed in the circuit. To remove the fragments, type CTRL-L to redraw the screen.

Click the *LEFT* mouse button to make a connection and then move the mouse away. You should now have the schematic shown below:

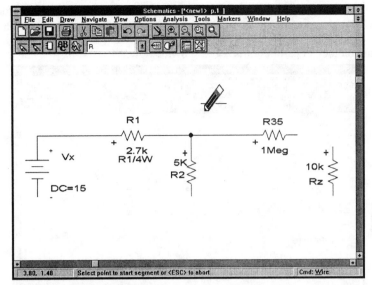

Continue wiring until you obtain the circuit shown below:

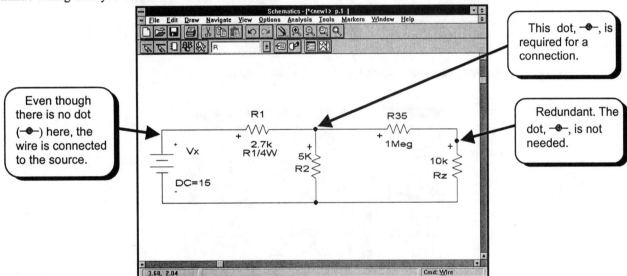

To stop drawing wires, click the *RIGHT* mouse button.

Note in the circuit above that some of the connections have a black dot, ─●─. The black dots indicate a connection. It is not necessary to have a dot present when a wire joins a pin, but having the dot emphasizes that a connection is present. To make a dot appear at a pin, make the wire overlap the pin when you are drawing wires. If you draw a wire to the tip of a pin, a dot will not be drawn. If you draw a wire that overlaps a pin, a dot will be drawn. Dots are always drawn when wires meet in a "T," ─┬─.

1.F. Correcting Wiring Mistakes

If you made a mistake drawing a wire, you can use the following procedure to delete unwanted wires, as well as components.

1. Make sure you are not in "wire" mode. If you are in "wire" mode, the mouse pointer is displayed as a pencil. To terminate "wire" mode, click the *RIGHT* mouse button so that the pencil disappears.

2. Move the mouse pointer to the segment of wire or the part that you wish to remove.

3. Click the *LEFT* mouse button on the wire or part you wish to remove. This will select the wire or part. When the wire or part has been selected, it will turn red.

4. Press the DELETE key to delete the selected wire or part.

1.G. Grounding Your Circuit

To run a circuit on PSpice, you must have at least one ground connection in your circuit. To ground your circuit, you must place a part called "AGND." To place this part, we will show a different method of obtaining the Part Browser dialog box.

Click the *LEFT* mouse button on the get part button as shown below:

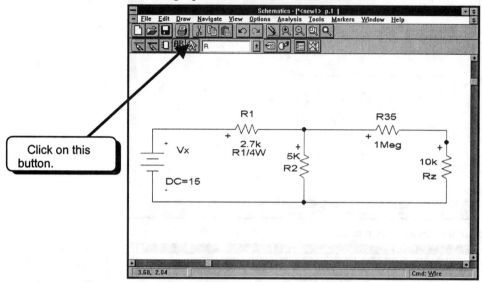

After clicking the button, the **Part Browser Advanced** dialog box should appear:

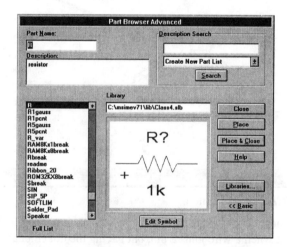

The name of the ground symbol is AGND so type the text **agnd** in the dialog box:

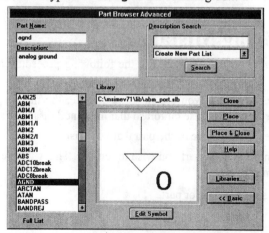

Click the **Place & Close** button to accept the part. The ground symbol will appear attached to the mouse pointer:

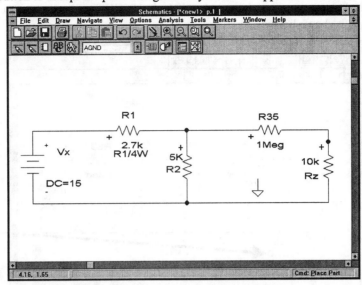

Note that the ground symbol moves with the mouse. Move the ground symbol to a wire as shown below:

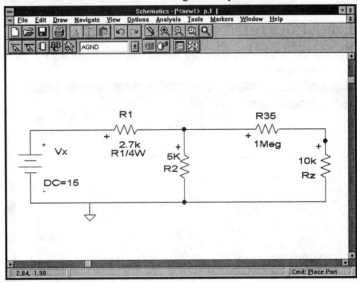

To place the part click the **LEFT** mouse button. The part is placed on the schematic and, when you move the mouse, a second ground symbol appears attached to the mouse pointer:

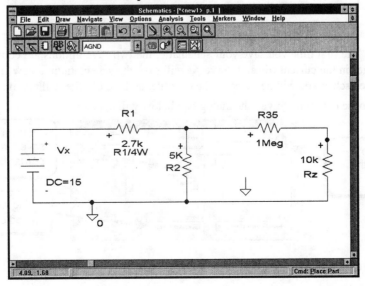

If you place the part close enough to the wire, the ground will automatically be connected to the wire. This is indicated in the schematic above because a dot is shown where the ground symbol joins the wire.

Since we do not need any more ground symbols, click the **RIGHT** mouse button to terminate drawing ground symbols. The completed circuit is shown below:

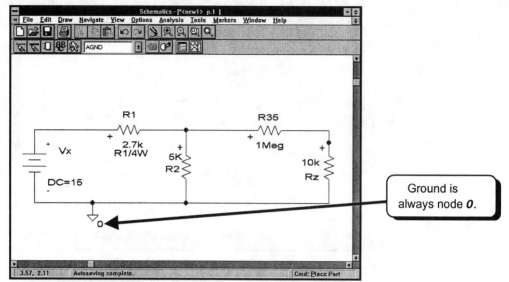

1.H. Saving Your Schematic

Now that we have finished the schematic, we need to save it. Click the **LEFT** mouse button on the **File** menu selection. The **File** pull-down menu will appear:

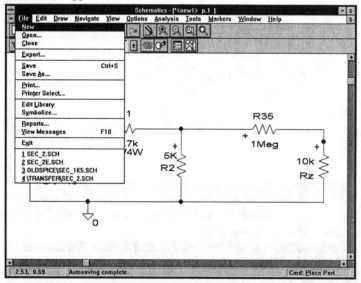

There are two save options. The first time you save your schematic, the two save options are equivalent. After the first time, **Save** will save all the changes in the current file and **Save As** will save the schematic in a new file. **Save As** will prompt for a new file name. The changed schematic will be saved in the new file, and the old file will remain unchanged.

Click the **LEFT** mouse button on **Save**. The dialog box below will appear:

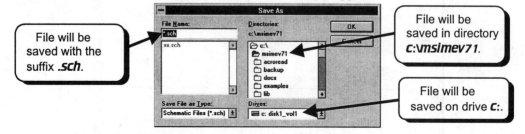

The current directory for saving a schematic is **C:\msimev71**. This is the default directory for saving files. The name of your directory may be slightly different depending on your installation. Type in the name of the schematic in the ***File Name*** text field. I will name the schematic **sec_1h** so I can remember with which section the file belongs:

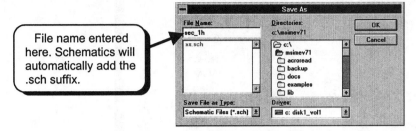

Click the ***OK*** button to accept the file name. When the file is saved the name of the circuit is displayed at the top of the schematic, and the status line at the bottom of the window indicates that the schematic has been saved.

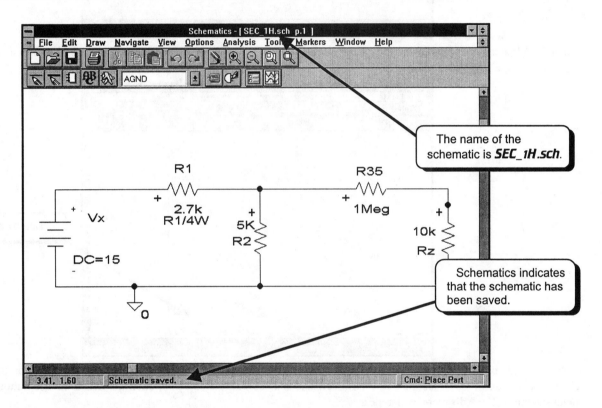

1.I. Editing the Title Block

The title block contains information about you, your company, and the schematic. We will be using the title block to identify the student, professor, class, homework number, and university where the schematic was created. There are two ways we can change the title block. The first way shows how to change the title block for the current schematic. This method must be repeated each time you create a new schematic. The second method is used to permanently change the title block so that all new schematics will reflect the changes. You would usually make permanent changes to things like the company name and the author of the schematic. Other information is usually changed frequently so permanent changes do not save time.

1.I.1. Modifying the Title Block in the Current Schematic

The first thing we want to do is place your name and class information on your schematic. In the bottom right corner of your schematic is a title block. To see this block we must switch to the page view of the schematic. To do this, click the ***LEFT*** mouse button once on the **View** menu selection:

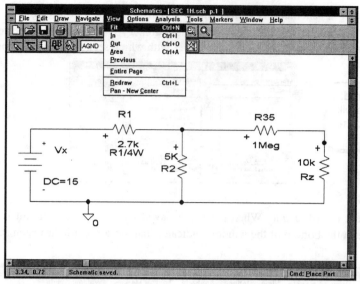

When the **View** menu appears, click the *LEFT* mouse button on the **Entire Page** option. The screen below will appear:

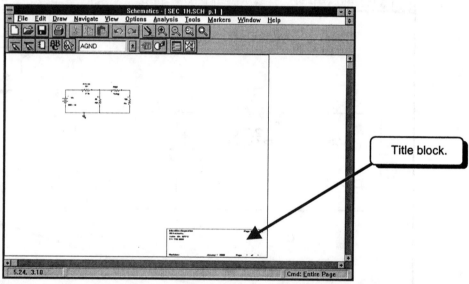

Note that a title block appears in the lower right corner. To edit its contents, click the *LEFT* mouse button on the title block. The block will turn red, indicating that it has been selected. To edit the contents, click the *LEFT* mouse button on the **Edit** menu selection. The **Edit** menu will appear:

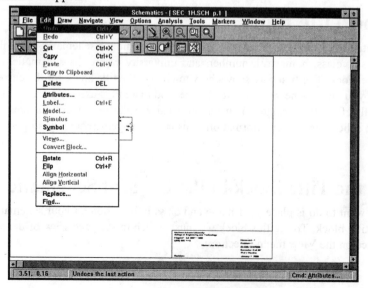

Click the **LEFT** mouse button on **Attributes**.* The dialog box below will appear. This box allows you to change most of the attributes in the title block.

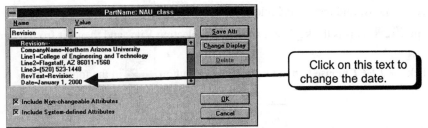

We will now illustrate how to change the date in the title block. This same procedure can be used to change all attributes in the title block. To change the date, click the **LEFT** mouse button on the line **Date=January 1, 2000**. The word **Date** will appear in the **Name** box, and the text **January 1, 2000** will appear in the **Value** box:

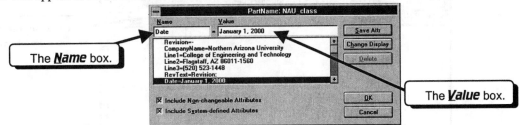

To change the date in the **Value** box:

1. Press and hold the **LEFT** mouse button to the right of the text **January 1, 2000**.
2. While holding down the **LEFT** mouse button, slide the cursor to the left, across the text **January 1, 2000,** by moving the mouse.
3. When the complete text **January 1, 2000** is highlighted, release the **LEFT** mouse button.
4. Type in a new date, `December 20, 1997`, for example.

The new date should be shown in the **Value** box:

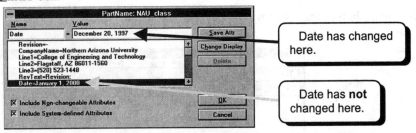

Note in the figure above that the date has not changed in the list of attributes. To permanently change the date attribute, click the **LEFT** mouse button on the **Save Attr** button. The date will now change in the list of attributes:

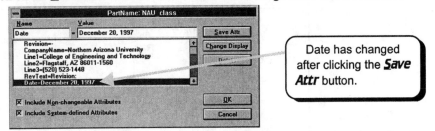

You may change all the attributes by the procedure described above. Never type anything in the **Name** box. This will add a new attribute to the list. Always click the **LEFT** mouse button on the attribute you want to change. When you have changed all of the attributes to your satisfaction, click the **OK** button with the **LEFT** mouse button. You will return to the page view of the schematic.

---

*If the text **Attributes** appears grayed out (Attributes), you have not properly selected the title block. Click the **LEFT** mouse button on the title block again until it becomes highlighted in red. When the block is highlighted in red, you may edit its attributes.

Usually schematics are edited while viewing a small area of a page. We would like to fit the schematic to the screen. There are three methods we can use to fit the schematic to the page:

1. Use the mouse and menus and select **View** and then **Fit**.

2. Type **CTRL-N**. This is the keyboard shortcut to selecting **View** and then **Fit** from the menus.

3. Click the *LEFT* mouse button on the View Fit button ![button] in the toolbar.

We will use the View Fit button ![button]. Click the *LEFT* mouse button on the View Fit button ![button] in the toolbar as shown below:

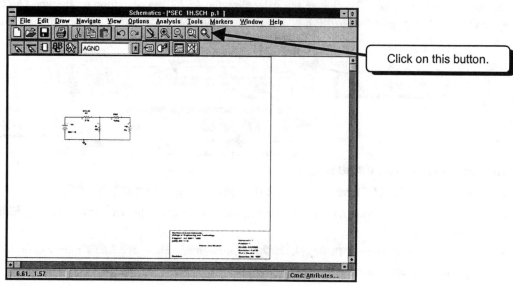

After clicking the button, the circuit will be enlarged to fit the screen:

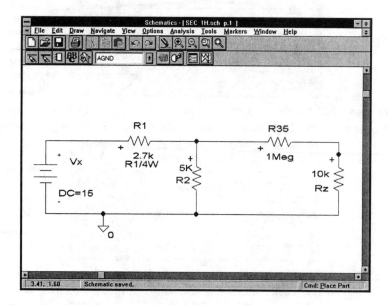

1.I.2. Permanently Changing the Title Block

If you have your own computer, you may wish to permanently change the title block rather than change the title block each time you create a new schematic. To see the title block we must view the entire page of the schematic. To do this, select **View** and then **Entire Page** from the Schematics menus:

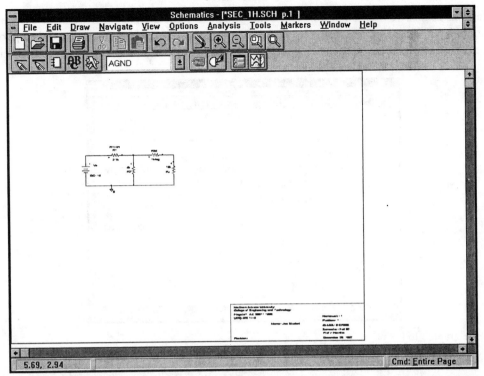

Note that a title block appears in the lower right corner. To select the title block, click the *LEFT* mouse button on the title block. It should turn red, indicating that it has been selected. When the title block is highlighted in red, click the *LEFT* mouse button on the **Edit** menu selection:

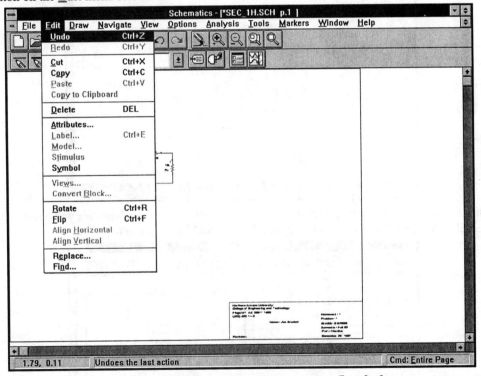

To permanently change the title block, click the *LEFT* mouse button on the **Symbol** menu selection. You may get the dialog box below:

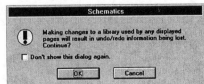

The dialog box informs us that we may not be able to use the **Undo** and **Redo** commands if we make changes. Also note that we can choose not to have this dialog box displayed in the future. If someone before you selected the option not to display the dialog box, you will not see the previous dialog box. Click the **OK** button. You will enter the Symbol Editor with the title block loaded:

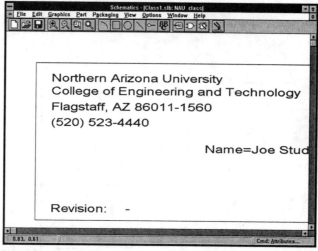

Type **CTRL-N** to fit the graphic to the page:

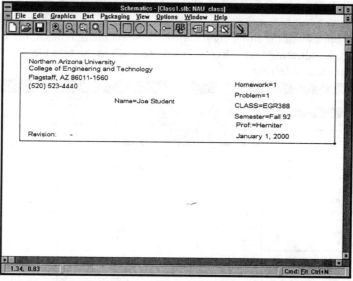

The Symbol Editor is used to modify existing graphic symbols and to create new symbols. We will modify the title block. To view the attributes, click the **LEFT** mouse button on **Part** in the Symbol Editor menu bar:

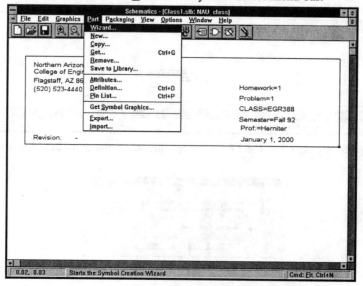

Click the *LEFT* mouse button on the **Attributes** menu selection to change the text in the title block:

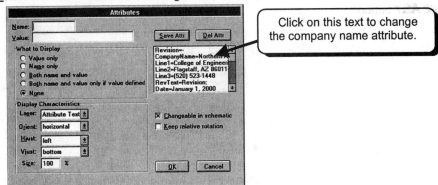

Click on this text to change the company name attribute.

This dialog box shows all of the attributes for the title block. You can change as many of these attributes as you wish. First, we will change the **CompanyName** attribute. Click the *LEFT* mouse button on the text **CompanyName**. The text will become highlighted, and the information will appear in the **Name:** and **Value:** text fields:

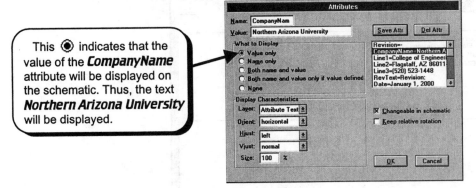

This ⊙ indicates that the value of the **CompanyName** attribute will be displayed on the schematic. Thus, the text *Northern Arizona University* will be displayed.

The text **CompanyName** is called an attribute. The value of the attribute is *Northern Arizona University*. Note that in the **What to Display** portion of the dialog box, the circle next to *Value Only* has a dot in it, ⊙. This tells Schematics to display the value of the attribute **CompanyName** on the schematic. Thus the text *Northern Arizona University* will be displayed on the schematic.

We wish to change the value of the **CompanyName** attribute from *Northern Arizona University* to your university or company name. To change the value of the attribute:

1. Press and hold the *LEFT* mouse button to the right of *Northern Arizona University*.
2. Slide the cursor to the left, across the text *Northern Arizona University*, by moving the mouse.
3. When the complete text *Northern Arizona University* is highlighted, release the *LEFT* mouse button.
4. Type in a new name, `Mountain High Voltage` for example.

The new name should appear in the **Value** text field:

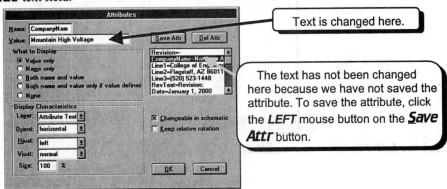

Text is changed here.

The text has not been changed here because we have not saved the attribute. To save the attribute, click the *LEFT* mouse button on the **Save Attr** button.

Note that the attribute has not changed in the list of attributes. To save the change, click the *LEFT* mouse button on the **Save Attr** button:

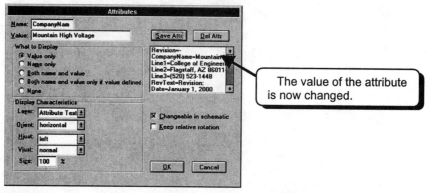

The value of the attribute is now changed.

You may make changes to as many attributes as you want. When you are finished, click the **OK** button. The changes in the dialog box will show on the screen:

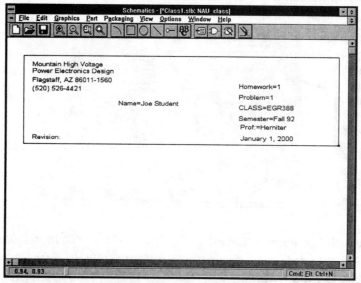

We can now save the changes and return to the original schematic. Select **File** from the Symbol Editor menu bar:

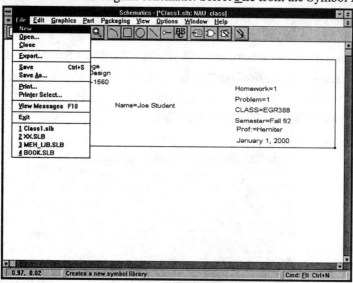

Click the *LEFT* mouse button on **Close**:

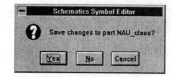

Click the **Yes** button. The changes will be saved and you will return to the schematic:

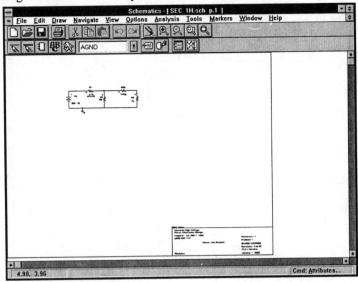

Each time you create a new schematic, the new title block will be displayed. When you return to the schematic, select **View** and then **Fit** to return to the normal view of the schematic:

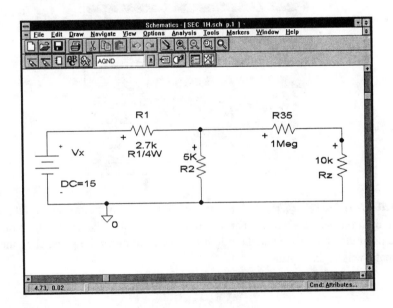

1.J. Printing Your Schematic

This section assumes that you have a correctly installed printer on your Windows system. If your printer is not installed under Windows, or if the printer is not working properly, refer to your Windows documentation. There are three ways to print your schematic:

1. Printing the entire page. This method fits the entire schematic page, including the title block, on one printer page. This method results in the circuit symbols appearing very small on the printed page.

2. Printing selected items with a specified zoom factor. The zoom factor can be chosen to make the symbols as large as desired. This method has the drawback that if the zoom factor is too large, the selected items will require more than one page.

3. Selecting an area and fitting the area to one printer page. This method usually works better than method 2 if you want to enlarge the symbols.

METHOD 1: Printing the entire page.

To print the schematic click the *LEFT* mouse button on **File** from the Schematics menu bar:

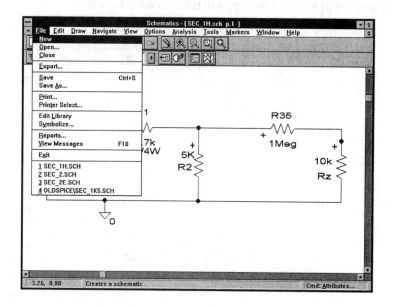

Click the *LEFT* mouse button on **Print**:

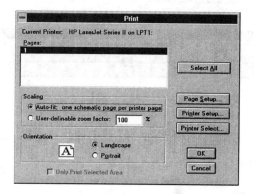

The default page *Orientation* is *Landscape*. Notice that the circle next to the text ***Auto-fit: one schematic page per printer page*** has a dot in it, ⊙. This tells us that the entire page of the schematic will fit on one page of paper. This option will print the contents of the entire page and the title block. To print the entire page, click the *LEFT* mouse button on the *OK* button. The schematic will begin to print:

The dialog box will display which printer is being used. Your printer may not be the same as the one indicated here. **Do not press the ENTER key.** Pressing the ENTER key will terminate printing. The printed page using this method is shown on page 45. Notice that the circuit elements come out fairly small.

METHOD 2: Printing a selected area with specified zoom factor.

When the entire page prints, you will notice that the electronic symbols print very small. This is because a large number of symbols can fit on a single page. Our circuit is very small, and we wish to make the symbols larger. It is possible to print only selected parts. We will select the entire circuit and print the selected parts. Move the cursor to the left and above the DC voltage source. Click and hold the *LEFT* mouse button. While holding down the *LEFT* mouse button, move the mouse down and to the right. A box should appear attached to the mouse cursor:

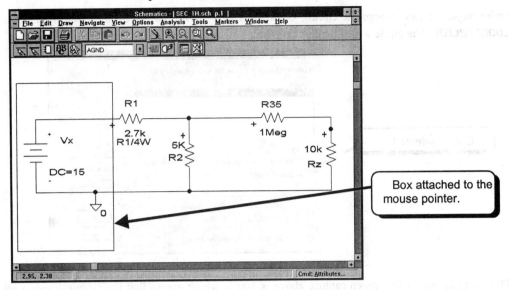

Move the cursor so that the box encloses the entire circuit and release the *LEFT* mouse button:

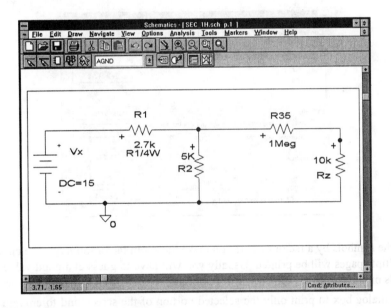

The entire circuit should be highlighted in red, indicating that the entire circuit has been selected. Select **File** and then **Print** from the Schematics menus:

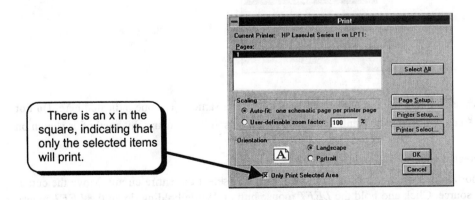

At the bottom of the dialog box is a square ☐ next to the text *Only Print Selected Area*. By default, the option is selected since the square is filled with an x ☒. This option will print only the selected items. Since we are not printing the entire page, we can enlarge the circuit. Click the *LEFT* mouse button on the circle ○ next to the text ***User-definable zoom factor***. The circle will fill with a dot, ◉:

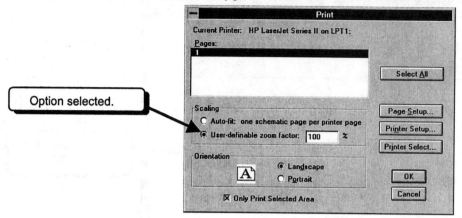

Option selected.

The zoom factor on the screen capture above is 100%, which means that the circuit drawing will not be enlarged. We will change the ***zoom factor*** to 200%:

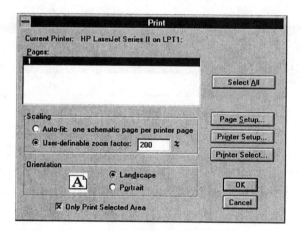

This setting will enlarge the symbols by a factor of 2. If you make the zoom factor too large, the circuit will be too large to fit on a single page, and multiple pages will be printed. Usually you will have to guess at a good value for the zoom factor. The factor you choose depends on the size of the circuit you are printing.

We have set the dialog box to print only the selected portion of the screen and to enlarge the circuit drawing by a factor of 2. Since the schematic is so small, it should fit on a single page. Click the ***OK*** button to print the circuit:

The dialog box will display which printer is being used. Your printer may not be the same as the one indicated here. **Do not press the ENTER key.** Pressing the ENTER key will terminate printing. A printout of the schematic using this method is shown on page 46.

METHOD 3: Printing a selected area to fit a printer page.

We first need to select the portion of the circuit we wish to print. We will select the entire circuit. Move the cursor to the left and above the DC voltage source. Click and hold the *LEFT* mouse button. While holding down the *LEFT* mouse button, move the mouse down and to the right. A box should appear attached to the mouse cursor:

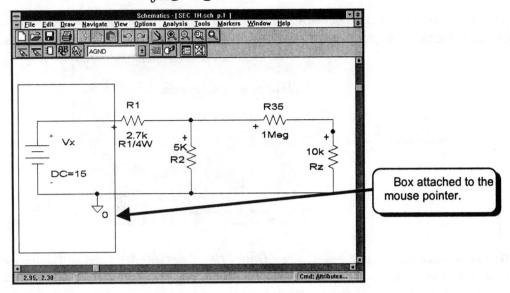

Move the cursor so that the box encloses the entire circuit:

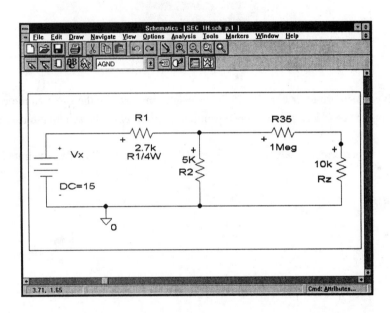

Release the **LEFT** mouse button. The entire circuit should be highlighted in red, indicating that the entire circuit has been selected. Select **File** and then **Print** from the Schematics menus:

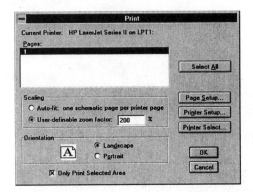

At the bottom of the dialog box is a square ⊠ next to the text ***Only Print Selected Area***. The square has an x in it, indicating that only the selected items will be printed. From the last example, the option ***User-definable zoom factor*** is selected. We need to select the ***Auto-fit*** option. Click the *LEFT* mouse button on the circle ○ next to the text ***Auto-fit: one schematic page per printer page***. The circle should fill with a dot ◉, indicating that the option is selected.

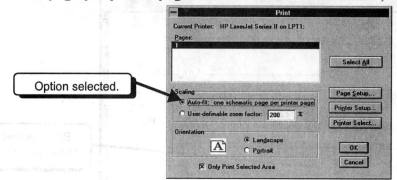

We have selected the ***Auto-fit*** option and the ***Only Print Selected Area*** option. These options have the effect of fitting the area of the selection box to a printer page. It is important to note that the selection box is fitted to the page and not the selected parts. Click the ***OK*** button to print the page:

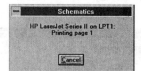

The dialog box will display which printer is being used. Your printer may not be the same as the one indicated here. **Do not press the ENTER key.** Pressing the ENTER key will terminate printing. A printout of the schematic using this method is shown on page 47.

      We are now ready to perform an analysis on this circuit. The analysis will be demonstrated in Part 3.

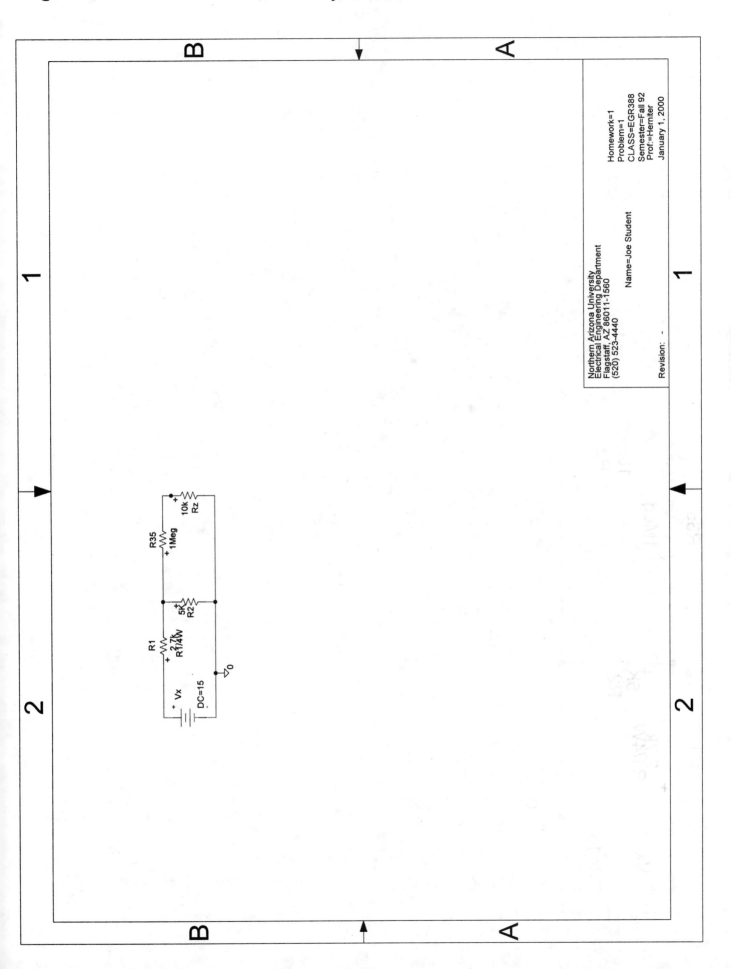

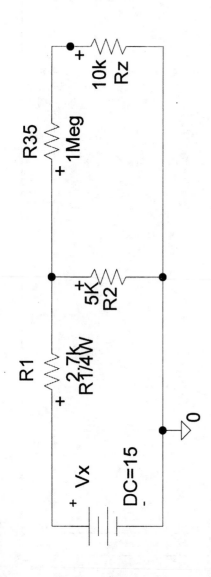

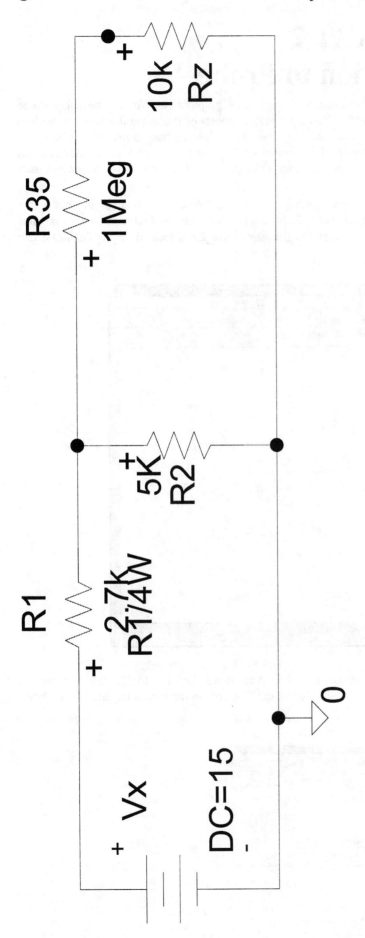

PART 2
Introduction to Probe

Probe is a program that will display the results obtained from PSpice graphically. We will be using Probe extensively throughout this manual to display the results of simulations. Various aspects of Probe are discussed in sections throughout this manual. However, if you pick specific examples you may miss those showing how to use some of the tools provided by Probe. The result may be that a section in the manual you are currently using refers to a tool in Probe that was discussed in a section that you did not cover. To avoid this problem, we will review in this section how to use the most frequently used tools.

To demonstrate Probe, we will simulate a power supply circuit with a Transient Analysis. Although this may be too complicated a circuit for beginning students, the methods discussed in Probe can be used with any analysis. We are using this simulation because it provides many interesting waveforms. Open the file named sec_2.sch located on the CD-ROM that accompanies this manual. Select **File** from the menu bar:

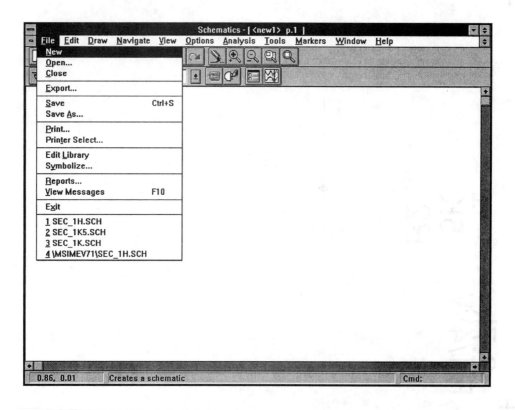

Select **Open**. We will assume that your CD-ROM drive is labeled as drive D:, so locate the file named D:\ctk_files\spice\sec_2.sch and select the file. My CD-ROM is labeled as the F: drive, so the screen capture below shows drive F:

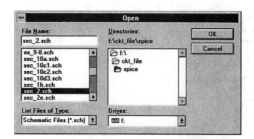

Click the **OK** button to open the file:

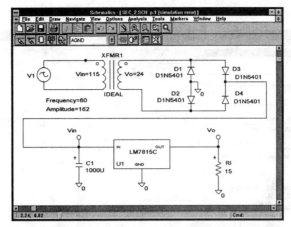

If your circuit has errors as displayed below, then you have not properly installed the libraries needed for this text:

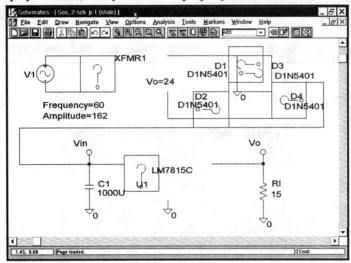

You may have forgotten to install the libraries as described in Section A.1.c on page 449. Attempt to install the libraries and then try Schematics again. If you still get the screen with errors, you should uninstall Schematics and then reinstall both Schematics and the libraries again. When installing the libraries, make sure that you choose the same installation directory for the libraries as you chose for installing the Schematics software.

If Schematics is working properly and the libraries for this manual are installed correctly, you should have the schematic below. We will continue this example starting at the screen below:

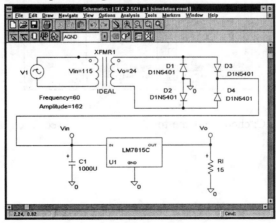

Notice that the file has two bubbles, ⟶○, one labeled Vo and the other labeled Vin.

This file is set up so that we can run it immediately. Before we can simulate the circuit we should save it to the local hard drive. When we simulate a circuit, several files will be created. The files will automatically be written to the location of the ".sch" file. Since we obtained this file from the CD-ROM, PSpice and Schematics will attempt to write files to the CD-ROM. Since a CD-ROM is read-only, an error will be generated. To avoid this problem we will save the file on the local hard drive. Select **File** from the Schematics menu bar:

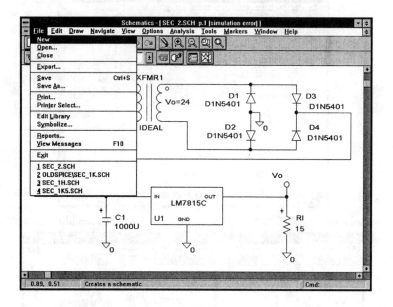

Select **Save As** to save the file with a different name or location:

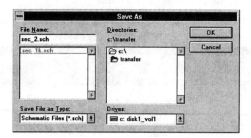

Save the file on drive C: in a directory specified for student files. I will save my file in directory c:\transfer. You may not have this directory and should choose a different directory:

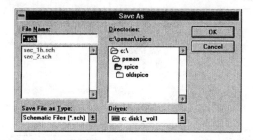

After specifying a directory on drive C: choose a name for the file (I chose sec_2.sch) and click the **OK** button to save the file:

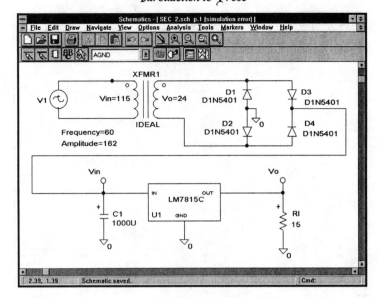

To simulate the circuit, select **Analysis** from the Schematics menu bar:

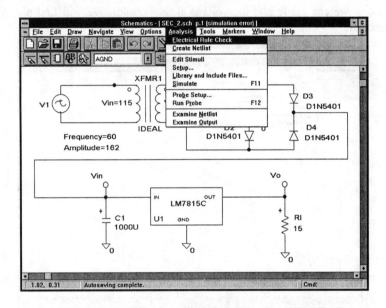

Select **Simulate**. The PSpice window will open and the simulation will begin:

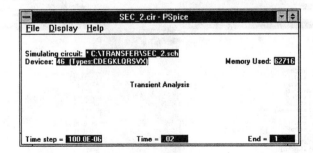

When the simulation is complete the Probe window will automatically open:

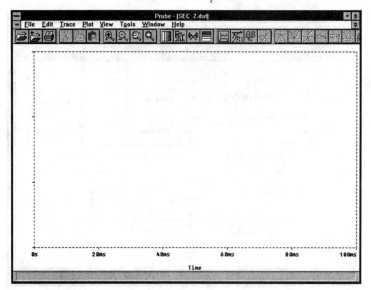

2.A. Plotting Traces

We will now add some traces. In the schematic we have labeled two nodes with bubbles, Vin and Vout. Since we know the names of the nodes, we can easily plot the voltage at the bubbles. We will first plot Vin. To add a trace, select **Trace** from the Probe menu bar:

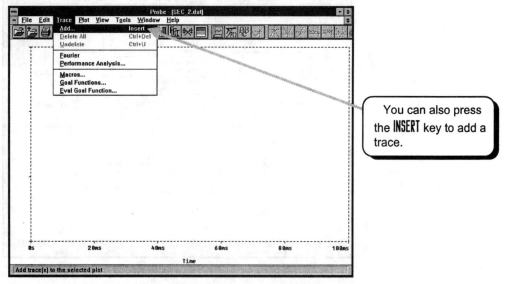

You can also press the **INSERT** key to add a trace.

Select **Add**:

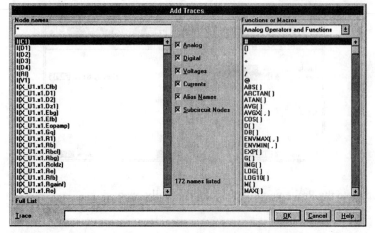

The left pane lists all of the traces we can plot. The right pane displays mathematical operations we can perform on the traces. It lists digital waveforms (we do not have any in our circuit), voltages, and currents. It also lists what are referred to as alias names. Aliases are different names that refer to the same thing. For example, in our circuit the cathodes of D3 and D4 are connected to the input of the voltage regulator and the capacitor. The node is also labeled as Vin. We can refer to the voltage of this node in several ways. This node could be addressed as pin 1 of D4, pin 1 of D3, pin 1 of C1, pin IN of the voltage regulator, or as Vin. Thus, to display the voltage at this node, we can use any of the aliases mentioned. Each node will have many aliases, and thus the list shown in the left pane is very large. Note that all trace types are displayed, **Analog**, **Digital**, **Voltages**, **Currents**, **Alias Names**, and **Subcircuit Nodes**:

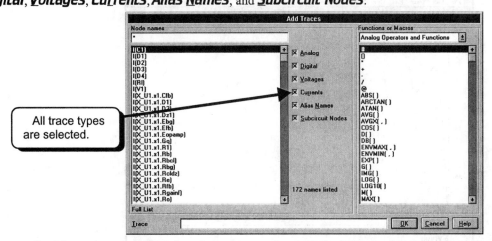

Because all of the options are selected, many names are shown in the left pane. Note that the voltage regulator is a subcircuit. A subcircuit is shown as a single block on the schematic, but there may be several circuit elements inside the subcircuit. The right pane is currently displaying all subcircuit nodes and the aliases for the subcircuit nodes. If you scroll through the list, you will see too many traces.

       We would first like to display Vin and Vout. The list of traces is too long and the traces are not easily spotted. Vin and Vout are analog voltages so we will select only the **Analog** box and the **Voltage** box:

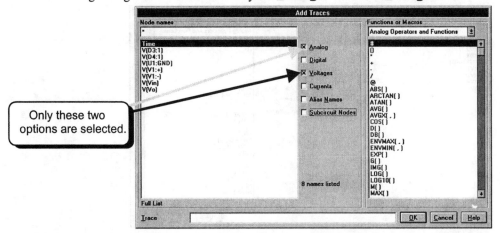

This list is now very short and we can easily spot the text **V(Vin)**. Click on the text **V(Vin)** to select it. It should become highlighted:

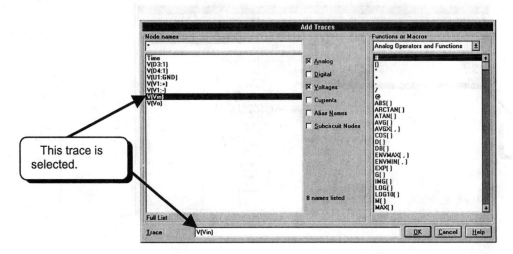

To plot the selected trace, click the **OK** button:

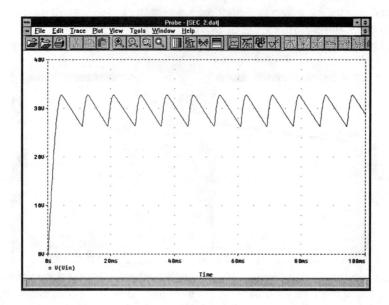

Next we will plot Vo. Press the INSERT key. This is a shortcut that will open the **Add Traces** dialog box:

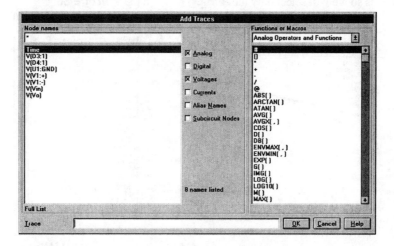

Click on the text **V(Vo)** to select the trace:

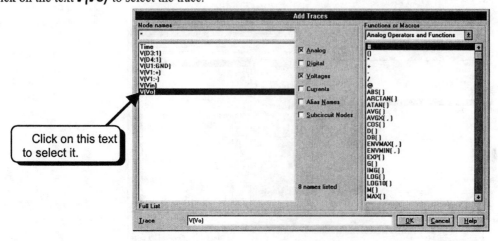

Click the **OK** button to plot the trace:

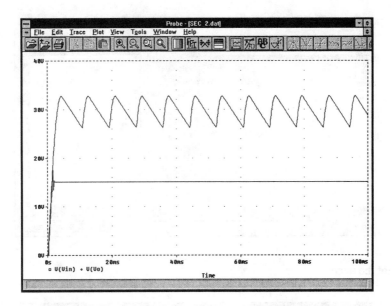

We can add many traces to the plot. Next we will display the current through D1. Press the INSERT key to obtain the **Add Traces** dialog box:

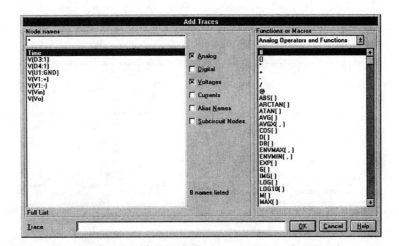

Presently the dialog box shows only voltages. We wish to plot a current, so specify the options as shown:

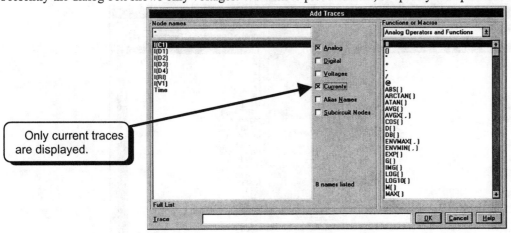

Only current traces are displayed.

We see that only the options **Analog** and **Currents** are selected. The left pane displays only the currents. Click on the text **I(D1)** to select the trace:

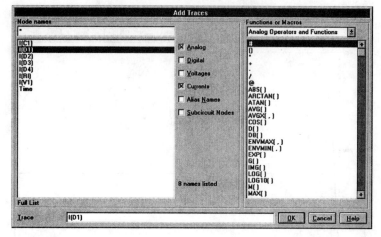

Click the **OK** button to plot the trace:

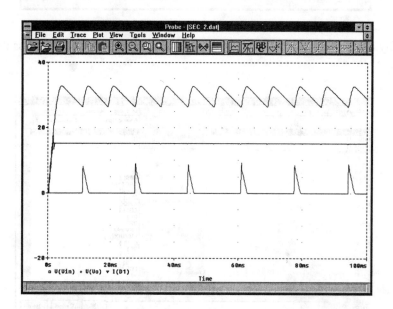

As a last example, we will show how to use one of the mathematical operations in the right pane of the **Add Traces** dialog box. Press the **INSERT** key to obtain the dialog box:

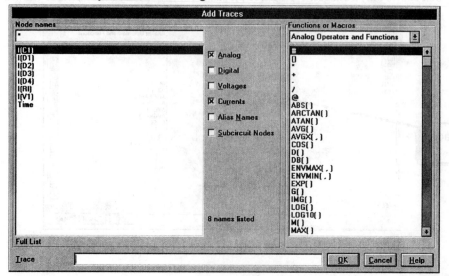

We will plot the time average current through D1. The AVG function will perform this function. Click the **LEFT** mouse button on the text **AVG()** to select the function:

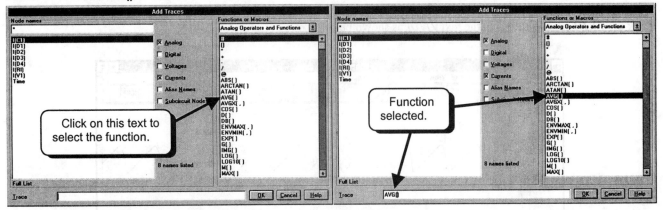

Notice that the text **AVG()** appears in the **Trace** text field and that the cursor is positioned between the parentheses waiting for a trace. Next, click the **LEFT** mouse button on the text **I(D1)**. This will select the trace and place it within the parentheses of the **AVG** function:

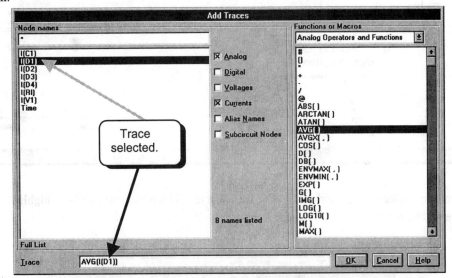

This trace is plotting the time average of the current through D1. Click the **OK** button to plot the trace:

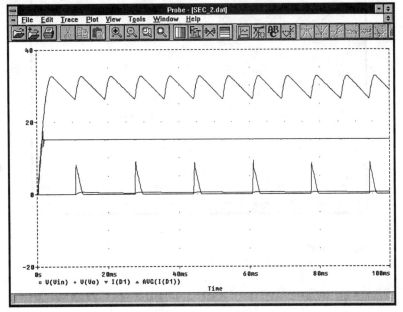

The trace starts out small but we can see the fourth trace.

2.B. Deleting Traces

We now have a number of traces on the plot. We can remove individual traces easily. We will remove trace **V(Vo)**. Click the *LEFT* mouse button on the text **V(Vo)** as shown below:

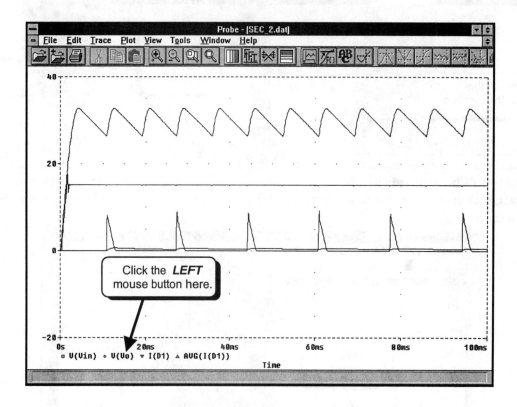

The text will become highlighted in red, indicating that it is selected. When the text **V(Vo)** is highlighted in red, press the DELETE key. The trace will be removed:

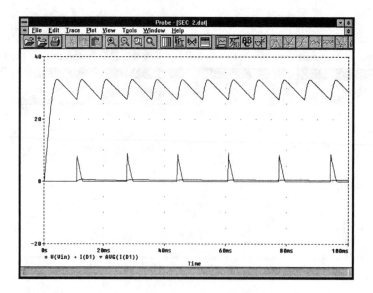

Delete all traces but V(Vin) using this method:

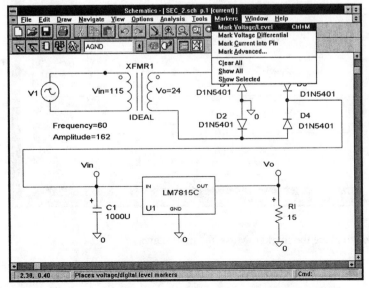

2.C. Using the Markers to Add Traces

In this example we are currently describing, both Probe and Schematics are running. Hold down the ALT key and press the TAB key twice to pop the Schematics window to the top:

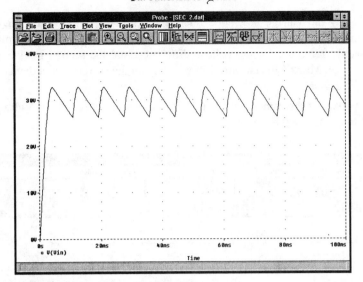

We can now use the markers to display the currents or voltages in the circuit. To obtain a voltage marker, select **Markers** from the Schematics menu bar:

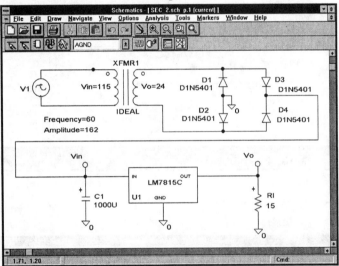

Mark Voltage/Level will display the voltage at a node relative to ground. **Mark Voltage Differential** will display the voltage between any two points. With **Mark Voltage Differential**, you will be required to place two markers. The first marker will designate the positive reference for the voltage difference, and the second marker will designate the negative reference for the voltage difference. **Mark Current into Pin** will display the current into a device. You will need to place the marker on one of the pins to the device.

We will look at the voltage at the cathode of D1 relative to ground. Select **Mark Voltage/Level** from the menu. A marker will become attached to the mouse pointer:

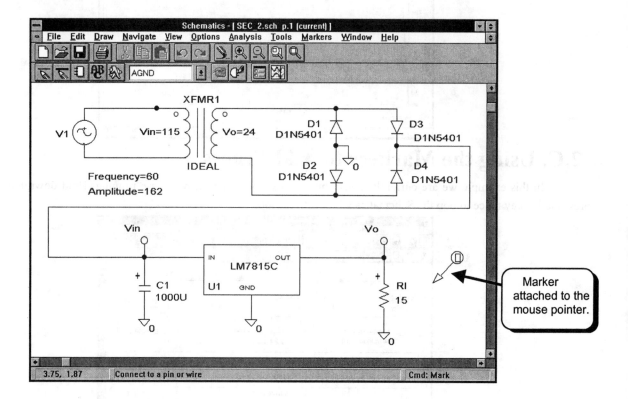

Move the mouse to position the pointer as shown:

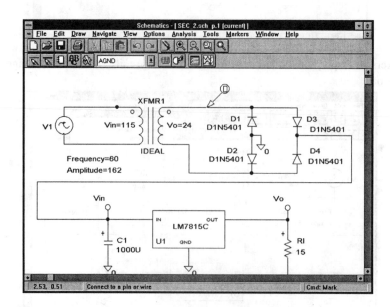

Click the *LEFT* mouse button to place the marker. Move the mouse away:

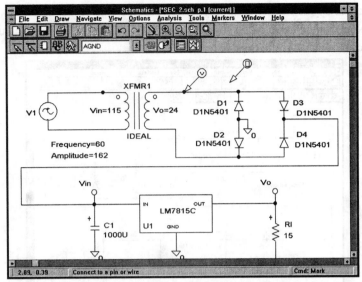

A marker is placed on the wire and a new marker is attached to the mouse pointer. We can place more markers if we want. Click the **RIGHT** mouse button to terminate placing markers:

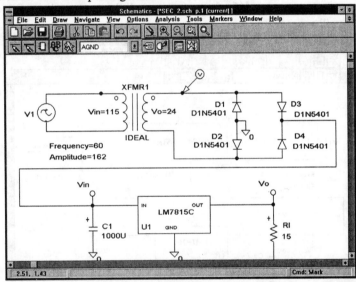

Hold down the ALT key and press the TAB key once to switch back to Probe. A new trace will be displayed:

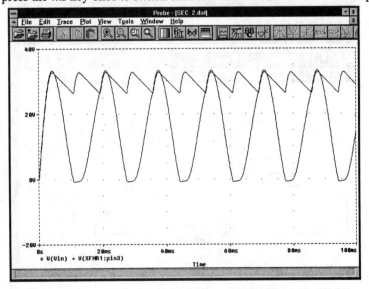

We see that the voltage specified by the marker is displayed. Markers are convenient because we do not need to know the node names to plot a trace. Note that the name of this trace is *V(XFMR1:pin3)*, hardly an obvious name. Hold down the ALT key and press the TAB key once to switch back to Schematics:

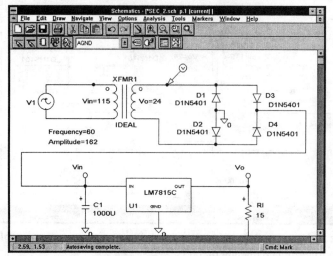

Next we will plot the voltage of source *V1* using the voltage differential markers. Select **Markers** and then **Mark Voltage Differential**. A marker will appear:

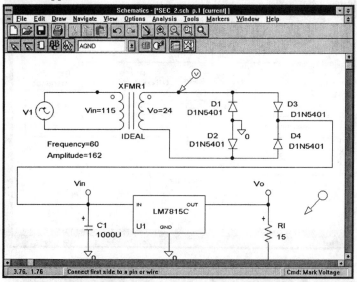

Move the mouse to position the marker as shown:

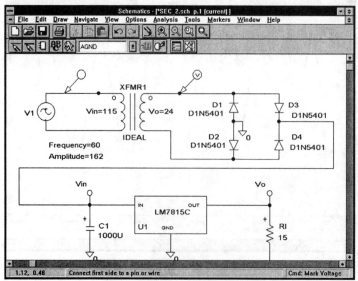

Click the *LEFT* mouse button to place the marker. As you move the mouse away, you will notice that a marker is placed on the wire and that a second marker is attached to the mouse pointer:

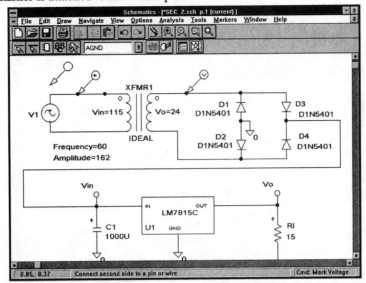

Move the mouse to position the marker as shown:

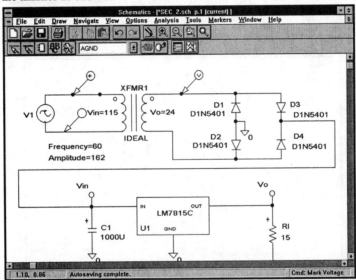

Click the *LEFT* mouse button to place the marker. Move the mouse pointer away:

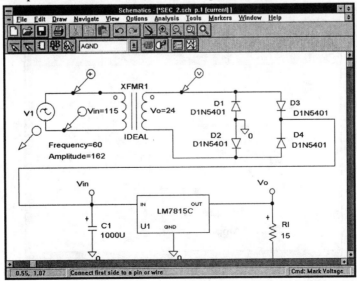

Click the **RIGHT** mouse button to stop placing markers. If you zoom in on the markers you will notice that one marker has a plus sign and the other has a minus sign:

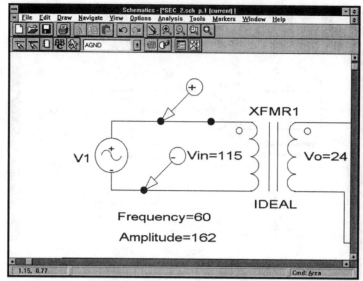

These markers display the voltage difference between the two markers. To view the trace, hold down the ALT key and press the TAB key once to switch back to Probe. The new trace will be displayed:

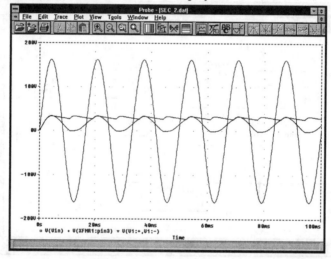

     Last, we will use a current marker to display the current through an element. Hold down the ALT key and press the TAB key once to switch back to Schematics:

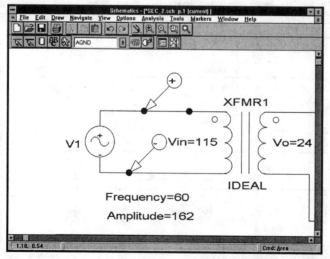

Hold down the CTRL key and type **n** to display the entire schematic:

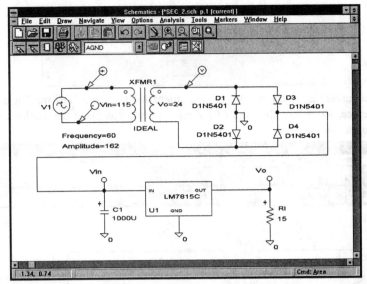

We would like to plot the capacitor current. Select **Markers** and then **Mark Current into Pin**. A marker with an I will become attached to the mouse pointer:

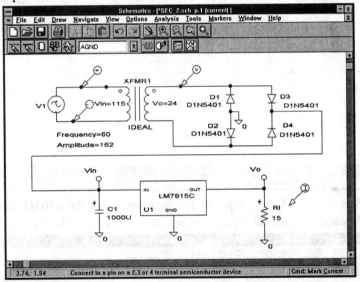

Position the marker as shown:

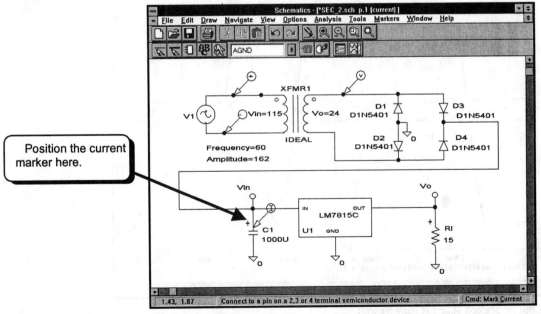

Position the current marker here.

Click the *LEFT* mouse button to place the marker. If you get the message:

you missed the pin. Move the marker up or down one grid increment and try again. The marker must be placed at the end of a blue pin connected to a device. After you click the *LEFT* mouse button to place the marker and do not receive an error message, move the mouse pointer away:

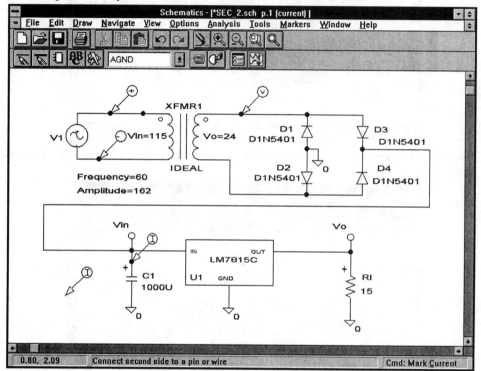

The marker is now attached to the end of the top pin of the capacitor. Click the *RIGHT* mouse button to stop placing markers. Hold down the ALT key and press the TAB key once to switch back to Probe. The current trace will be displayed:

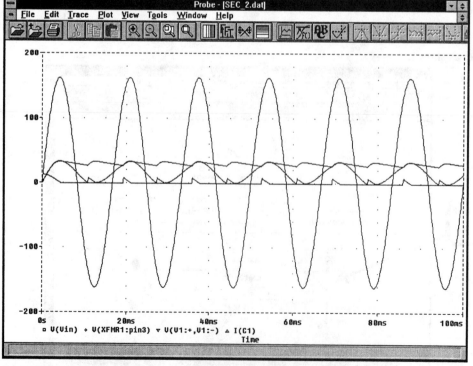

The current trace is small but its peaked shape is easy to spot.

2.D. Zooming In and Out

      We now have a number of traces displayed. However, the current trace is small. Suppose that we would like to look a little closer at a peak in the current waveform. We can do this by using some of the zoom features provided by Probe. Select **View** from the Probe menu bar:

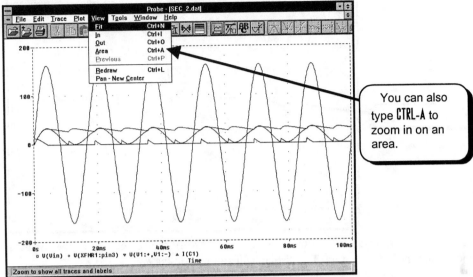

You can also type **CTRL-A** to zoom in on an area.

The menu lists 5 ways to zoom in and out. Select **Area**. The cursor will be replaced by a cross hair:

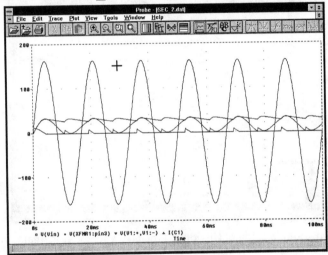

Position the cross hair as shown below:

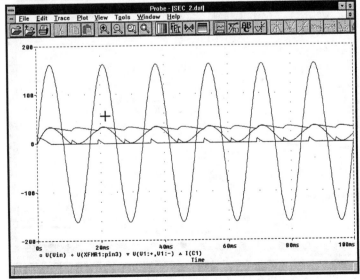

Click and **HOLD** the *LEFT* mouse button. While continuing to hold down the mouse button, move the mouse down and to the right. An outline will appear:

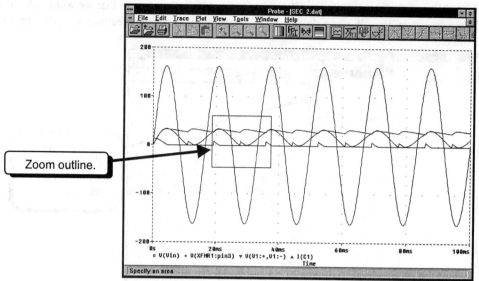

The portion of the plot inside the outline will be enlarged to fit the screen. Move the mouse to make an outline as shown above and release the mouse button. The display will zoom in on the area:

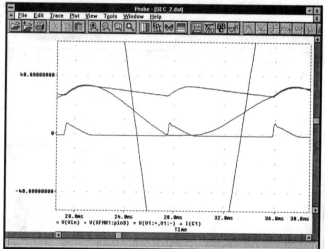

We can use the same technique to zoom in further. This time, type **CTRL-A** to zoom in again with the same method. Draw an outline as shown:

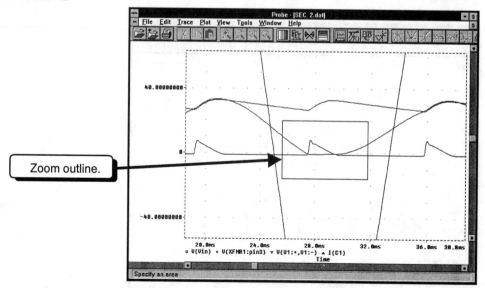

When you release the mouse button, Probe will zoom into the specified area:

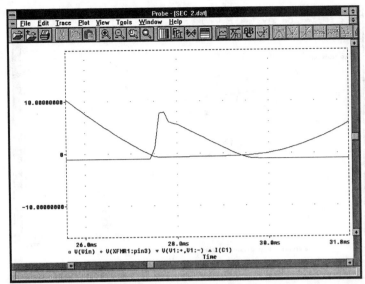

Suppose that we do not like the present view. To return to the previous view, select **View** and then **Previous**:

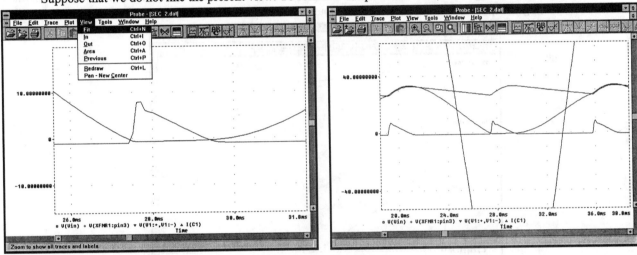

A second method for zooming in is selecting **View** and then **In**. This will zoom in around the cursor by a fixed percent. Select **View** and then **In**. The cursor will be replaced by a cross hair:

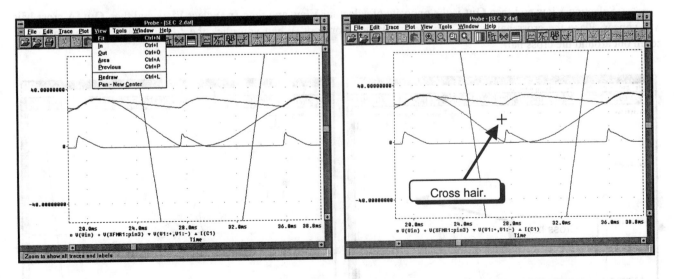

Cross hair.

Place the cursor as shown above and click the **LEFT** mouse button. Probe will zoom in around the location of the cross hair:

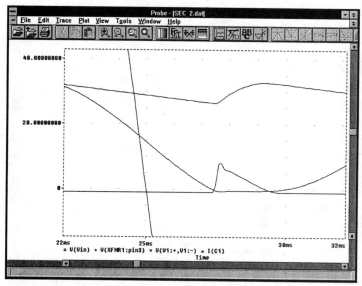

We can repeat the procedure and zoom in further. Select **View** and then **In** to obtain the cross hair. Place the cross hair where you would like to enlarge the plot and click the *LEFT* mouse button:

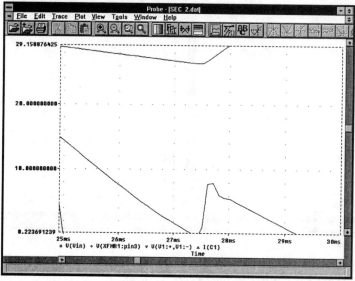

Selecting **View** and then **Out** is the opposite of selecting **View** and then **In**. Select **View** and then **Out**. A cross hair will appear. Place the cross hair where you would like to zoom out. When you click the *LEFT* mouse button, Probe will zoom out around the cross hair:

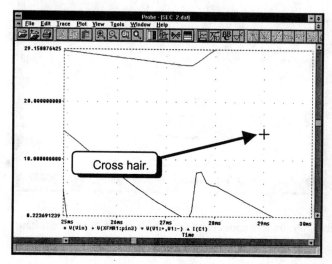

Cross hair.

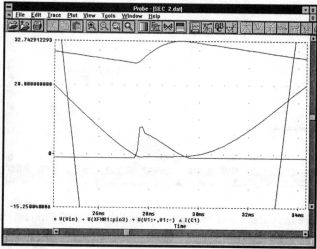

You can zoom out more if you wish.

To return the plot to its original view, select **View** and then **Fit**. This will fit the entire plot to the screen:

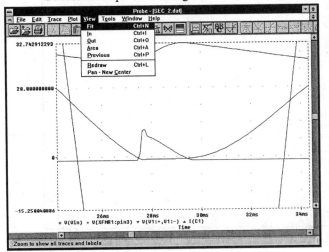 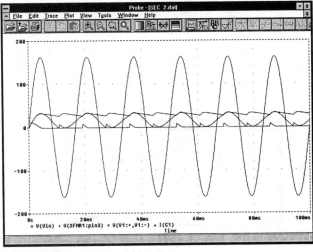

2.E. Adding a Second Y-Axis

In the previous example, we had several traces on a single plot. Some of the traces became hard to see when the magnitude of the numerical values of the traces differed by large amounts. A typical example would be plotting a voltage trace and a current trace on the same plot. Typically voltage traces may be in volts and currents may be in milliamperes or microamperes. When Probe plots traces with different units on the same plot, it plots the magnitude of the numerical values on the plot. If we plot a voltage that ranges from 0 to 5 volts and currents that range from 0 to 5 mA, both traces will be plotted with a y-axis that can accommodate numerical values from 0 to 5. With this scale, the current trace will be displayed close to zero and it will be hard to see any detail. We will use the circuit below to illustrate. The name of the circuit is sec_2e.sch. Open the file from the CD-ROM and save the file in the same manner as we did with file sec_2.sch at the beginning of this chapter. The schematic is shown below:

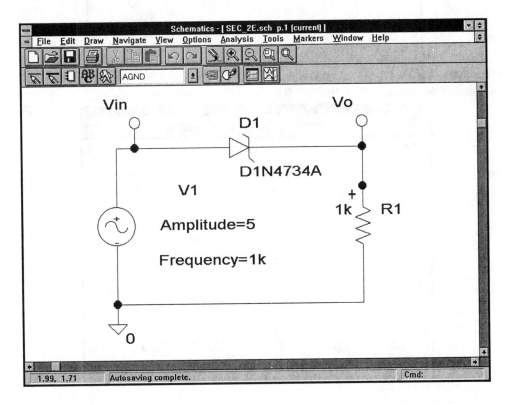

Press the F11 key to simulate the circuit and run Probe :

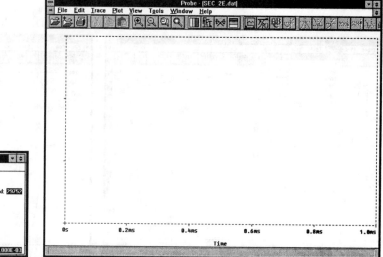

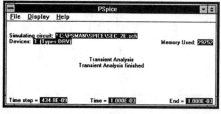

Add traces V(Vin) and I(R1):

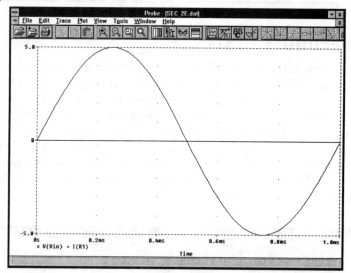

We see that the trace of I(R1) looks like it is constant at zero. This is because the numerical values of the current are 1000 times smaller than those of the voltage. Delete the trace I(R1)[*]:

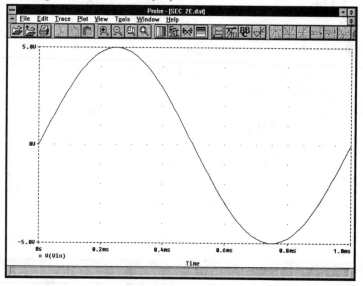

---

[*]See page 58 for deleting traces.

To fix this display problem, we will add a second y-axis. One y-axis will be used for the voltage trace and the second y-axis will be used for the current trace. The advantage is that the two y-axes can have different scales. To add another y-axis, select **Plot** and then **Add Y Axis**:

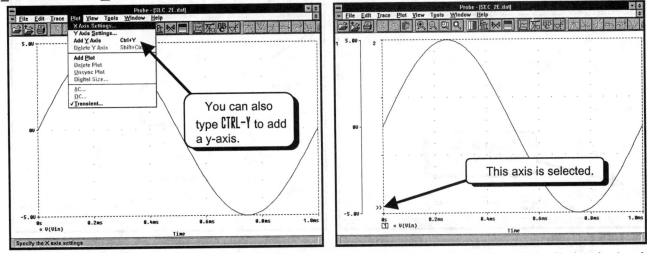

Note in the screen capture above that the axis we just added is selected. The next trace we add will be displayed using the selected axis, in this case, the new axis. Add the trace I(R1):

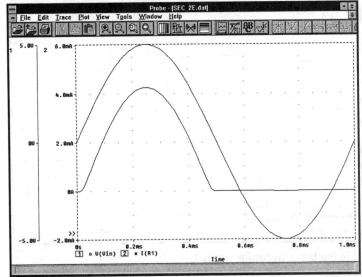

Traces that are added are placed on the selected y-axis. Below, the **>>** symbol indicates that the second y-axis is selected:

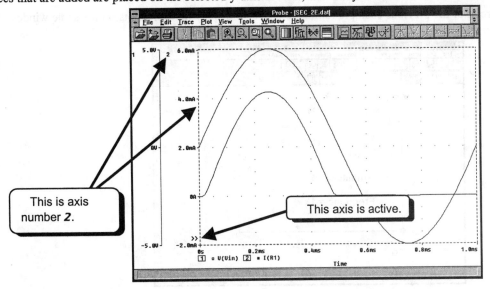

Suppose we want to plot V(Vo). This plot would fit best on the first y-axis. To select a y-axis, click the **LEFT** mouse button on the axis:

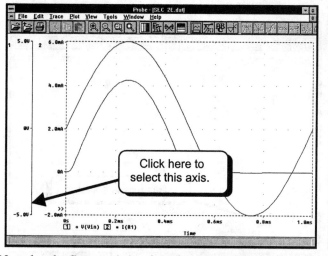
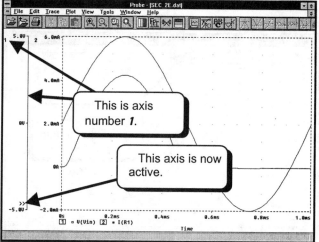

Now that the first y-axis is selected, we can add trace V(Vo):

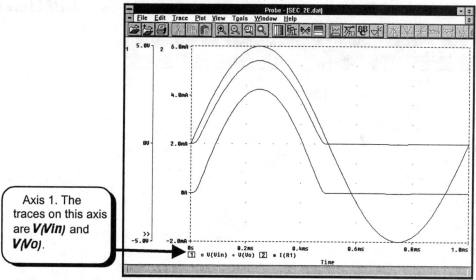

Note that trace **V(Vo)** is placed in the list of traces for y-axis 1.

2.F. Adding Plots

We will use the circuit of the previous example to illustrate how to display multiple plots on the same window. We will start with the schematic:

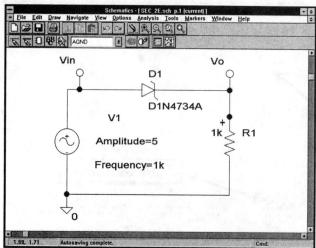

Press the F11 key to simulate the circuit and run Probe :

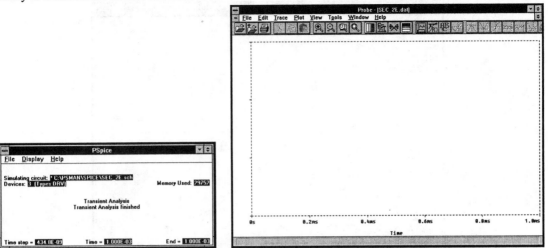

Add the trace V(Vin):

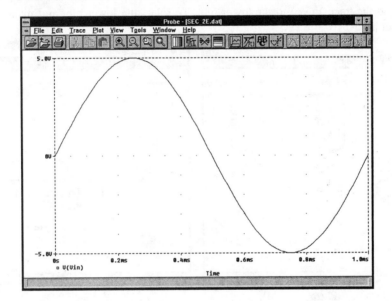

We will now create a second plot. Select **Plot** and then **Add Plot**:

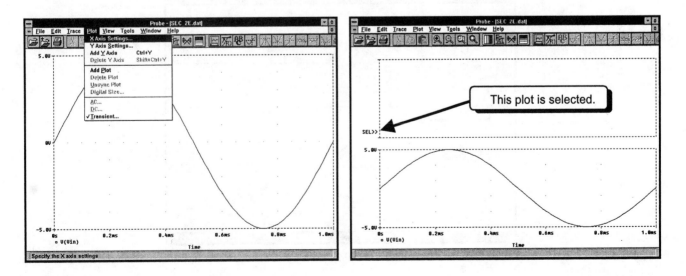

Notice that the top plot is selected. All new traces are added to the selected plot. Add the trace I(R1):

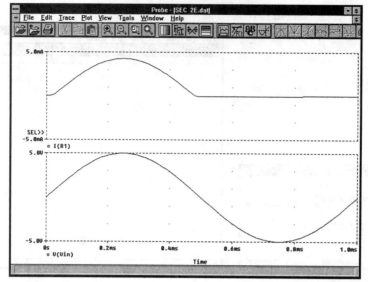

We would now like to add the trace V(Vo) to the lower plot. To select the lower plot and make it active, click the *LEFT* mouse button on the lower plot:

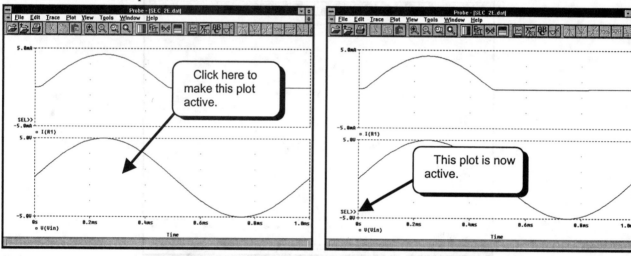

The bottom plot is now selected. Add trace V(Vo):

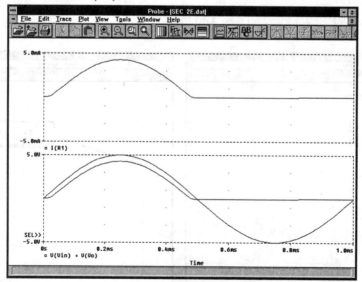

We see that the newly added trace is placed on the selected plot. We can add more plots to this page if we wish. Select **Plot** and then **Add Plot**:

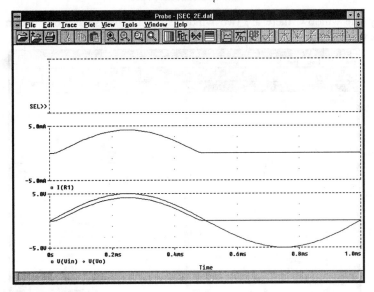

2.G. Adding a Window

Probe has the ability to display multiple windows. Each window can display different traces. We will continue now, starting at the end of the previous example. Select **Window** and then **New** to create a new window:

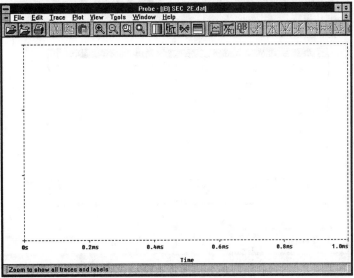

Add the trace V(Vin):

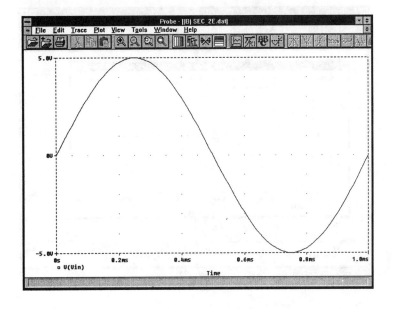

Although we can see only one window, there are two windows open. To display both windows at the same time, select **Window** and then **Tile Vertical**:

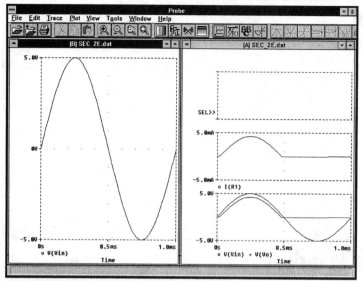

We will add a third window. Select **Window** and then **New**:

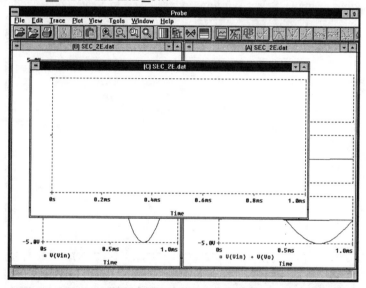

Add the trace I(R1):

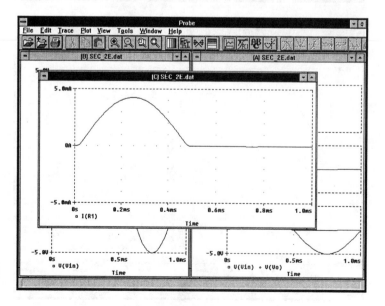

We would now like to display all windows at the same time. Select **Window** and then **Tile <u>H</u>orizontal**:

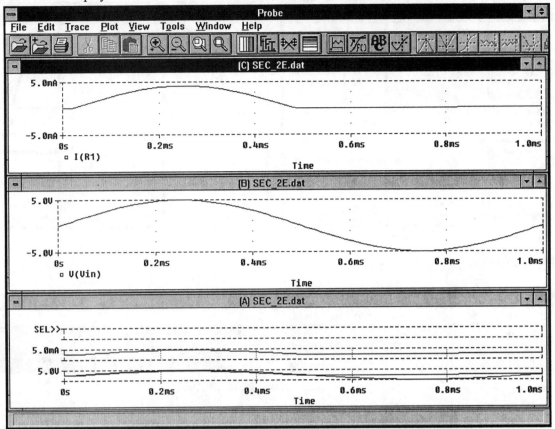

These plots do not look good when displayed in this manner. Select **Window** and then **Tile <u>V</u>ertical**:

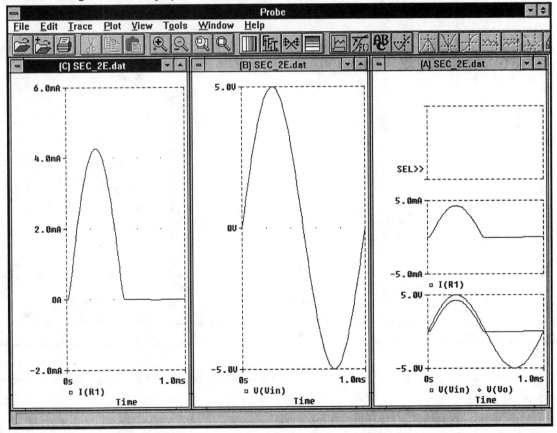

New traces are added to the selected window. To select a window, click the *LEFT* mouse button on the window you wish to use. For example, click the *LEFT* mouse button on the rightmost window. It will become selected:

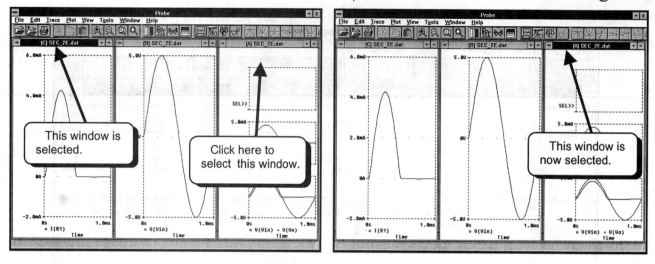

Add the trace V(Vo):

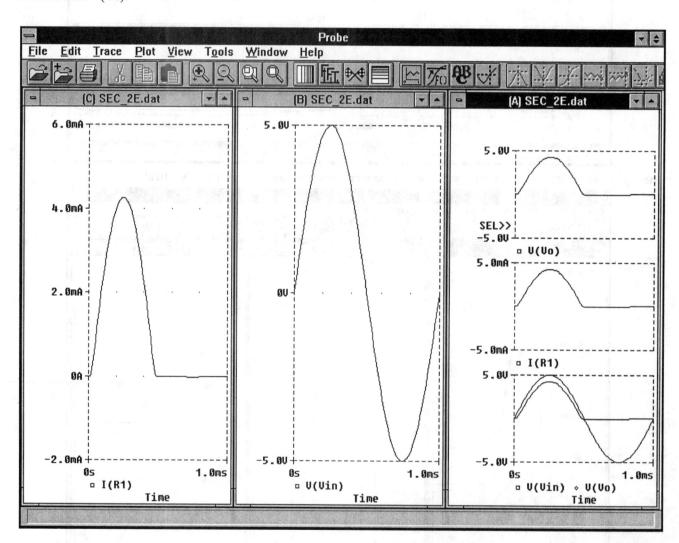

Notice that the trace is added to the active window. To enlarge a window, click the **LEFT** mouse button on the up triangle as shown below:

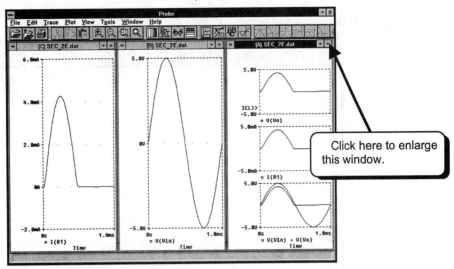

After clicking on the up triangle, the window will be enlarged:

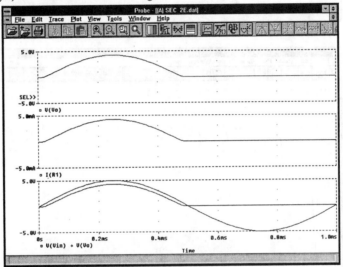

2.H. Placing Text on Probe's Screen

We will use the circuit named sec_2.sch. Open this file[*]:

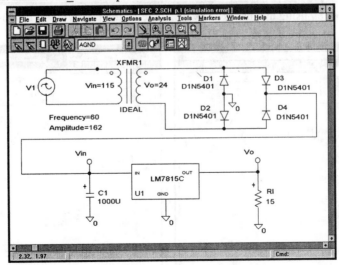

---

[*]We used this circuit at the beginning of the chapter and the file may be on your hard disk. If it is, open the file from the hard disk. If the file is not on your hard disk, follow the instructions on pages 48–50 to open the file from the CD-ROM and save it on your hard disk.

Press the **F11** key to simulate the circuit and then run Probe :

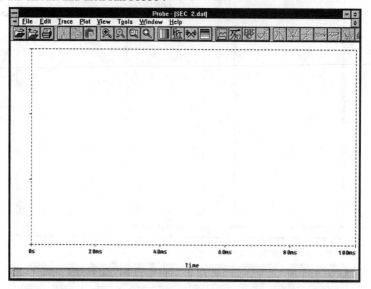

Add the trace V(Vin):

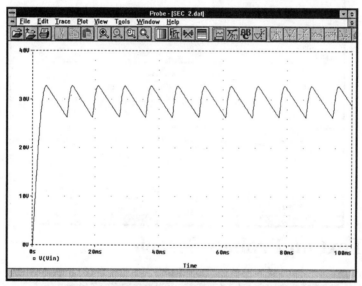

We can place text on the screen by using the menus or by using the button bar. We will first use the menus. Select **Tools** and then select **Label**:

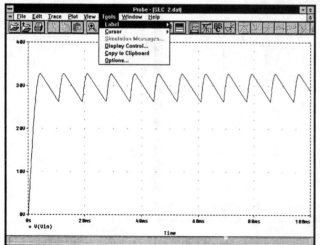 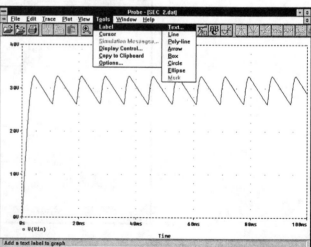

Select **Text**:

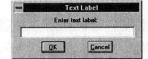

Type the text string you would like to display. Type **Input Voltage to Regulator**:

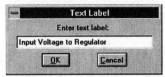

Click the **OK** button. The text string will become attached to the mouse pointer and move with the pointer:

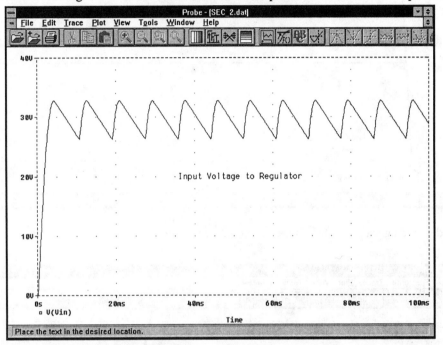

Position the text as shown below and click the **LEFT** mouse button. The text will be placed and the normal mouse pointer will return:

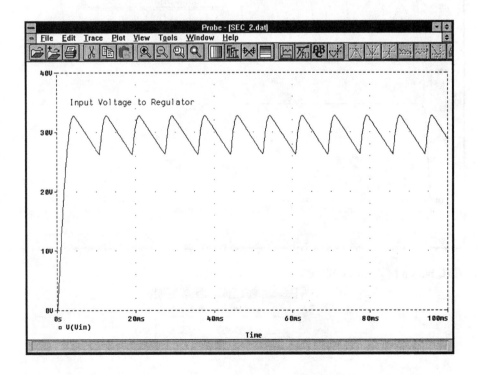

Next, add the trace V(Vo) to the plot:

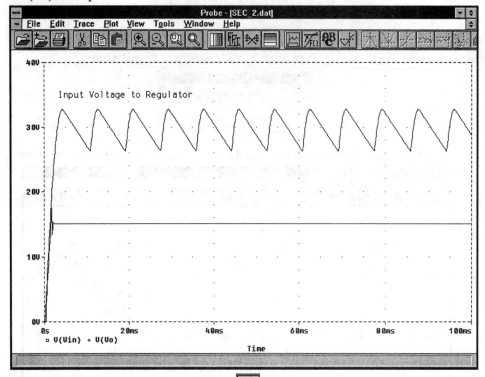

The second method of adding text is to click the **ABC** button 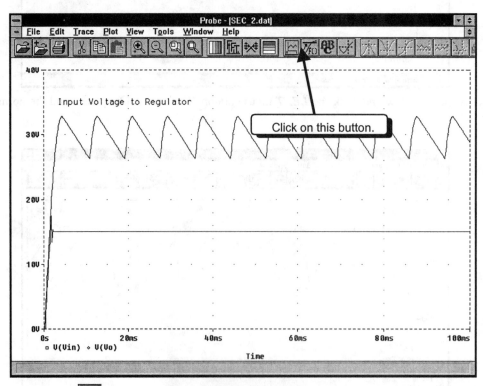 as shown below:

After clicking the **ABC** button  the **Text Label** dialog box will open:

Type the text **Regulator Output Voltage** and press the **ENTER** key. Position the text as shown and click the *LEFT* mouse button:

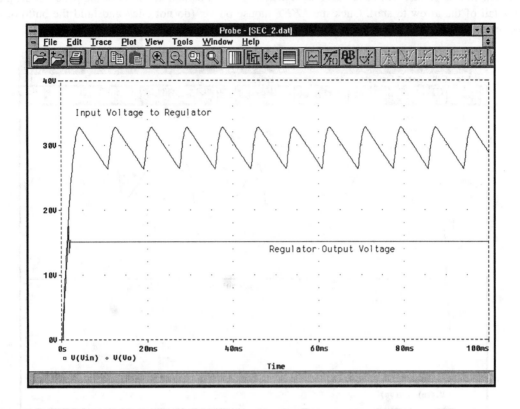

2.I. Placing Arrows on the Screen

We will now place arrows on the previous plot to point from the text to the appropriate traces. To add an arrow select **Tools** and then **Label**:

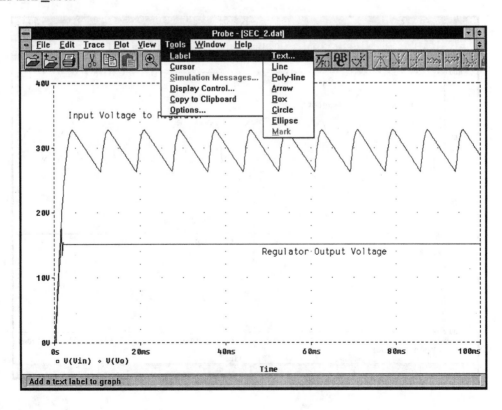

Select **Arrow**. The mouse pointer will be replaced by a pencil ✏️ . Position the point of the pencil at the place where you would like the tail of the arrow to start. Click the *LEFT* mouse button (do not click and hold the button). An arrow will appear on the screen and change its size as you move the mouse:

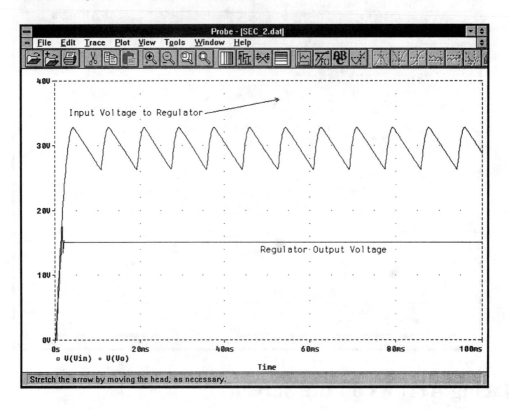

Position the head of the arrow as shown below and click the *LEFT* mouse button to place the arrow:

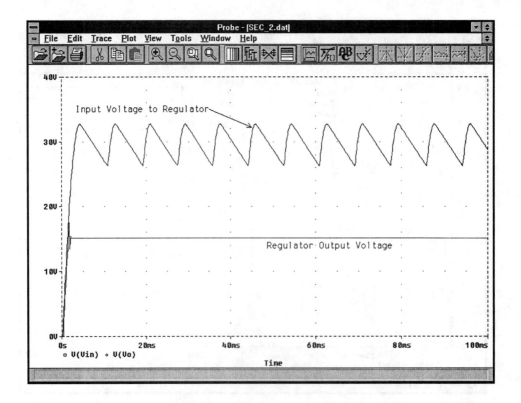

2.J. Moving Items on the Screen

Items that we place on the Probe screen can be moved with the same techniques we use to move items in Schematics. We will first move the text ***Regulator Output Voltage***. Click the *LEFT* mouse button on the text ***Regulator Output Voltage***. The text should turn red, indicating that it is selected. Next, place the mouse pointer on the red text and then click and **HOLD** the *LEFT* mouse button. While continuing to hold down the mouse button, move the mouse. The text should move with the mouse pointer:

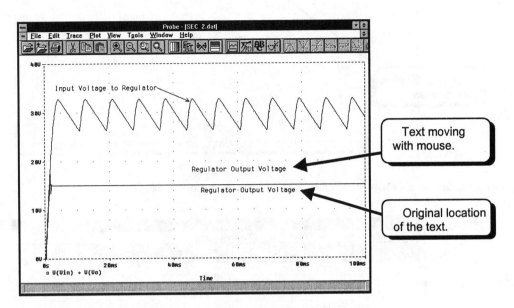

Position the text in a convenient location and release the mouse button to place the text:

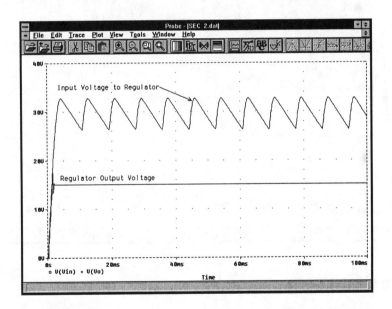

We can use the same technique to move the arrow. Click the *LEFT* mouse button on the arrow to select the arrow. It should turn red, indicating that it is selected. Next, place the mouse pointer at the center of the red arrow and click and **HOLD** the *LEFT* mouse button. While continuing to hold down the mouse button, move the mouse. The arrow should move with the mouse pointer:

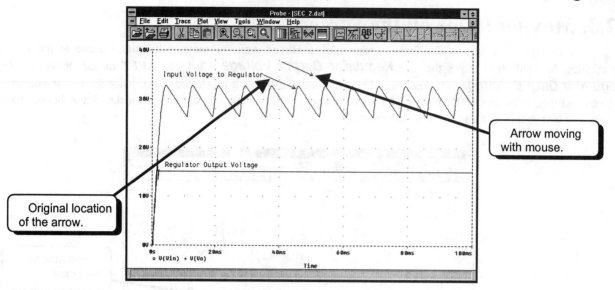

Position the arrow as shown below and release the mouse button to place the arrow:

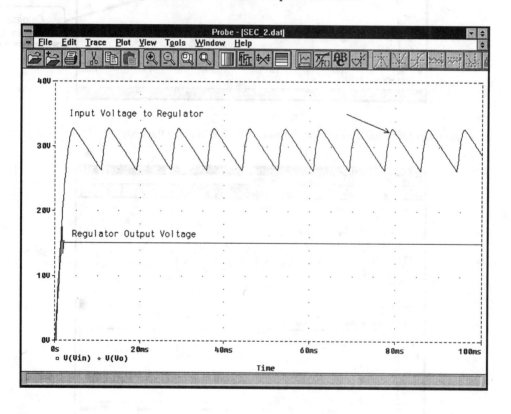

2.K. Using the Cursors

The cursors can be used to obtain numerical values from traces. To display the cursors, click the cursor button as shown below:

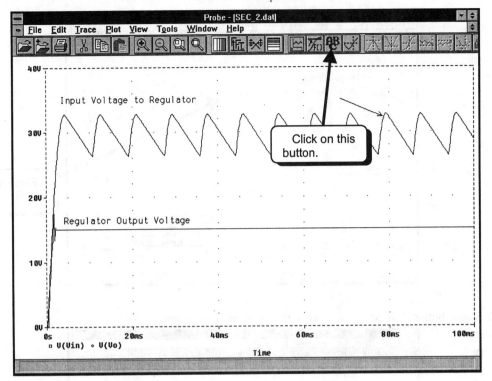

After clicking the button, the cursors will be displayed:

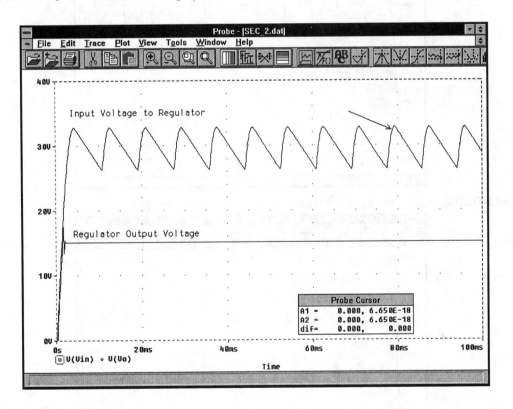

The cursors are positioned at the leftmost data point of the first trace so they cannot be easily seen. A new dialog box is displayed on the Probe screen. This dialog box displays the coordinates of each cursor and the difference between the two cursors. (There are two cursors.) The cursors can be controlled using the mouse buttons or the keyboard. The left mouse button moves cursor 1 and the right mouse button moves cursor 2. Also, the left and right arrow keys (⬅➡) move cursor 1, and the SHIFT key plus the left and right arrow keys (⬅➡) move cursor 2. Place the mouse pointer as shown below:

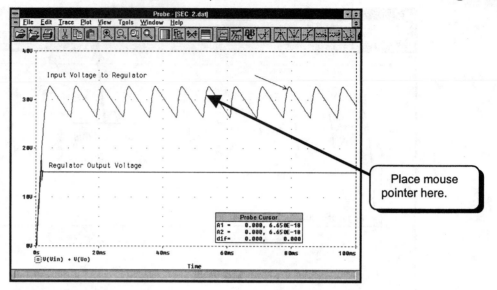

Click the *LEFT* mouse button. Cursor 1 will move to the location of the pointer:

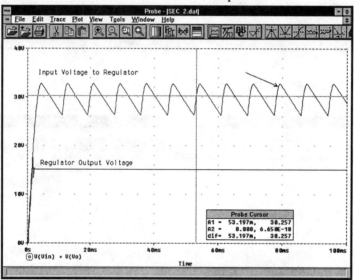

Next, press and **HOLD** the right arrow key (⬛). The cursor should move to the right:

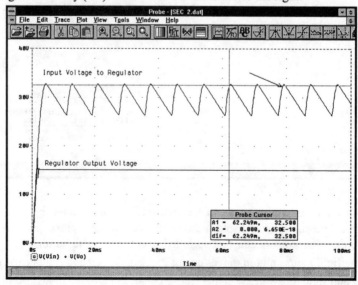

If you press the left arrow key (⬛), the cursor will move to the left. Note that as you move the cursor, the values in the **Probe Cursor** dialog box change.

Next, we will move cursor 2. Place the mouse pointer as shown:

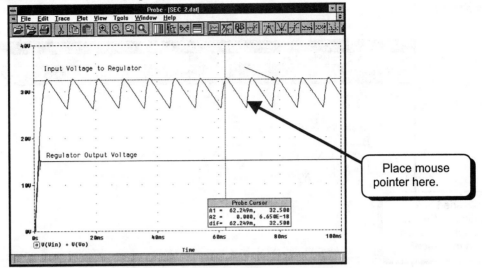

Click the **RIGHT** mouse button. Cursor 2 will move to the location of the pointer:

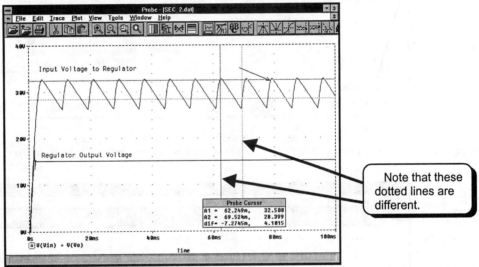

Notice that the dotted lines of the cursor are slightly different for the two cursors. Next, press and **HOLD** the SHIFT key and press and **HOLD** the right arrow key (SHIFT-→). Cursor 2 should move to the right:

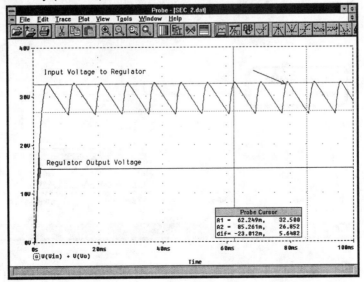

If you hold down the SHIFT key and press the left arrow key (SHIFT-←), cursor 2 will move to the left.

Presently, both cursors are displayed on the trace V(Vin). An indication of this is given by the dashed box around the symbol for V(Vin), as shown below:

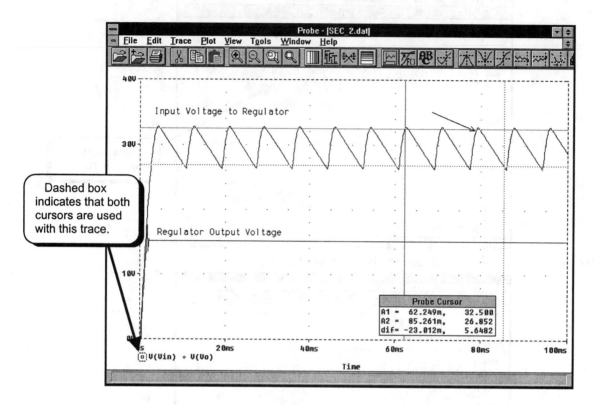

To place cursor 1 on trace V(Vo), click the **LEFT** mouse button on the marker for trace V(Vo):

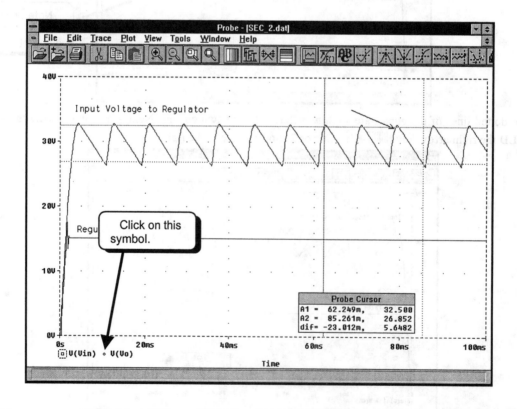

Cursor 1 will jump to trace V(Vo). Notice that a new dotted box encircles the symbol for trace V(Vin) and a different dotted box encircles the symbol for V(Vo).

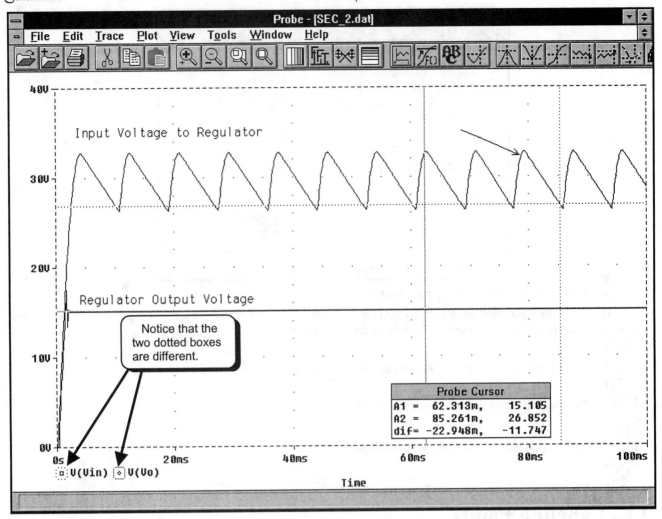

The dot patterns of the boxes match the dot patterns of the lines of two cursors. These dotted boxes indicate which cursor is attached to which trace.

To place cursor 2 on trace V(Vo), click the **RIGHT** mouse button on the marker for trace V(Vo):

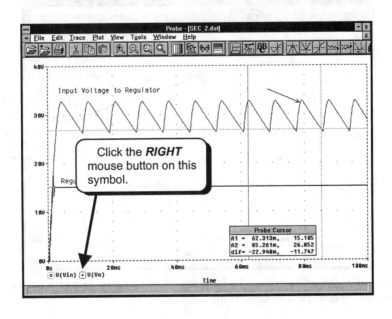

The second cursor will jump to the V(Vo) trace.

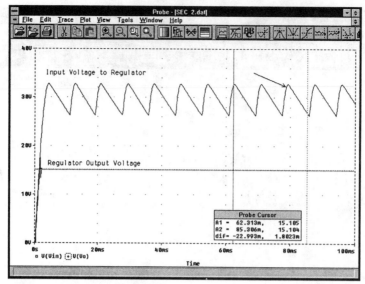

We can use the same method to move the cursors back to trace V(Vin).

The cursor that you used most recently (cursor 1 or cursor 2) is the active cursor. There are several tools in the button bar for positioning the active cursor. The buttons are described below:

- - Positions the cursor to the absolute maximum of the trace.

- - Positions the cursor to the absolute minimum of the trace.

- - Positions the cursor to the next local maximum.

- - Positions the cursor to the next local minimum.

2.L. Labeling Points

The cursors are used to find numerical values of a trace. Once important points are found, the coordinates of those points can be placed on the plot. The cursor that you used most recently is the active cursor. Place cursor 1 on trace V(Vin) at the point shown:

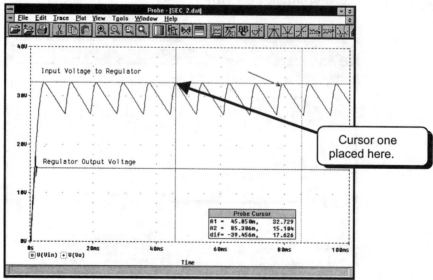

Since we just moved cursor 1, it is the active cursor. To label the coordinates of the active cursor, select **Tools**, **Label**, and then **Mark**. The coordinates will be displayed on the screen:

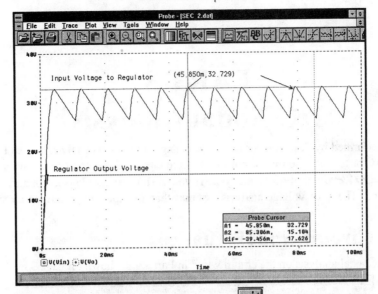

To hide the cursors, click the **LEFT** mouse button on the cursor button :

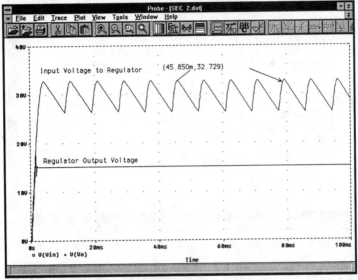

We can now move any of the items on the screen, including the coordinates of the point just added. The steps for moving any item are the same as shown in Section 2.J.

PART 3
DC Nodal Analysis

The node voltage analysis performed by PSpice is for DC node voltages only. This analysis solves for the DC voltage at each node of the circuit. If any AC or transient sources are present in the circuit, those sources are set to zero. Only sources with an attribute of the form **DC=value** are used in the analysis. If you wish to find AC node voltages, you will need to run the AC Sweep described in Part 5. The node voltage analysis assumes that all capacitors are open circuits and that all inductors are short circuits.

3.A. Resistive Circuit Nodal Analysis

We will perform the nodal analysis on the circuit wired in Part 1 and shown on page 31.

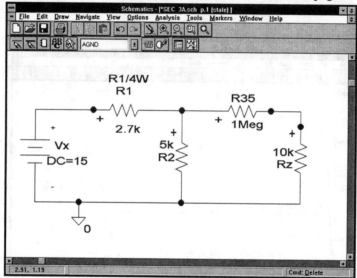

We will add two parts to this schematic for viewing DC voltages and currents. Add the parts shown below:

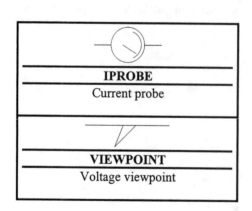

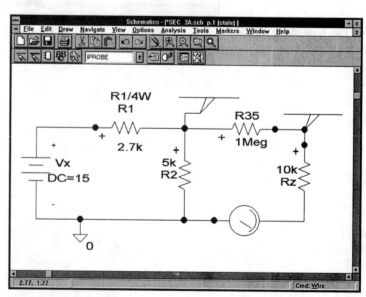

We will now create a netlist. Click the *LEFT* mouse button on **Analysis**:

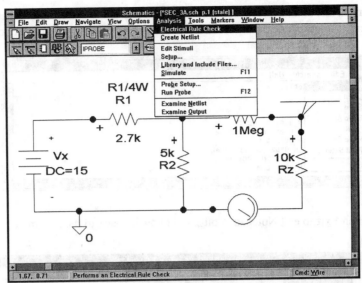

Next, click the *LEFT* mouse button on **Create Netlist**. The netlist has now been created, and we wish to examine it. If your schematic has errors, refer to Appendix D on page 461. Click the *LEFT* mouse button on the **Analysis** menu selection again.

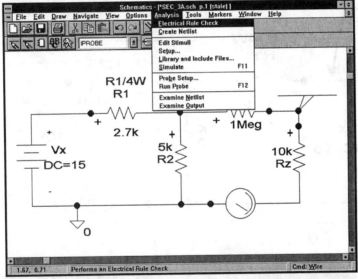

Click the *LEFT* mouse button on **Examine Netlist**. The Windows program Notepad will open the netlist file:

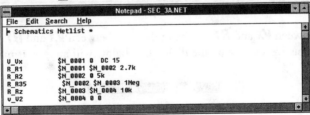

This netlist shows the following circuit connections:

- The text **V_VX** refers to the voltage source we named Vx. Vx is connected between nodes $N_0001 and 0. It is a DC source. Its value is 15 volts. Node 0 is ground.

- The text **R_R1** refers to the resistor we named R1. R1 is connected between nodes $N_0001 and $N_0002. Its value is 2.7 kΩ.

- Resistor R2 is connected between nodes $N_0002 and 0. Its value is 5 kΩ.

- Resistor R35 is connected between nodes $N_0002 and $N_0003. Its value is 1 MΩ.

- Resistor Rz is connected between nodes $N_0003 and $N_0004. Its value is 10 kΩ.

We also see a part labeled **V_V2**. This source was created by the IPROBE part and is used to sense the current. It is connected between nodes $N_0004 and 0 (between Rz and ground) and has a value of zero volts. Since the source has a value of zero volts, it does not affect the circuit.

The problem with this netlist is that, except for ground, we do not know what the node names are unless we look at the netlist. Next, we will discuss how to name particular nodes. To exit the Notepad program, click the *LEFT* mouse button on the **File** menu selection in the Notepad menu bar.

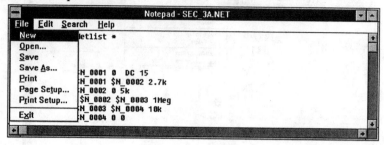

Click the *LEFT* mouse button on **Exit** to exit Notepad. You should now be back to the schematic:

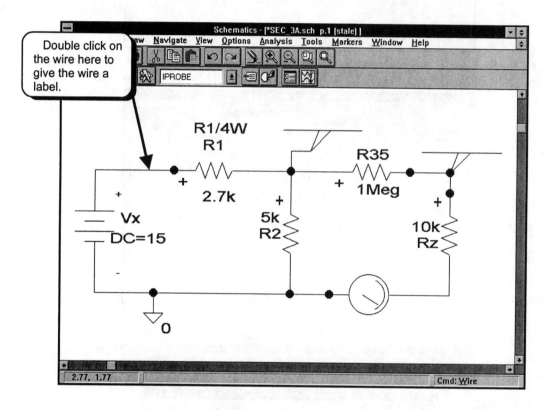

We now wish to label the node between *Vx* and *R1* by giving the wire between *Vx* and *R1* a name. Double click the *LEFT* mouse button on the wire. When you are successful, the dialog box below will appear. It may take a few tries to obtain this dialog box.

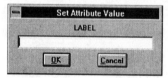

Enter a name for this wire. The name of the wire will become the name of the node. Names may be text or numbers, **but the label must not contain any spaces.** I will label the wire **1** as shown below.

After entering the label, click the *LEFT* mouse button on the *OK* button. The label *1* will appear next to the wire:

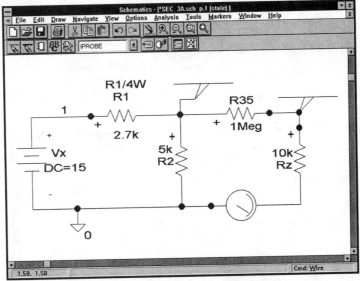

We will now use a different method to name the node between *R35* and *Rz*. We will get a part called BUBBLE and place it in the circuit. Type **CTRL-G** to get a part. The *Part Browser* dialog box will appear:

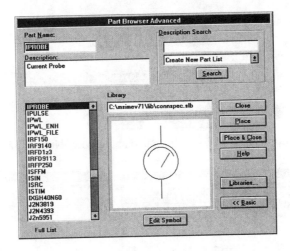

The BUBBLE part is in the abm_port.slb library. Since we know the name of the part, we will not browse for it. Type the name **bubble** in the *Part Browser* dialog box.

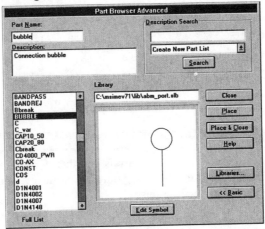

Click the *Place & Close* button to accept the part. Place the bubble next to the wire between *R35* and *Rz*. To place the part, click the *LEFT* mouse button. To stop placing parts, click the *RIGHT* mouse button. To rotate the bubble, type **CTRL-R**.

When you have placed the BUBBLE, you should have a schematic similar to the one following. If no dot appears between the BUBBLE and the wire, you will need to draw a wire to connect the BUBBLE to your circuit.

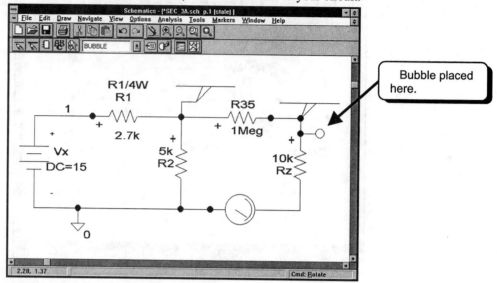

To name the BUBBLE we need to edit its attributes. You may use any of the methods described above for changing attributes. I will double click the *LEFT* mouse button on the BUBBLE graphic, ──◯. The dialog box below will appear:

Type in the label for the BUBBLE. I will call it **Vo**.

To accept the label click the **OK** button. You will see the schematic below. Notice that the label **Vo** appears next to the BUBBLE.

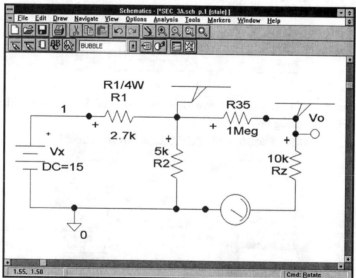

We are now ready to create a new netlist. Select **Analysis** and then **Create Netlist** from the Schematics menus. When the netlist is complete, click the *LEFT* mouse button on **Analysis** and then **Examine Netlist** to view the new netlist. The Windows Notepad program will run and display the netlist:

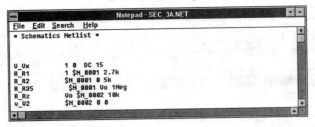

This netlist shows the following connections:

- Vx is a 15 volt DC source connected between nodes 1 and 0. Node 0 is ground.
- R1 is connected between nodes 1 and $N_0001. Its value is 2.7 kΩ.
- R2 is connected between nodes $N_0001 and 0. Its value is 5 kΩ.
- R35 is connected between nodes $N_0001 and Vo. Its value is 1 MΩ.
- Rz is connected between nodes Vo and $N_0002. Its value is 10 kΩ.

We are now done with the netlist, so we can close Notepad. To exit the Notepad program, select **File** and then **Exit** from the Notepad menus. You will return to the schematic:

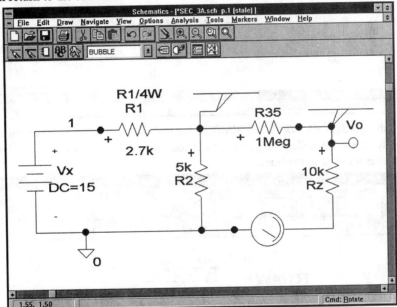

We now wish to simulate the circuit. Click the *LEFT* mouse button on **Analysis**.

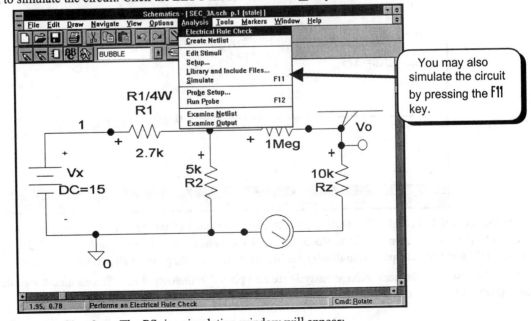

Click the *LEFT* mouse button on **Simulate**. The PSpice simulation window will appear:

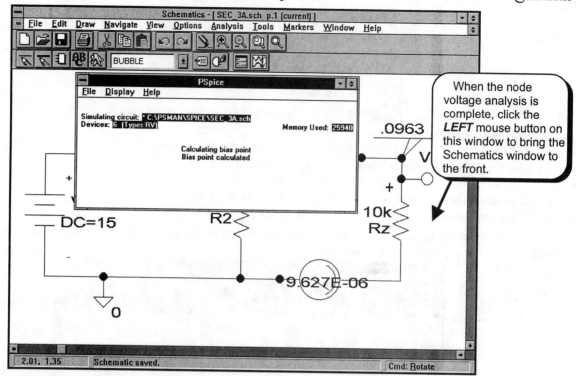

When the window displays the words **Bias point calculated**, the node voltage analysis is complete. This message will be displayed almost immediately. When the node voltage analysis is complete, click the *LEFT* mouse button on the Schematics window to bring the Schematics window to the front:

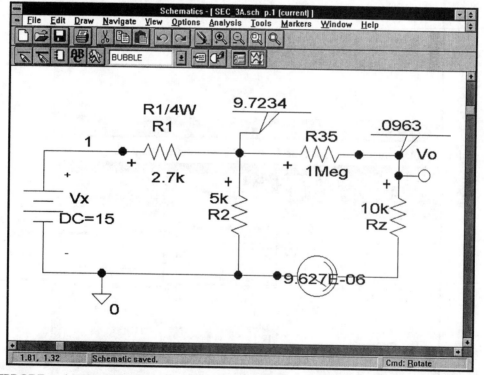

We see that the IPROBE and VIEWPOINT parts display the results of the node voltage analysis. The current through **Rz** is **9.627** μA, the voltage at node **Vo** is **96.3** mV, and the voltage at the center node is **9.7234** V. The purpose of the VIEWPOINT and IPROBE parts is to display the bias point (node voltage analysis) on the schematic.

The results of the node voltage analysis are also placed in the output file. To examine the contents of the output file, select **Analysis** from the menu.

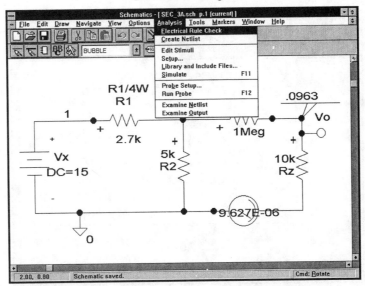

Select **Examine Output**. The Notepad program will display the contents of the output file:

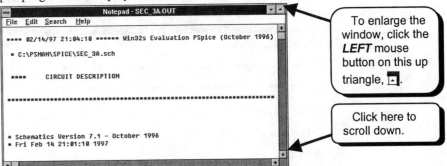

To scroll down through the output file, click the down arrow. If you go down far enough in the file, you will see the following information:

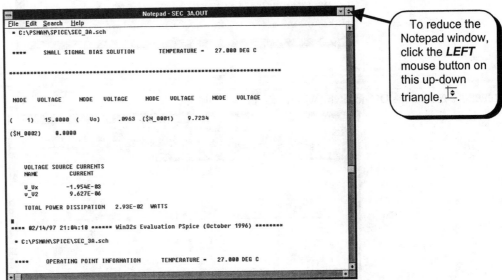

This screen shows the DC node voltages for the circuit. Node *1* is at *15* volts DC relative to ground, node *Vo* is at *.0963* volts DC relative to ground, and node *$N_0001* is at *9.7234* volts DC relative to ground. The screen also shows that *Vx* is sourcing *1.954* mA of current, and the current through V2 (our current probe) is *9.627* μA. If you wish to print these results, select **File** and then **Print** from the Notepad menus. To exit the Notepad program, select **File** and then **Exit** from the Notepad menus.

EXERCISE 3-1: Find the DC node voltages for the circuit below:

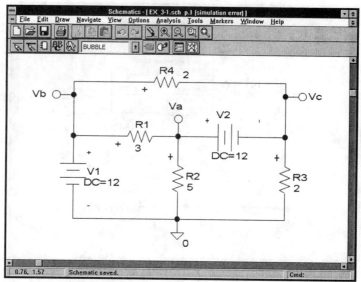

SOLUTION: The results of the DC node voltage analysis can be viewed using the VIEWPOINT part or by examining the contents of the output file. The output file is shown below:

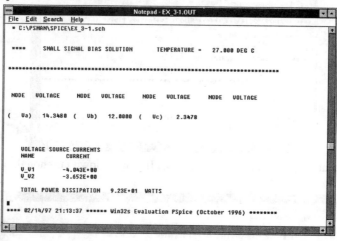

EXERCISE 3-2: Find the DC node voltages for the circuit below:

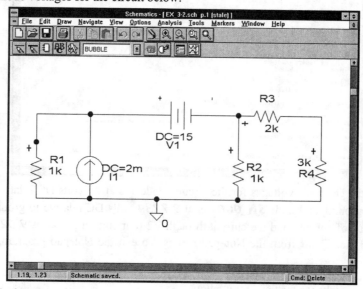

HINT: A DC current source is the part called IDC.

SOLUTION: The results of the DC node voltage analysis can be viewed using the VIEWPOINT part or by examining the contents of the output file. The results below are displayed using the VIEWPOINT part:

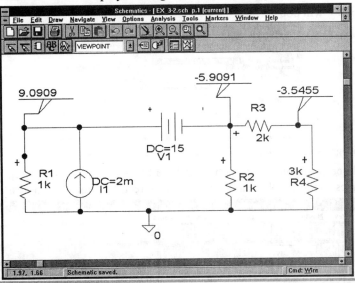

3.B. Nodal Analysis with Dependent Sources

To illustrate an example with dependent sources, we will perform a node voltage analysis on the circuit below. If you are unfamiliar with wiring a circuit, review Part 1. Create the circuit below.

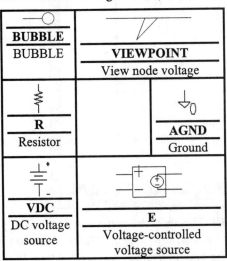

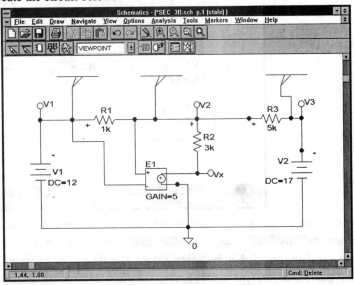

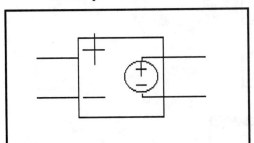

Figure 3-1: Voltage dependent voltage source.

The new part in this circuit is the voltage-controlled voltage source. The way this element is wired, the voltage at node **Vx** is 5 times the voltage across **R1**: $V_x=5(V_2-V_1)$. If you zoom in* on the voltage-controlled voltage source, you will see the graphic shown in Figure 3-1. Notice that the plus (+) and minus (−) terminals are open circuited. These connections draw no current and only sense the voltage of the nodes to which they are connected. The right half of the graphic contains the dependent source. The voltage of this source is the gain times the voltage at the sensing nodes.

When you have wired the circuit, run PSpice by selecting **Analysis** and then **Simulate** from the Schematics menus. Schematics will first create an updated netlist and then run PSpice. Schematics will inform

---

*To zoom in on a particular spot on the screen, select **View** and then **In** from the Schematics main menu. The cursor will be replaced by a cross hair (+). Move the cross hair to the spot on the screen where you want to zoom in. Click the **LEFT** mouse button. Repeat the steps to make the drawing larger if necessary.

you if there are any errors in your netlist. See Appendix D on page 461 if you have errors. At this point one of two things will happen.

1. The PSpice simulation window will pop up:

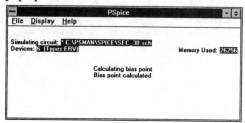

    This screen tells you the progress of the simulation.

2. If a simulation has already been run once and you reduced the PSpice window to an icon, the PSpice window will not open. Instead, the Schematics window will remain in front. This screen tells you the progress of the simulation:

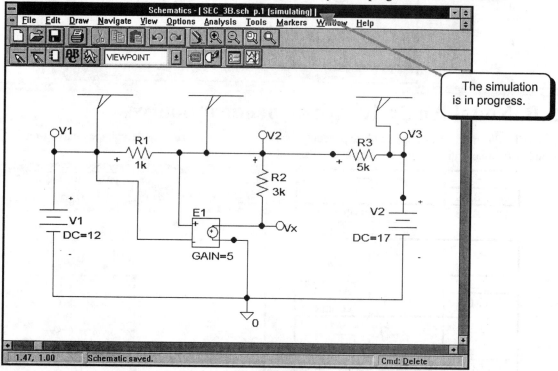

Notice that at the top of the screen Schematics says that it is *(simulating)*. This indicates that a simulation is in progress. When *(simulating)* changes to *(current)*, the simulation is complete. If you wish to see the PSpice simulation window, you can find the window by holding down the **ALT** key and pressing the **TAB** key. Remember to keep holding down the **ALT** key. Each time you type the **TAB** key, an icon will show on the screen:

Continue pressing the **TAB** key until you see the PSpice icon:

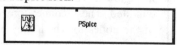

Release the **ALT** key. The PSpice window will open:

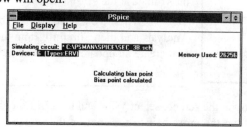

    When the simulation is complete, the results are displayed on the schematic:

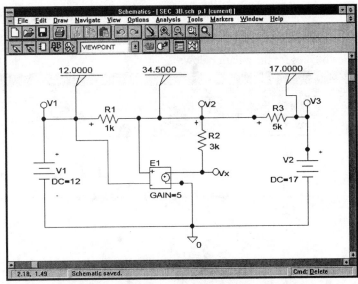

The results are also contained in the output file. To examine the output file, select **Analysis** and then **Examine Output** from the Schematics menu bar. The Windows Notepad program will run:

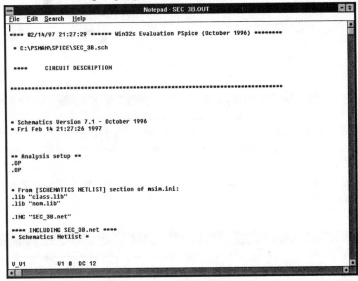

The node voltage results are contained near the bottom of the file. Click the *LEFT* mouse button on the down arrow 🔽 until you see this text:

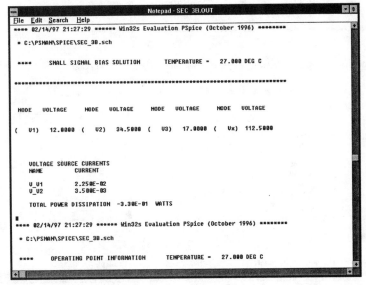

The results show the voltages at the specified nodes relative to ground. To close the Notepad program, select **File** and then **Exit** from the Notepad menu bar.

EXERCISE 3-3: Find the DC node voltages for the circuit below:

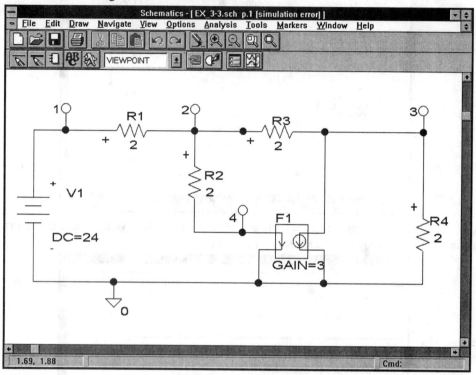

HINT: F1 is a current-controlled current source. Note that node 4 is connected to ground with a wire. Thus, the voltage of node 4 should be zero volts. Node 4 is necessary because it joins the lower terminal of R2 to the current-sensing terminal of F1.

SOLUTION: The results of the DC node voltage analysis can be viewed using the VIEWPOINT part or by examining the contents of the output file. The output file is shown below:

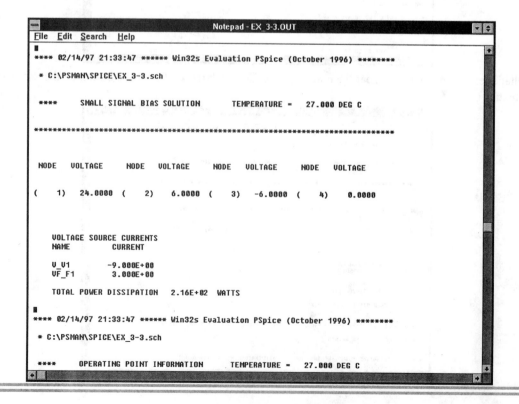

EXERCISE 3-4: Find the DC node voltages for the circuit below:

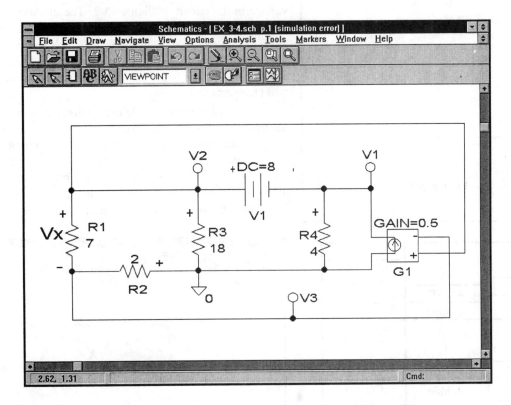

HINT: **G1** is a voltage-controlled current source. The current through **G1** is 0.5 times the voltage Vx. Note that Vx is the voltage at node V2 minus the voltage at node V3. It is not necessary to add the text Vx to your circuit.

SOLUTION: The results of the DC node voltage analysis can be viewed using the VIEWPOINT part or by examining the contents of the output file. The output file is shown below:

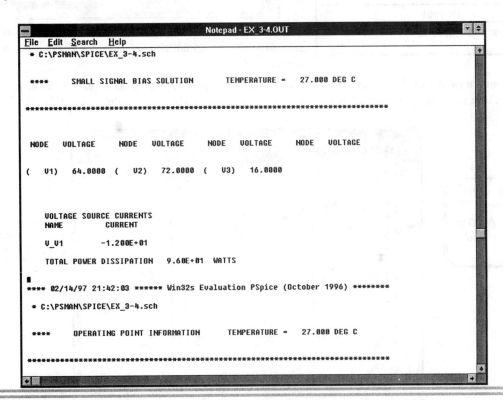

3.C. Diode DC Current and Voltage

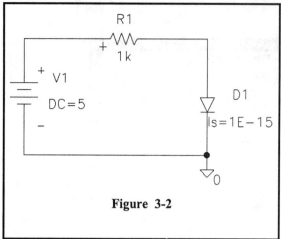

Figure 3-2

We will now use PSpice to find the diode current and voltage in the circuit of Figure 3-2. The diode current is given as $I_D=I_s[\exp(V_D/\eta V_T)-1]$. I_S is the diode saturation current and is 10^{-15} amps for this example. V_T is the thermal voltage and is equal to 25.8 mV at room temperature. η is the emission coefficient for the diode and its default value is 1. PSpice automatically runs all simulations at room temperature by default.

When you use a diode in a circuit, you will have to specify a model for the diode. In our case, the model will tell PSpice the value of I_S for our diode. The class libraries have a number of predefined models that are usually used in a classroom environment. However, the model for this diode is not in our libraries so we will have to define a new model for it. The part for the diode you should use is **Dbreak**. This diode is used to define your own model. Draw the circuit below:

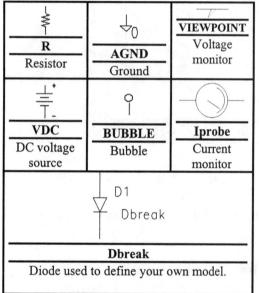

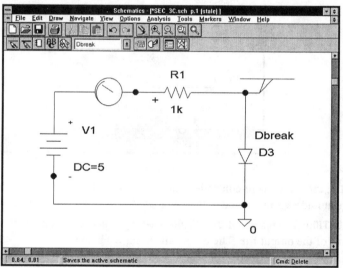

Before we can define a model for the diode, we must save the schematic. If you do not remember how to save the schematic, refer to Section 1.H.

We must now define the model for the diode. Click the **LEFT** mouse button on the diode graphic, ▷├. The graphic should turn red, indicating that it has been selected. Click the **LEFT** mouse button on **Edit** in the Schematics menu bar:

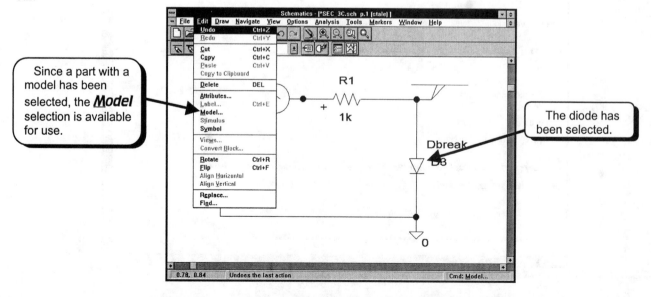

If you have selected a part with a model, the menu selection **Model** should be in a dark font as shown above. If a part with a model is not selected, the text **Model** would appear as Model. If the text **Model** is not in a dark font, return to the schematic and select the diode again by clicking the *LEFT* mouse button on it.

Click the *LEFT* mouse button on the **Model** menu selection:

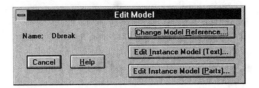

The button ***Change Model Reference*** will allow us to change the model from ***Dbreak*** to another model that is already defined in our libraries. The button ***Edit Instance Model (Parts)*** will allow us to create a model using the MicroSim Parts program. The button ***Edit Instance Model (Text)*** will allow us to edit a previously defined model with the intent of creating a new model. Click the *LEFT* mouse button on the ***Edit Instance Model (Text)*** button. When you create a new model for a schematic, you will create a new library file. When you create a new model for a part in a schematic, a library file will be created with the same name as the schematic but with the extension "lib." In the screen capture below, the name of the new file will be ***C:\PSMAN\SPICE\SEC_3C.lib***:

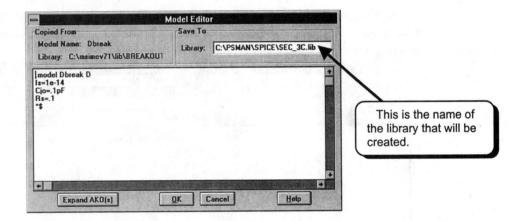

This screen shows us the present definition of model Dbreak. In this model, ***Rs*** is the series resistance of the diode and ***Cjo*** is the junction capacitance. The only parameter that we will specify for this example is the saturation current ***Is***. The window is a text editor and is used to modify the model. Change the text as follows:

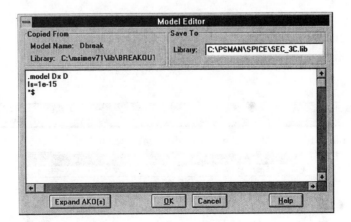

We have defined a new model called ***Dx*** with I_s specified as 10^{-15} amps. Click the ***OK*** button to accept the model. You will return to the schematic. Notice that the diode model name has changed from Dbreak to ***Dx***:

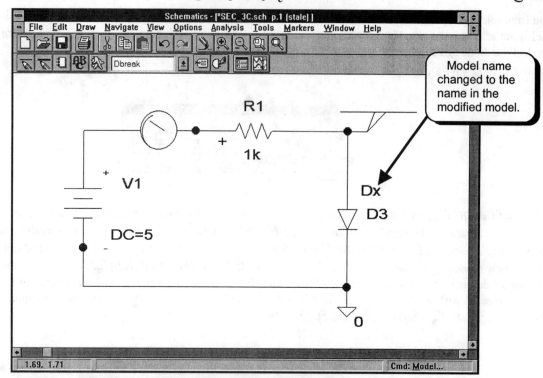

All we have to do now is run the simulation. Select **Analysis** from the menu bar:

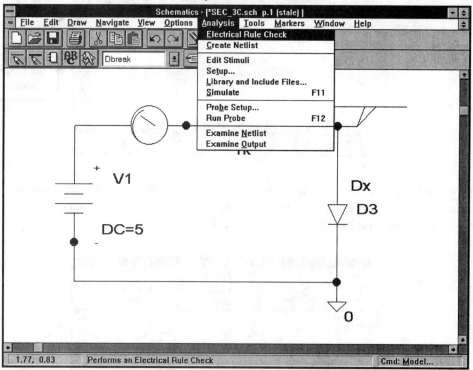

Select **Simulate**. The simulation will run and the PSpice window will appear:

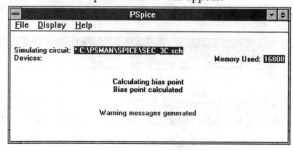

When the simulation is complete the PSpice monitor will remain on top:

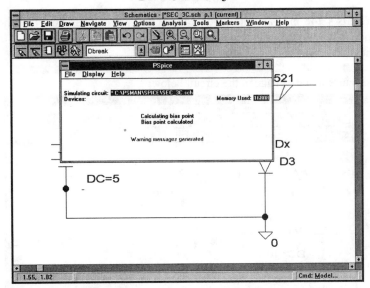

To view the results, we must bring the Schematics window to the front. Click the *LEFT* mouse button on the schematics window:

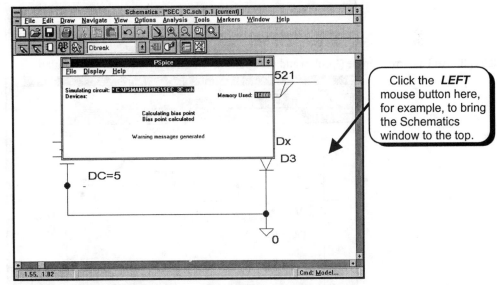

The Schematics window should come to the front and a value should appear next to the voltage and current monitors:

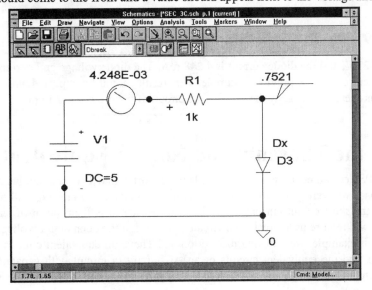

We see that the diode current is 4.248 mA, and the diode voltage is 0.7521 volts. Note that the viewpoint voltage monitor gives the voltage of the node to which it is connected relative to ground. Since the other side of the diode is grounded, the voltage displayed is also the diode voltage.

EXERCISE 3-5: Find the diode current and voltage in the circuit below:

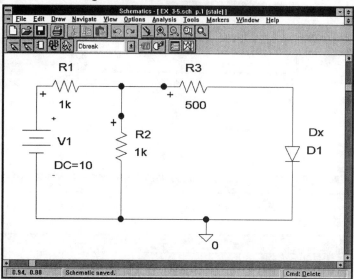

SOLUTION: Add a current probe (Iprobe) and viewpoint to the circuit. Then run the simulation:

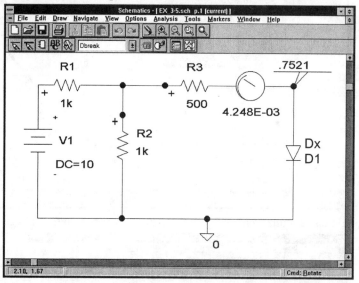

The diode voltage is 0.7521 V, and the diode current is 4.248 mA. The diode voltage and current of this circuit are the same as the diode voltage and current of the previous example. This result should be expected since the Thevenin equivalent of V1, R1, R2, and R3 in this exercise is exactly the same as the circuit of the previous example.

3.D. Finding the Thevenin and Norton Equivalents of a Circuit

Schematics and PSpice can be used to easily calculate the Norton and Thevenin equivalents of a circuit. The method we will use is the same as if we were going to find the equivalent circuits in the lab. We will make two measurements, the open circuit voltage and the short circuit current. The Thevenin resistance is then the open circuit voltage divided by the short circuit current. This will require us to create two circuits, one to find the open circuit voltage, and the second to find the short circuit current. In this example, we will find the Norton and Thevenin equivalent circuits for a DC circuit. This same procedure can be used to find the equivalent circuits of an AC circuit (a circuit with capacitors or inductors). However, instead of finding the open circuit voltage and short circuit current using the DC Nodal Analysis, we would need to use the AC analysis.

For this example, we will find the Thevenin and Norton equivalent circuits for the circuit attached to the diode in **EXERCISE 3-5**. The circuit is repeated below:

Schematics

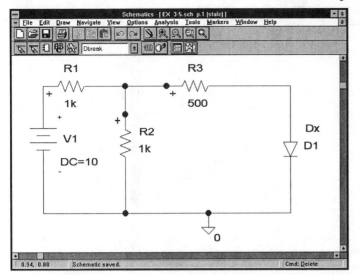

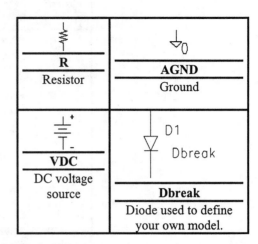

This circuit is difficult because it contains a nonlinear element (the diode) and a complex linear circuit. If we could replace V1, R1, R2, and R3 by a simpler circuit, the analysis of the nonlinear element would be much easier. To simplify the analysis of the diode, we will find the Thevenin and Norton equivalent circuits of the circuit connected to the diode; that is, we will find the Thevenin and Norton equivalents of the circuit below:

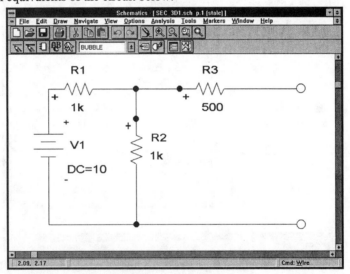

We will convert this circuit into the Thevenin equivalent :

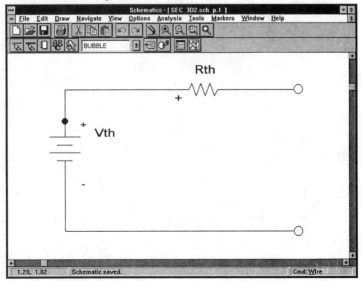

Once we find numerical values for Vth and Rth the entire circuit of **EXERCISE 3-5** reduces to:

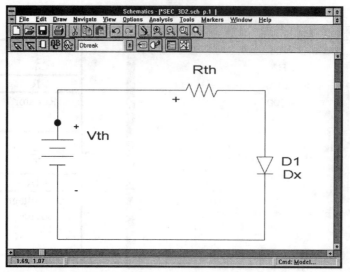

For determining the diode voltage and current, this circuit is much easier to work with than the original. This example is concerned with finding the numerical values of the equivalent circuit. The analysis of the circuit above was covered in Section 3.C. We will now find the Thevenin and Norton equivalent circuits of the circuit shown below:

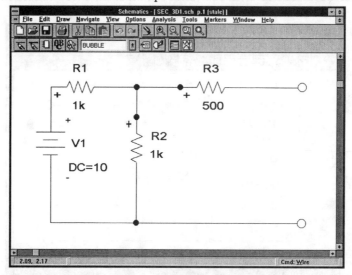

We will first find the open circuit voltage. This is just the voltage across the two terminals in the circuit shown above. First we must add a ground to the circuit. This is necessary because PSpice requires all circuits to have a ground reference:

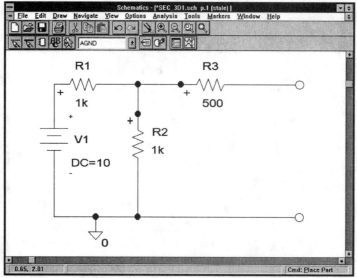

The lower terminal is now at ground potential, zero volts. We need to find the voltage of the upper terminal. To view the voltage on the screen, we will add a viewpoint part:

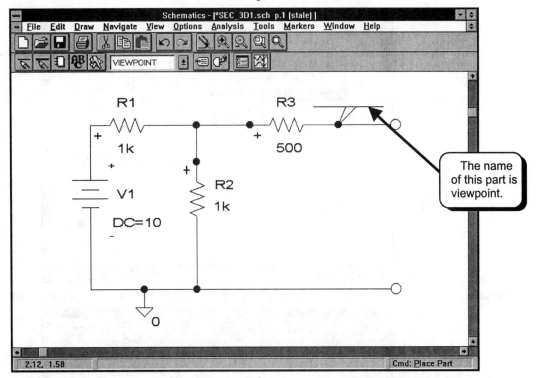

There are several errors in the circuit above. One is that there are two bubbles in the circuit and neither of them have labels. All bubbles must be labeled or errors will be generated and the simulation will not run. The second error is that there is only one element connected to the upper right node. That is, nothing is connected to the right terminal of the 500 Ω resistor. PSpice requires that all nodes have at least two elements connected to them. To fix this problem, we must add another element to the circuit that does not affect the operation of the circuit. To simulate an open circuit, I will add a resistor of value 100T. The suffix T in PSpice is a multiplier with a value of 10^{12}. Thus, a resistor with the value 100T will have a resistance of 100×10^{12} Ω. This value is significantly larger than all other resistors in the circuit and is an open circuit for all practical purposes. Thus, we will simulate the circuit below:

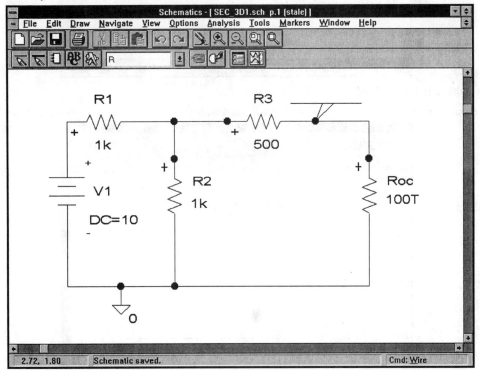

We can now simulate the circuit (select **Analysis** and then **Simulate**). PSpice will run and simulate the circuit:

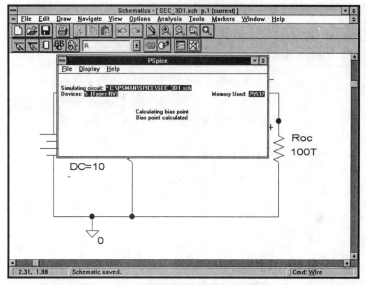

When the simulation is complete, bring the Schematics window to the top. The open circuit voltage will be displayed:

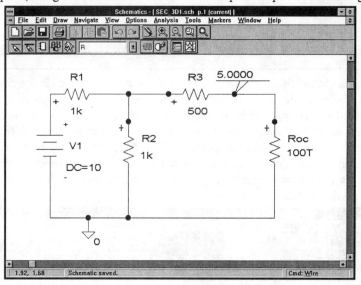

The open circuit voltage is **5.0000** volts.

Next we must find the short circuit current. We will start with the original circuit as shown below:

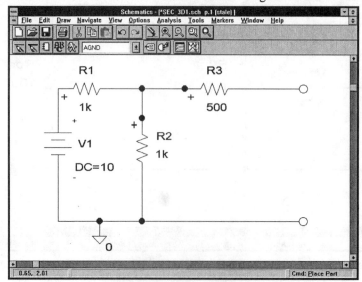

A current probe has no resistance. We can short the two terminals and measure the current at the same time by placing a current probe between the two terminals. The current probe part is named IProbe. Place the part as shown:

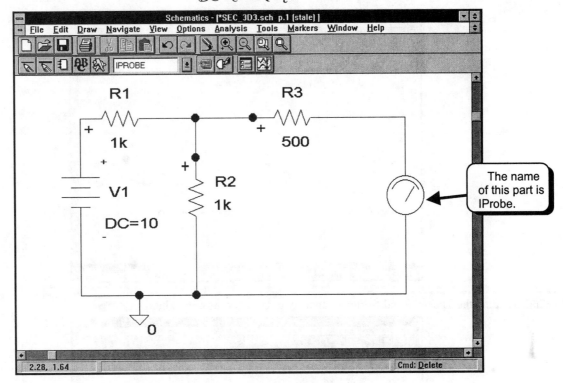

Simulate the circuit and display the Schematics window when the simulation is complete:

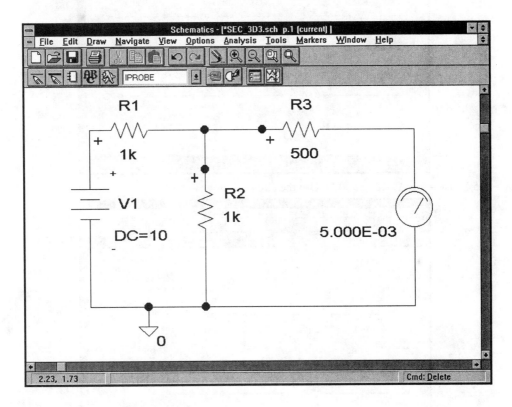

We see that the short circuit current is **5.000** mA.

We can now find the Thevenin resistance by dividing the open circuit voltage by the short circuit current:

$$R_{th} = \frac{Voc}{Isc} = \frac{5.000V}{5.00\,mA} = 1000\,\Omega$$

Our Thevenin and Norton equivalent circuits are shown below:

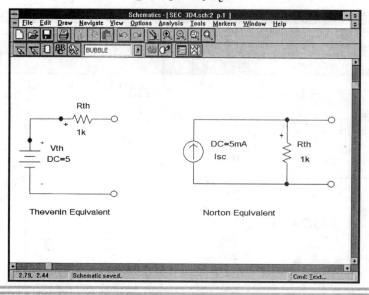

EXERCISE 3-6: Find the Thevenin and Norton equivalent circuits for the circuit below:

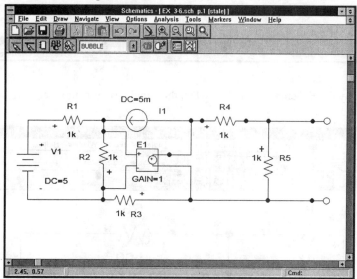

SOLUTION: Voc = 2.5 V, Isc = 5 mA, Rth = 500 Ω. Use the circuits below:

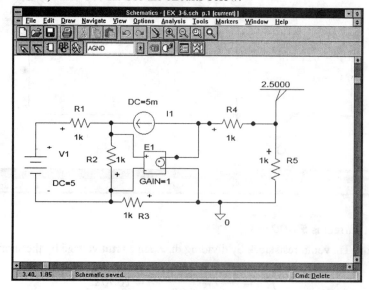

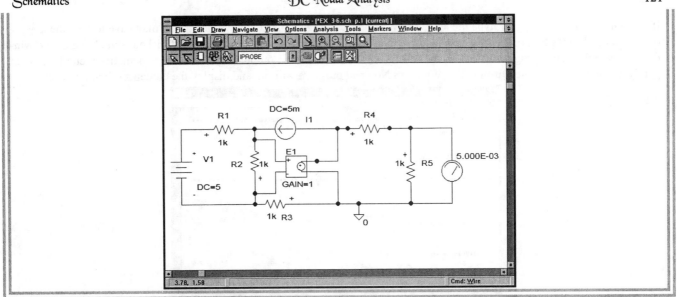

3.E. Transistor Bias Point Detail

One of the first things you should do when you are simulating an amplifier circuit is to check the transistor operating point. If the transistor bias is incorrect, none of the other analyses will be valid. If another analysis does not make sense, check the operating point. When PSpice finds the bias point, it assumes that all capacitors are open circuits and that all inductors are short circuits.

For a BJT, the Bias Point Detail gives the collector current, the collector-emitter voltage, and some small-signal parameters for the BJT at the bias point. For a jFET, the Bias Point Detail gives the drain current, the drain-source voltage, and some small-signal model parameters at the bias point. The results of the Bias Point Detail are contained in the output file. We will illustrate the Bias Point Detail analysis with the circuit below:

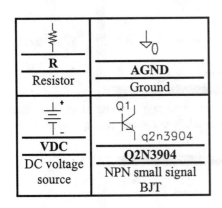

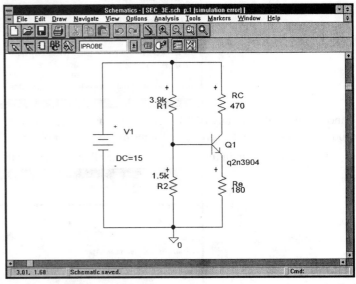

By default, the Bias Point Detail analysis is set to run automatically. We can now run the simulation. Select **Analysis** and then **Simulate** from the Schematics menu bar. PSpice will run:

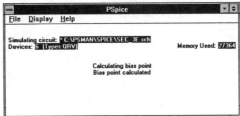

If this window does not appear, type the ALT-ESC key sequence to toggle the active window. You may have to type the key sequence several times before the PSpice window comes to the front. When the simulation is finished, switch the active window to the Schematics program. To view the results we must look at the output file. From the Schematics menu bar, select **Analysis** and then **Examine Output**. The Windows Notepad program will run and display the contents of the output file:

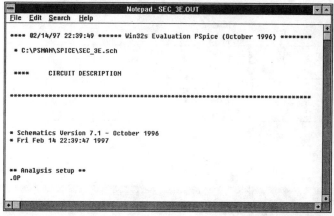

To see the results, click the *LEFT* mouse button on the down arrow, ⊞. You will first see the node voltages:

```
                                   Notepad - SEC_3E.OUT
 File  Edit  Search  Help

   * C:\PSMAN\SPICE\SEC_3E.sch

   ****     SMALL SIGNAL BIAS SOLUTION       TEMPERATURE =   27.000 DEG C


   ***************************************************************************


   NODE   VOLTAGE     NODE   VOLTAGE     NODE   VOLTAGE     NODE   VOLTAGE

  ($N_0001)   4.0445            ($N_0002)   6.4423

  ($N_0003)   3.2977            ($N_0004)  15.0000


     VOLTAGE SOURCE CURRENTS
     NAME          CURRENT

     V_V1         -2.102E-02

     TOTAL POWER DISSIPATION   3.15E-01   WATTS

 ****  02/14/97 22:39:49  ****** Win32s Evaluation PSpice (October 1996) ********

   * C:\PSMAN\SPICE\SEC_3E.sch

   ****     OPERATING POINT INFORMATION    TEMPERATURE =   27.000 DEG C
```

Since we did not place any BUBBLEs or name any wires, the node names are a bit cryptic to us. If we were interested in the node voltages, we could have named some of the wires or placed a few BUBBLEs at the nodes in question.

Further down in the file we see the Bias Point Detail results:

```
                                   Notepad - SEC_3E.OUT
 File  Edit  Search  Help


   ***************************************************************************



   ****  BIPOLAR JUNCTION TRANSISTORS

   NAME        Q_Q1
   MODEL       Q2N3904
   IB          1.13E-04
   IC          1.82E-02
   VBE         7.47E-01
   VBC        -2.40E+00
   VCE         3.14E+00
   BETADC      1.61E+02
   GM          5.82E-01
   RPI         2.57E+02
   RX          1.00E+01
   RO          4.20E+03
   CBE         1.84E-10
   CBC         2.34E-12
   CJS         0.00E+00
   BETAAC      1.50E+02
   CBX         0.00E+00
   FT          4.96E+08


      JOB CONCLUDED

      TOTAL JOB TIME       .77
```

These results show several parameters for the BJT. Had there been more than one BJT in the circuit, the operating point information would be displayed for all BJTs. To close the Notepad program select **File** and then **Exit** from the Notepad menu bar.

EXERCISE 3-7: Find the bias point for the transistor circuit:

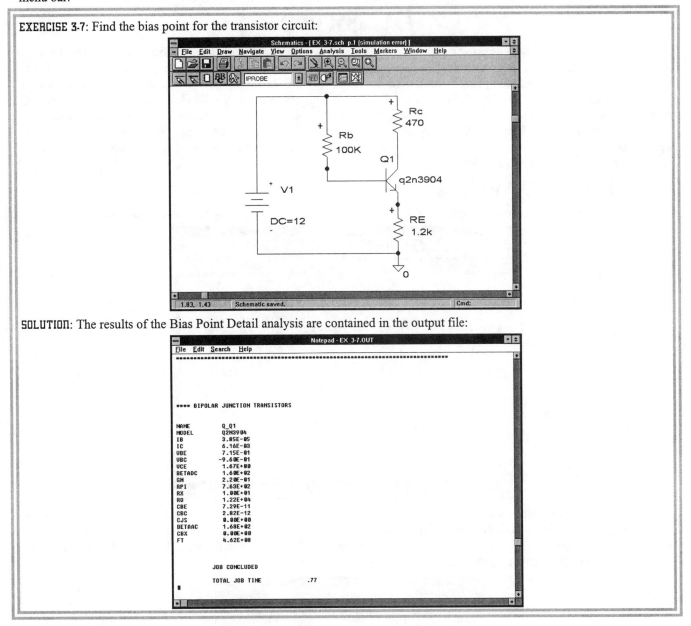

SOLUTION: The results of the Bias Point Detail analysis are contained in the output file:

3.F. Summary

- The DC Nodal Analysis finds the DC voltage at every node in the circuit. The voltages are relative to ground.
- The results are given in the output file. Use **Examine Output** from the **Analysis** menu to view the results.
- All capacitors are replaced by open circuits.
- All inductors are replaced by short circuits.
- All AC and time-varying sources are set to zero (IAC, VAC, Vsin, Isin, Vpulse, etc.).
- Use the part called "BUBBLE" to label the nodes of interest. If you do not label nodes with BUBBLEs, Schematics will label the nodes for you and you will not know which node is which.
- The IPROBE and VIEWPOINT parts can be used to view the results of the Nodal Analysis on the schematic.

PART 4
DC Sweep

The DC Sweep can be used to find all DC voltages and currents of a circuit. The DC Sweep is similar to the node voltage analysis, but adds more flexibility. The added flexibility is the ability to allow DC sources to change voltages or currents. For example, the circuit on page 102 will give us results only for the single value of Vx=15 V if the node voltage analysis is used. For each different value of Vx we are interested in, we must run the simulation again. If we use the DC Sweep, we can simulate the circuit for several different values of Vx in the same simulation. How node voltages vary for changing source voltages or how a BJT's bias collector current changes for different DC supply voltages would be example applications of the DC Sweep. As in the node voltage analysis, all capacitors are assumed to be open circuits, and all inductors are assumed to be short circuits.

4.A. Basic DC Analysis

We will first start with a modification of the circuit discussed in Part 1. Using Schematics, create the schematic:

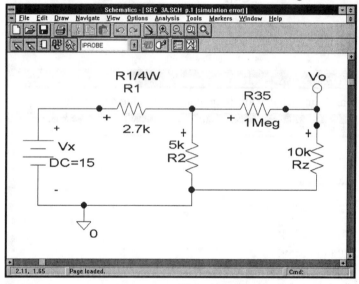

The question we will ask is: How does the voltage at **Vo** vary as **Vx** is raised from 0 to 25 volts? We will also view some of the currents through the components. Since this is a DC Sweep, all capacitors are assumed to be open circuits, and all inductors are assumed to be short circuits. We will now set up the DC Sweep. From the menu bar select **Analysis** and then **Setup**. The **Analysis Setup** dialog box will appear:

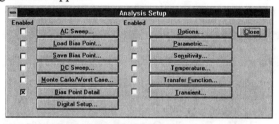

Click the **LEFT** mouse button on the **DC Sweep** button. The dialog box below will appear:

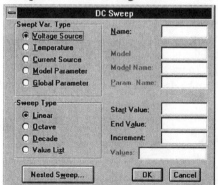

For this example we will demonstrate a linear sweep. The voltage source **Vx** will be swept from 0 volts to 25 volts in 1 volt increments. A linear sweep means that points between the beginning and ending values are equally spaced. Fill out the dialog box as shown below:

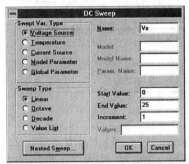

When you have set all the parameters as shown above, click the *LEFT* mouse button on the *OK* button:

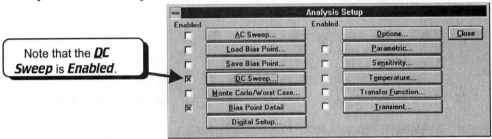

Note that the *DC Sweep* is *Enabled*.

Click the *Close* button.

We are now ready to run PSpice. Select **Analysis** from the menu bar and then select **Simulate**. Schematics will first create an updated netlist and then run PSpice. At this point one of two things will happen.

1. The PSpice simulation window will pop up:

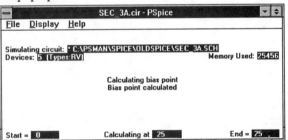

This screen tells you the progress of the simulation.

2. If a simulation has already been run once and you reduced the PSpice window to an icon, the PSpice window will not open. Instead, the Schematics window will remain in front. This screen tells you the progress of the simulation:

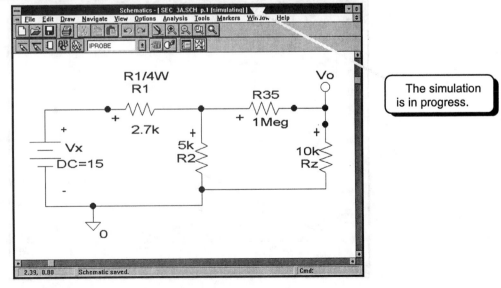

The simulation is in progress.

Notice that at the top of the screen Schematics says that it is *(simulating)*. This indicates that a simulation is in progress. When *(simulating)* changes to *(current)*, the simulation is complete. If you wish to see the PSpice simulation window, you can find the window by holding down the ALT key and pressing the TAB key. Remember to keep holding down the ALT key. Each time you type the TAB key, an icon will show on the screen:

Continue pressing the TAB key until you see the PSpice icon:

Release the ALT key. The PSpice window will open:

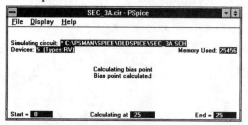

        When PSpice has finished simulating the circuit, the Probe program will run automatically. The first time Probe runs, the window below will appear:

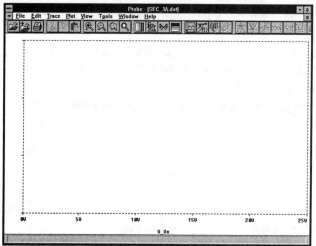

If you do not see this window, toggle the windows by pressing the ALT-ESC keys until this window is displayed. Press the INSERT key to add a trace. Add the trace V(Vo).

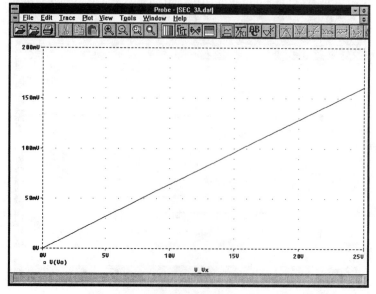

You can add as many traces to this graph as you wish. We will now add a second trace to show the capabilities of Probe. Select **Trace** and then **Add** to get the ***Add Traces*** dialog box. Fill in the dialog box as shown below:

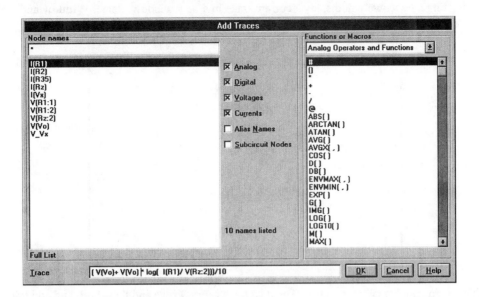

The trace command was entered by a combination of typing operators and parentheses in the text field and clicking the **LEFT** mouse button on the desired voltage.* Although the displayed waveform may not have much meaning to us, it does show what can be displayed with Probe. Click the **OK** button to display the trace:

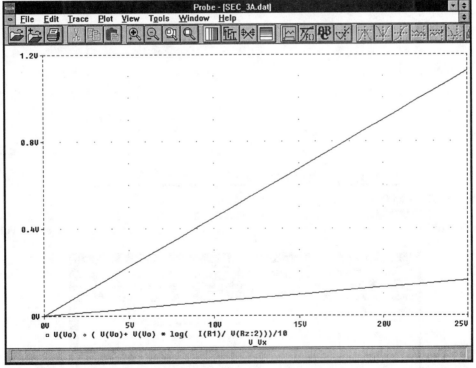

Currents through any device can also be displayed. Be careful, since the magnitude of currents is usually much smaller than that of voltages. Typically, currents are in milliamperes and voltages are in volts or tens of volts. If voltage and current traces are shown on the same screen, you will probably not see the current trace due to its small magnitude relative to that of the voltage. It is usually better to display currents on a different plot, or to delete some of the traces. To select a trace for deletion, click the **LEFT** mouse button on the name or expression of the trace you wish to delete. When a trace is

---

*The box next to ***Trace*** is a text field. Select the text field by clicking the **LEFT** mouse button at the desired point in the text field. A vertical cursor will appear.

selected, its expression or name should turn red, indicating that it has been selected. Delete the selected trace by selecting **Edit** and then **Delete** from the Probe menu bar, or by pressing the **DELETE** key. The trace should disappear from the screen.

Instead of deleting a trace, we will display a current trace in a new window. Select **Window** and then **New** from the Probe menus. A new empty window will open:

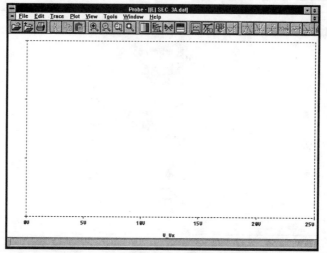

We can now use this window to display more traces. To add a trace select **Trace** and then **Add**, and add trace I(R1):

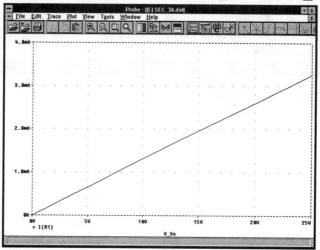

You can open several windows and add as many traces as you want.

EXERCISE 4-1: Find the voltages V1, V2, and V3 if the source voltage Vs is swept from a DC voltage of 6 volts to a DC voltage of 36 volts.

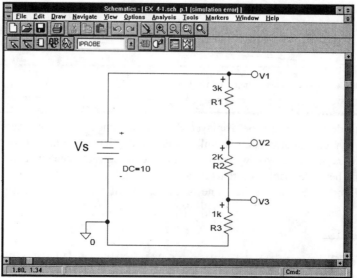

SOLUTION: The results of the DC Sweep are shown graphically using Probe :

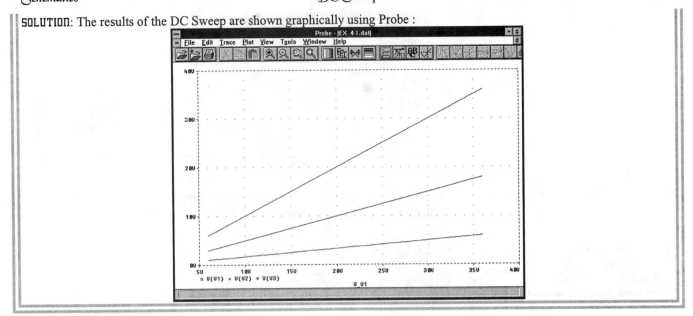

4.B. Diode I-V Characteristic

We would now like to use PSpice to obtain the I-V characteristic of a semiconductor diode. Wire the circuit shown below:

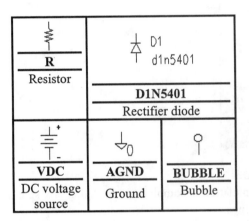

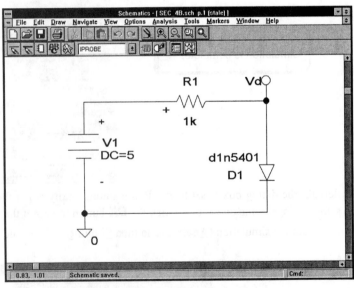

A DC Sweep will be run to sweep **V1** from −15 volts to +15 volts. To obtain the DC Sweep dialog box select **Analysis** and **Setup** from the Schematics menus:

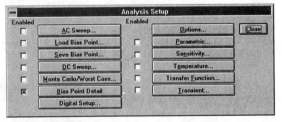

Click the **LEFT** mouse button on the **DC Sweep** button. Fill out the dialog box as shown on the following screen capture:

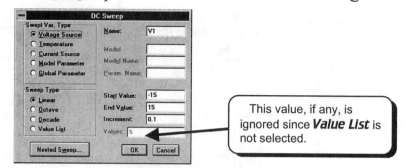

This dialog box is set up to sweep the voltage source named **V1** from −15 V to +15 V in 0.1 V increments. The sweep type is linear, which means that the voltage points are equally spaced. Click the **OK** button to accept the settings:

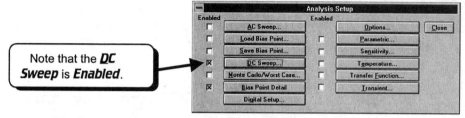

Click the **Close** button to return to the schematic.

For this simulation we would like to view the results using Probe. We would like Probe to automatically run when the simulation is complete. To check the Auto-run feature, select **Analysis** and then **Probe Setup**:

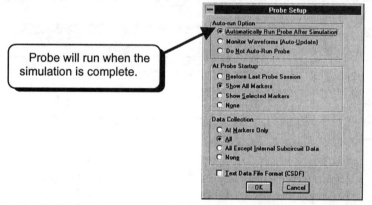

By default, the dialog box is set to run Probe automatically when the simulation is complete. Your dialog box should have the same settings as the one above. Click the **OK** button to accept the settings.

Run the simulation (**Analysis** and then **Simulate**). When the simulation is complete, Probe should run:

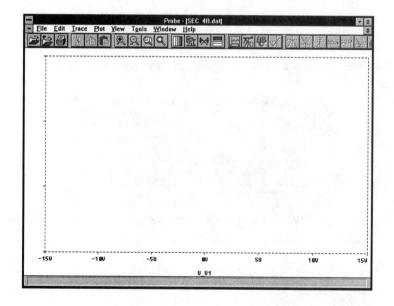

We now need to add a trace displaying the diode current. Select **Trace** and then **Add** from the Probe menus and add the trace I(D1):

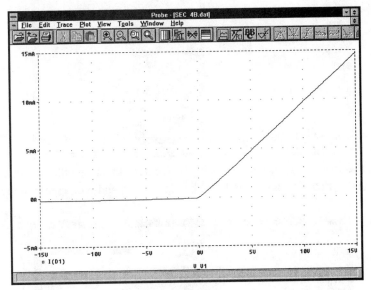

This is a trace of the diode current versus the voltage **V1**. We now need to change the x-axis from the voltage **V1** to the diode voltage. Click the *LEFT* mouse button on **Plot** in the Probe menu bar. The **Plot** pull-down menu will appear:

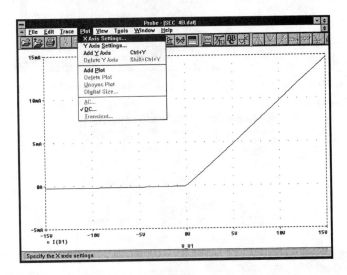

Click the *LEFT* mouse button on **X** Axis Settings. The **X Axis Settings** dialog box below will appear:

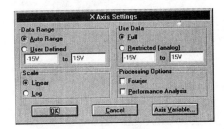

Click the *LEFT* mouse button on the **Axis Variable** button:

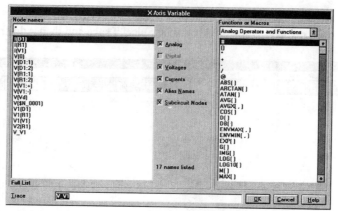

We see that the present **Trace** is the voltage of our source **V_V1**. We need to change the axis to the diode voltage. Click the **LEFT** mouse button on the text **V(Vd)**. The text **V(Vd)** will be highlighted and will appear in the text field next to the text **Trace**:

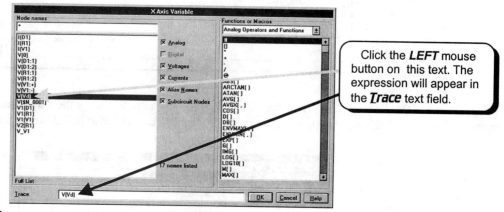

Click the **LEFT** mouse button on this text. The expression will appear in the **Trace** text field.

This is the voltage at node **Vd** relative to ground. Since one side of the diode is grounded, **Vd** is also the diode voltage. Click the **OK** button to accept the changes.

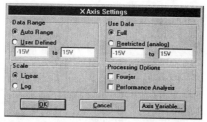

Click the **OK** button to display the I-V characteristic:

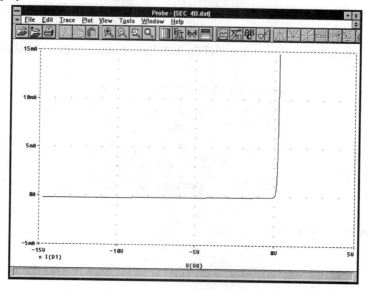

EXERCISE 4-2: Display the I-V characteristic of the Zener diode in the circuit below:

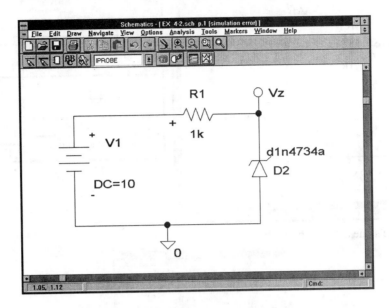

SOLUTION: Use the DC Sweep to plot the current through the resistor versus the Zener voltage, **Vz**. Note that the plus sign on **R1** indicates the positive voltage reference for the resistor.

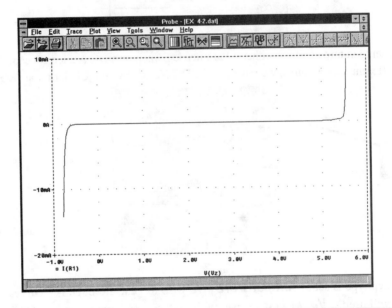

4.B.1. Temperature Sweep — Diode I-V Characteristic

Most of the devices used by PSpice can include temperature effects in the model. Most of the semiconductor models provided by MicroSim include temperature dependence. By default, the passive devices such as resistors, capacitors, and inductors do not include temperature dependence. To make these items include temperature effects, you will need to create models that include temperature effects. The temperature dependence of resistors in discussed in Section 4.G.1. In this section, we will show only how the I-V characteristic of a 1n5401 diode is affected by temperature. The D1n5401 diode model already includes temperature effects so we will not need to modify the model. We will use the standard resistor, which does not include temperature effects. We will continue with the circuit of Section 4.B:

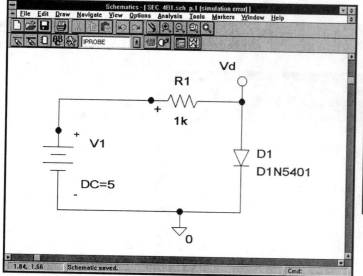

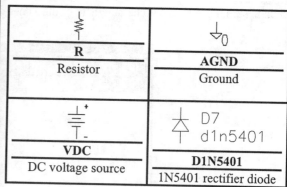

The DC Sweep is set up the same as in Section 4.B:

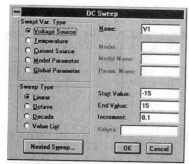

In general, we can run the temperature sweep as either a Parametric Sweep or a DC Nested Sweep. However, to generate I-V characteristic curves for different temperatures, we must use the DC Nested Sweep. Click on the **Nested Sweep** button:

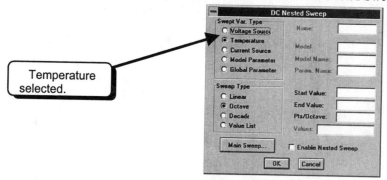

We see that the default sweep type is set to **Temperature**. We wish to generate three I-V curves, one at −25 °C, one at 0 °C, and one at 50 °C. We will use the **Value List** to specify these three temperature values. Click on the circle ○ next to the text **Value List** to select the option:

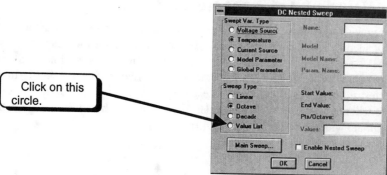

The circle ○ will fill with a dot ◉, indicating that the option is selected:

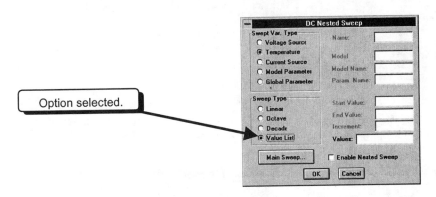

In the **Values** field, place the values **-25, 0, 50**

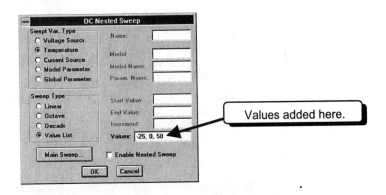

Last, we need to enable the Nested Sweep. Click the **LEFT** mouse button on the square ☐ next to the text **Enable Nested Sweep**:

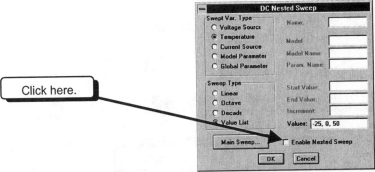

The square ☐ should fill with an x ☒, indicating that the option is selected:

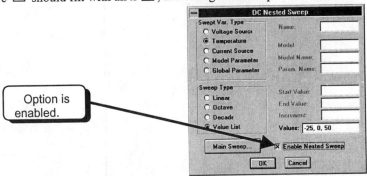

Click the **OK** button to accept the settings:

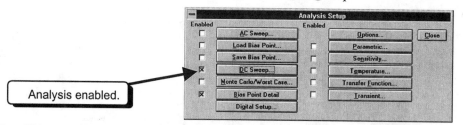

Analysis enabled.

We will be running both the DC Sweep and the DC Nested Sweep. Logically the DC Sweep executes inside the DC Nested Sweep. That is, for our settings, the temperature is set to –25 °C and V1 is swept from –15 to 15. Then the temperature is set to 0 °C and V1 is swept from –15 to 15. Last, the temperature is set to 50 °C and V1 is swept from –15 to 15.

Click the **Close** button to return to the schematic:

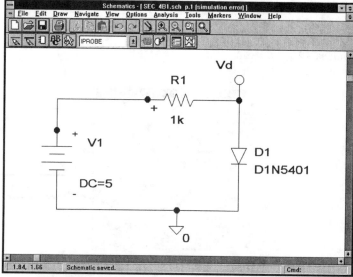

Simulate the circuit (**Analysis** and then **Simulate**):

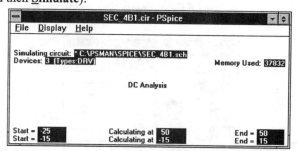

When the simulation is complete, Probe will run.

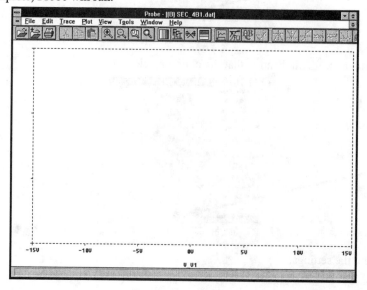

Follow the instructions on pages 130–132 to display the I-V characteristic of the diode for the three temperatures:

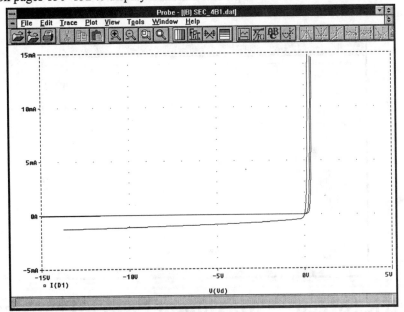

We can zoom in on the trace to see more detail. See page 67 for instructions on zooming in on a trace.

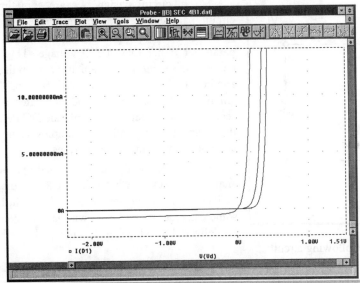

EXERCISE 4-3: Display the I-V characteristic of the Zener diode in the circuit below for temperatures −25°C, 0°C, and 50 °C.

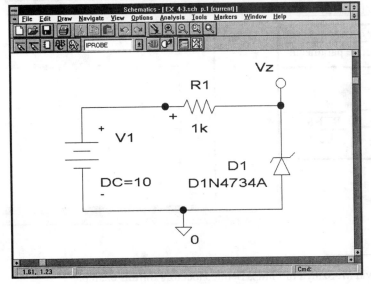

SOLUTION: Use the DC Sweep and DC Nested Sweep to plot the current through the resistor versus the Zener voltage, **Vz**. Note that the plus sign on **R1** indicates the positive voltage reference for the resistor.

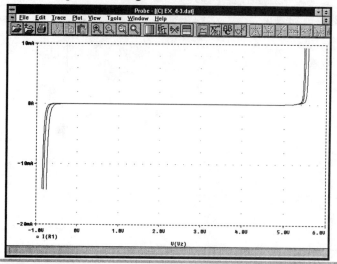

4.C. Parametric Sweep — Maximum Power Transfer

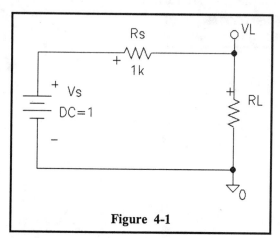

Figure 4-1

result using PSpice. Wire the following circuit:

In circuit design, we are sometimes concerned with how a circuit parameter affects performance. There are two ways to vary parameters in PSpice. The first is the DC Sweep, where we vary a parameter rather than a DC voltage. This method generates a single curve. The second is a Parametric Sweep that is run in conjunction with another analysis such as an AC Sweep, DC Sweep, or a Transient Analysis. The second method generates a family of curves. In this section we will demonstrate only the DC Parametric Sweep. Throughout this manual there will be examples using the Parametric Sweep in conjunction with the other analyses.

A frequently demonstrated problem in beginning circuit analysis courses is, what value of RL in the circuit of Figure 4-1 will deliver maximum power to RL?* With a little bit of circuit analysis and some calculus, it can be shown that for fixed Rs, maximum power will be delivered to RL when RL is equal to Rs. We will demonstrate this

	PARAMETERS:	
R Resistor	**PARAM** Part to define parameters	
VDC DC voltage source	**AGND** Ground	**BUBBLE** Bubble

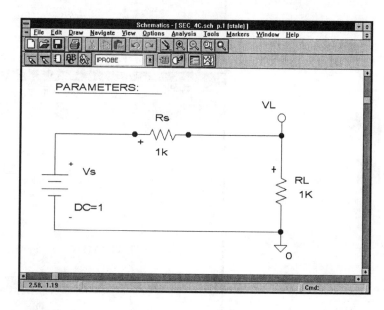

*Thanks to Dr. David M. Szmyd of Philips Semiconductors for this example.

We would like to vary the value of **RL**. To do this we need to define the value of **RL** as a parameter. Double click on the text **1K** for resistor **RL**:

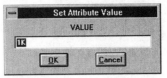

Type the text {**RL_val**}:

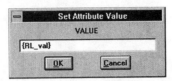

Click the **OK** button to accept the value:

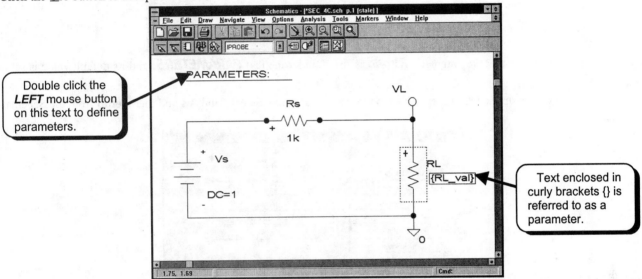

The value of **RL** is now the value of the parameter **{RL_val}**. We must now define the parameter and assign it a default value. Double click the **LEFT** mouse button on the text **PARAMETERS**:

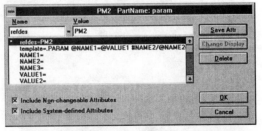

This attribute box is used to define parameters. We can define up to three parameters with this part. We will define **RL_val** as a parameter and give it a default value of 1k. Fill in the attributes as shown:

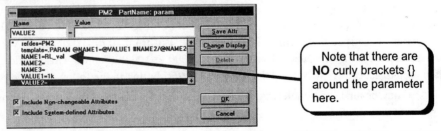

For every parameter defined (**NAME1**, **NAME2**, **NAME3**), a value must also be assigned (**VALUE1**, **VALUE2**, **VALUE3**). Click the **OK** button to accept the changes:

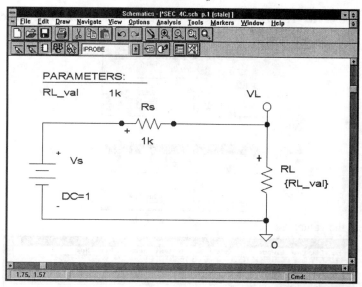

In the schematic we notice that the parameter **RL_val** is listed under the text ***PARAMETERS*:** and its default value is set to
1k.

We will now set up the DC Sweep to sweep the parameter value. Select **Analysis** and then **Setup** from the Schematics menus:

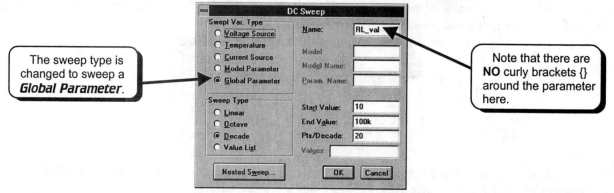

Click the ***LEFT*** mouse button on the ***DC Sweep*** button and fill in the dialog box as shown:

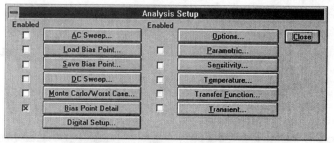

We are sweeping the global parameter ***RL_val***. The sweep will be in decades. The starting value is 10 Ω and the ending value is 100 kΩ. A decade is a factor of 10 and would be 1 to 10, 10 to 100, 100 to 1000, and so on. The sweep is set up for 20 points per decade so there will be 20 values of ***RL_val*** from 10 to 100 ohms, 20 values from 100 to 1000 ohms, 20 values from 1000 to 10,000 ohms, and 20 values from 10,000 to 100,000 ohms. This should give us a fairly detailed plot. Click the ***OK*** button to accept the sweep and then click the ***Close*** button in the following dialog box to return to the schematic.

Run the analysis by selecting **Analysis** and then **Simulate**. If the simulation runs successfully, Probe should run:

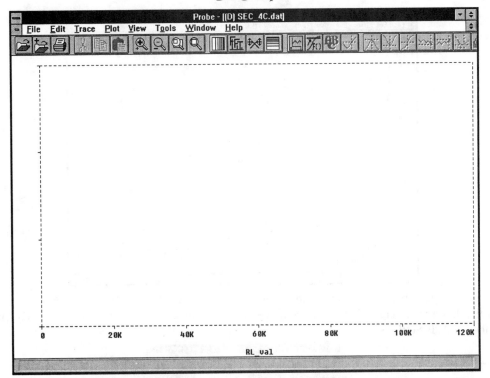

Notice that the x-axis is automatically set to **RL_val**. We would like to plot the power absorbed by **RL** versus the value of **RL**, **RL_val**. Select **Trace** and then **Add** to add a trace:

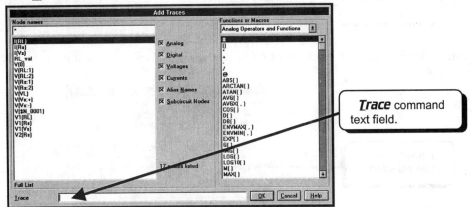

Trace command text field.

The power absorbed by **RL** can be expressed many ways, but I will use V^2/R. Type the text **V(VL)*V(VL)/RL_val** in the **Trace** text field:

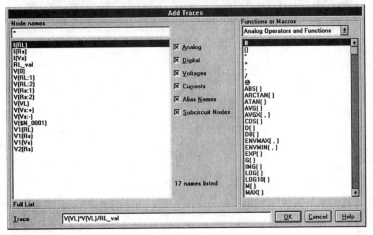

Click the **OK** button to plot the trace:

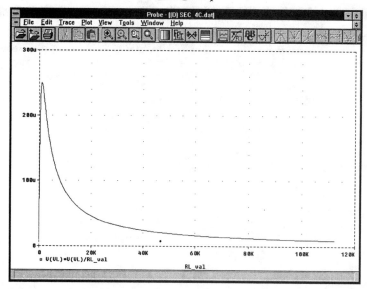

This is not an attractive graph because the maximum value should occur at RL_val = 1000 Ω. 1000 is a very small number on a linear axis that has a maximum value of 100,000. We need to change the x-axis to a log scale. To change the x-axis, select **P**lot and then **X** Axis Settings:

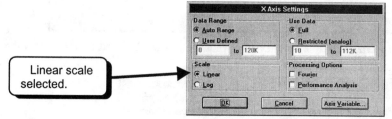

The current **Scale** is **Linear**. Click the *LEFT* mouse button on the circle ○ next to the text **Log**. The circle should fill with a black dot, ⦿:

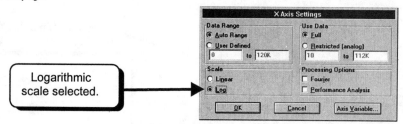

Click the **OK** button to accept the changes.

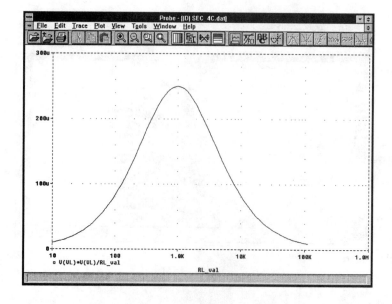

We see that the maximum power to the load occurs when the load is 1000 Ω, equal to the source resistance.

EXERCISE 4-4: Repeat the above example, but this time assume that the load is fixed and the source resistance is free to vary. With RL = 1 kΩ, find the value of Rs that delivers maximum power to RL.

SOLUTION: Set up the circuit the same as in the previous example, except let the value of Rs be the value of the parameter Rs_val:

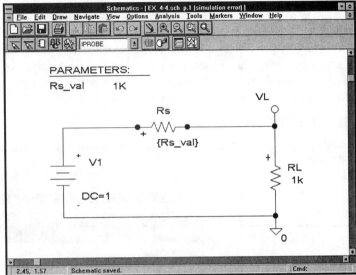

Set up the DC Sweep to sweep the global parameter Rs_val from 0.001 Ω to 10 kΩ:

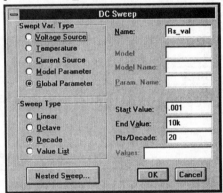

Simulate the circuit and then run Probe. Plot the trace V(VL)*V(VL)/1000 on a log scale.

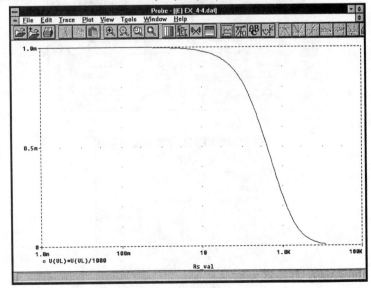

V(VL)*V(VL)/1000 is the power delivered to RL. We see that when Rs is free to vary, maximum power is delivered to the load when the source resistance is zero, or as close to zero as possible.

4.D. DC Transfer Curves

One of the more useful functions of the DC Sweep is to plot transfer curves. A transfer curve usually plots an input versus an output. A DC transfer curve plots an input versus an output, assuming all capacitors are open circuits and all inductors are short circuits. In a DC Sweep, all capacitors are replaced by open circuits and all inductors are replaced by short circuits. Thus the DC Sweep is ideal for DC transfer curves. The Transient Analysis can also be used for DC transfer curves, but you must run the analysis with low-frequency waveforms to eliminate the effects of capacitance and inductance. Usually a DC Sweep works better for a transfer curve. The one place where a transient analysis works better is plotting a hysteresis curve for a Schmitt Trigger. For a Schmitt Trigger, the input must go from positive to negative, and then from negative to positive to trace out the entire hysteresis loop. This is not possible with a DC Sweep.

4.D.1. Zener Clipping Circuit

The circuit below should clip positive voltages at the Zener breakdown voltage and negative voltages at the diode cut-in voltage, approximately 0.6 volts:

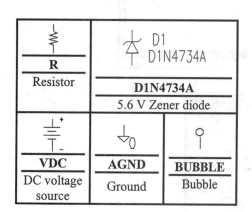

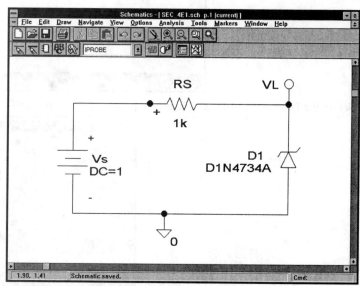

This circuit is also simulated in Section 6.F.1. We would like to sweep **Vs** from −15 volts to +15 volts and plot the output. Select **Analysis** and then **Setup** from the Schematics menus:

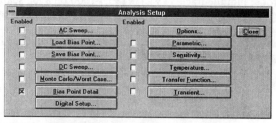

Click the **LEFT** mouse button on the **DC Sweep** button:

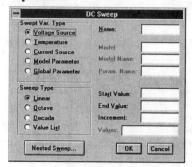

Fill in the dialog box as shown:

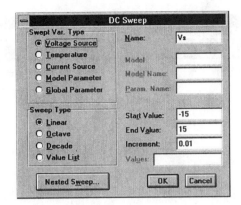

The dialog box is set to sweep **Vs** from **-15** volts to **+15** volts in **0.01** volt steps. Click the **OK** button:

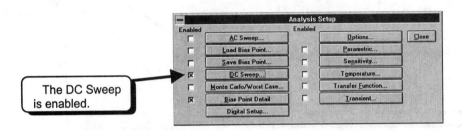

The DC Sweep is enabled.

Before returning to the schematic, make sure the box next to the **DC Sweep** button has an **X** in it, ⊠, as in the dialog box above. Click the **Close** button to return to the schematic. Run the analysis by selecting **Analysis** and then **Simulate**. When the simulation is complete, Probe should run automatically:

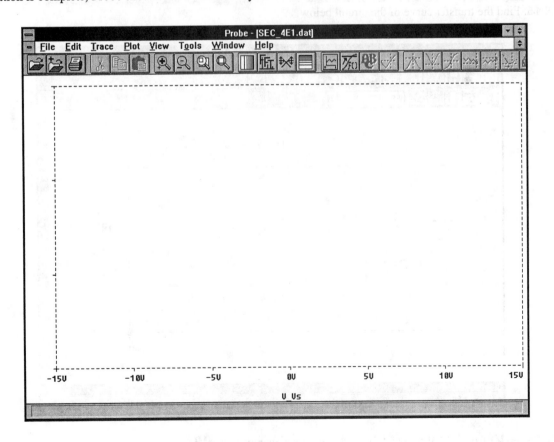

Add the trace **V (VL)** (select **Trace** and then **Add**):

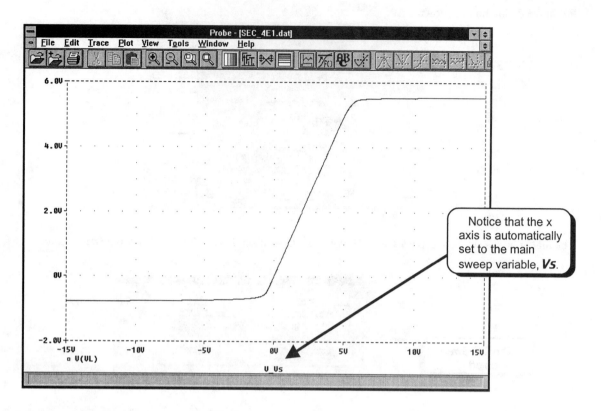

The screen capture above is the transfer curve for this circuit.

EXERCISE 4-5: Find the transfer curve of the circuit below:

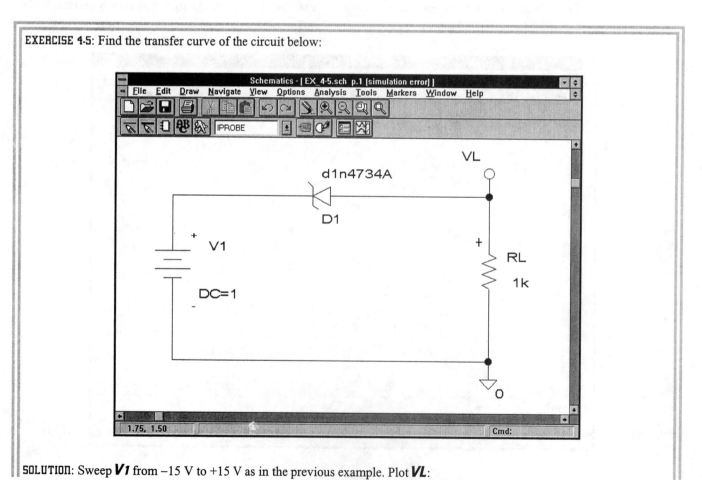

SOLUTION: Sweep *V1* from −15 V to +15 V as in the previous example. Plot *VL*:

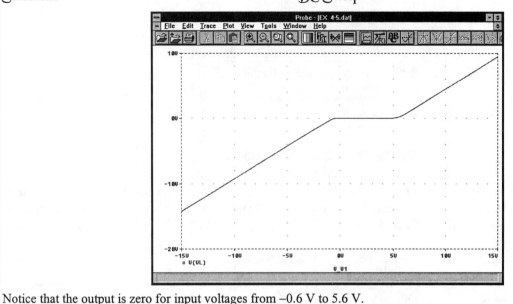

Notice that the output is zero for input voltages from −0.6 V to 5.6 V.

4.D.2. DC Nested Sweep — Family of Transfer Curves

In the previous section, the question may arise as to how the Zener breakdown voltage affects the transfer curve. We will assume that you have followed the procedure of the previous section and have already set up the DC Sweep.

We will now set up a Nested DC Sweep to sweep the breakdown voltage of the Zener diode. First we must look at the model for the Zener diode. Click the *LEFT* mouse button on the graphic symbol for the Zener, ⊸⊢ . It should turn red, indicating that it has been selected. Select **Edit** and then **Model** from the Schematics menus*:

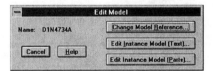

Click the *LEFT* mouse button on the **Edit Instance Model (Text)** button. This will allow us to edit the model. We will not change the model; we just want to look at it:

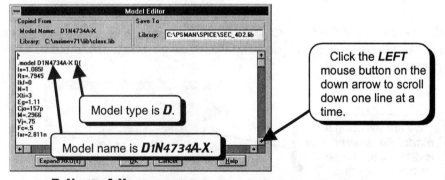

This screen shows that the model name is **D1N4734A-X**. The model we are editing is actually the D1N4734A model. However, Schematics is assuming that we are going to change the D1N4734A model so it automatically changes the name to distinguish it from the original model. The parameter for breakdown voltage is BV and cannot be seen on the screen. Click the down arrow 🔽 to see the rest of the model:

---

*If the menu selection **Model** appears grayed out (Model), you will not be able to choose the **Model** menu selection.

Attempt to select the Zener graphic again until its graphic, ⊸⊢ , is highlighted in red.

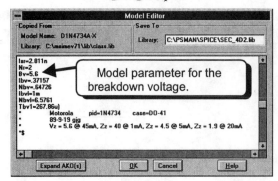

We see that the breakdown voltage is **5.6** volts. Click the **Cancel** button to return to the schematic. Since we canceled the operation, the model for the diode will be unchanged. Select **Analysis** and **Setup** from the Schematics menus, and then click the **DC Sweep** button to obtain the **DC Sweep** dialog box. The dialog box should have the parameters from the previous section:

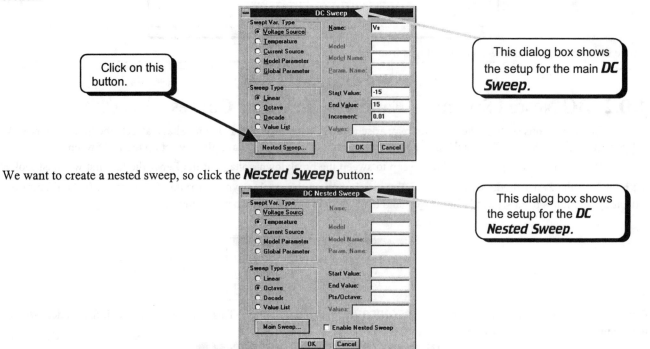

We want to create a nested sweep, so click the **Nested Sweep** button:

Notice that this box is titled **DC Nested Sweep**. The first DC Sweep dialog box is referred to as the **Main Sweep**. We would like to sweep the model parameter BV. Click the **LEFT** mouse button on the circle ○ next to the text **Model Parameter**:

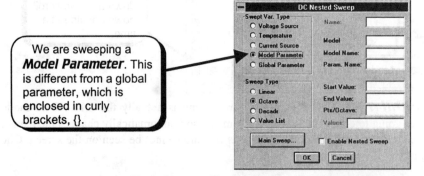

New text appears for the **Model Type**, **Model Name**, and **Param Name**. As in the model shown on page 147, the model type was a diode (D), the model name was D1N4734A, and the parameter we wish to vary is BV. Fill in the dialog box as shown:

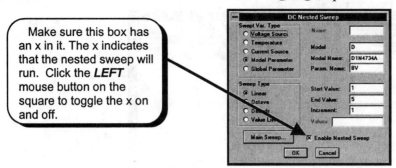

Make sure this box has an x in it. The x indicates that the nested sweep will run. Click the *LEFT* mouse button on the square to toggle the x on and off.

The box is set to sweep the model parameter *BV* from *1* volt to *5* volts in *1* volt steps. Notice that the box next to the text **Enable Nested Sweep** has an *X* in it, ⊠. The *X* means that the nested sweep is enabled. Click the *OK* button and then click the *Close* button to return to the schematic. Run PSpice (**A**nalysis and then **S**imulate):

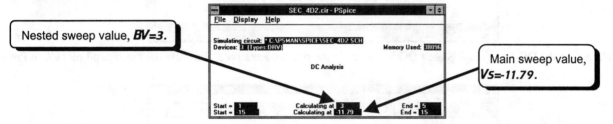

Nested sweep value, *BV=3.*

Main sweep value, *Vs=-11.79.*

Notice that two values are displayed in the PSpice window. The bottom line changes quickly and is the main sweep variable, Vs. The line second to the bottom is the nested sweep variable, BV. Notice that for each value of BV, Vs is swept from −15 to +15. Logically, the main sweep loop executes inside the nested sweep loop. That is, BV is set to 1 and Vs is swept from −15 to +15. Then BV is set to 2, and Vs is swept from −15 to +15. Then BV is set to 3, and Vs is swept from −15 to +15, and so on. When the analysis is finished, Probe will run. Add the trace **V(VL)** (select **T**race and then **A**dd):

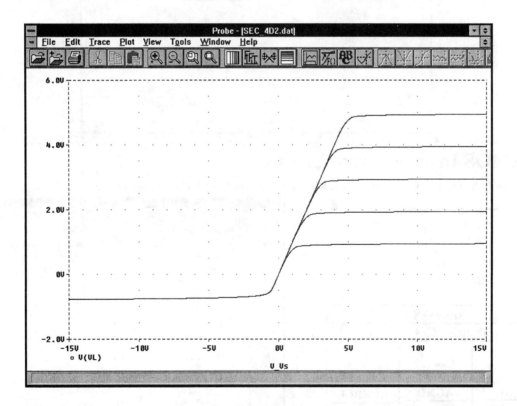

EXERCISE 4-6: For the circuit below, find the nested family of transfer curves if the breakdown voltage of the Zener is swept from 1 V to 6 V.

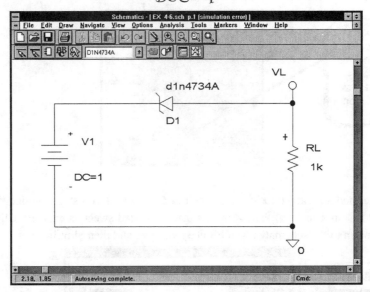

SOLUTION: Set up the DC Sweep and Nested DC Sweep as in the previous example. Simulate the circuit and plot **VL** in Probe :

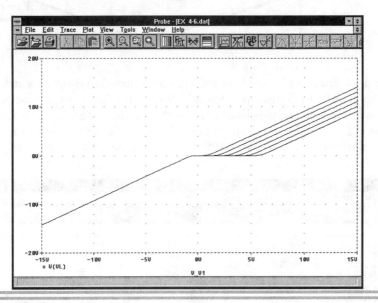

4.D.3. NMOS Inverter Transfer Curve

We would like to plot the transfer curve **Vo** versus **Vin** for the NMOS inverter below:

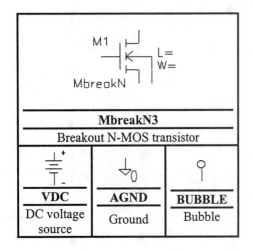

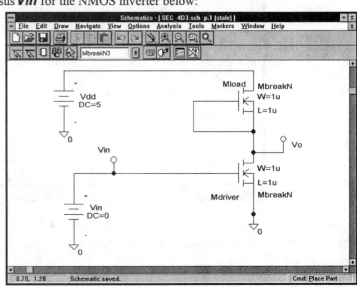

This circuit is an NMOS inverter with a depletion load. Many textbooks describe MOSFET operation by the following equations:

$$I_D = \left(\frac{W}{L}\right) K \left(V_{GS} - V_T\right)^2 \left(1 + \lambda V_{DS}\right) \qquad \text{saturation region}$$

$$I_D = \left(\frac{W}{L}\right) K \left[2\left(V_{GS} - V_T\right)V_{DS} - V_{DS}^2\right]\left(1 + \lambda V_{DS}\right) \qquad \text{linear region}$$

For our example, the load MOSFET is a depletion mode NMOS transistor. Its parameters are K = 20 μA/V², V_T = −1.5 V, and λ = 0.05 V⁻¹. The driver is an enhancement mode NMOS transistor. Its parameters are K = 20 μA/V², V_T = +1.5 V, and λ = 0.05 V⁻¹.

First, we need to create models to describe these transistors. When you look up the MOSFET models in the MicroSim PSpice reference manual on the CD-ROM that accompanies this text, you will find the following equations to describe MOSFET operation:

$$I_D = \left(\frac{W}{L}\right) \frac{Kp}{2} \left(V_{GS} - Vto\right)^2 \left(1 + \lambda V_{DS}\right) \qquad \text{saturation region}$$

$$I_D = \left(\frac{W}{L}\right) \frac{Kp}{2} \left[2\left(V_{GS} - Vto\right)V_{DS} - V_{DS}^2\right]\left(1 + \lambda V_{DS}\right) \qquad \text{linear region}$$

The conversion between the two sets of equations is shown in Table 4-1. The PSpice model parameter names are close to the standard names used to represent MOSFET operation in many textbooks. One difference is that the PSpice model parameter Kp is twice the value of K. Thus, in our model we should set K_P=40μA/V². All other model parameters will be the same.

Table 4-1	
PSpice Model Parameter	**Equation Variables**
Kp	2K
VTO	V_T
lambda	λ

We must now define the models for these two MOSFETs. Two separate models are required. Click the **LEFT** mouse button on the graphic symbol of the driver MOSFET, ⊥. It should turn red, indicating that it has been selected. Select **Edit** and then **Model***:

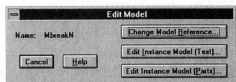

We would like to create a new model, so click the **LEFT** mouse button on the **Edit Instance Model (Text)** button:

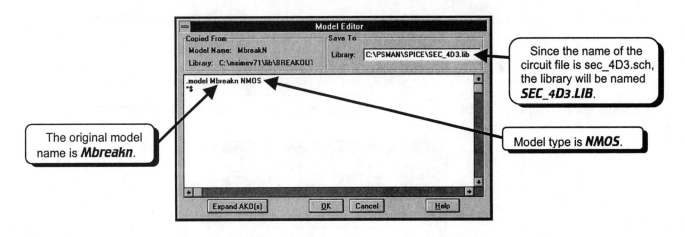

---

*If the menu selection **Model** appears grayed out (Model), you will not be able to choose the **Model** menu selection.

Attempt to select the MOSFET graphic again until its graphic, ⊥, is highlighted in red.

This is the default model for all n-channel MOSFETs (no model parameters specified). This window is a text editor. Change the model as shown:

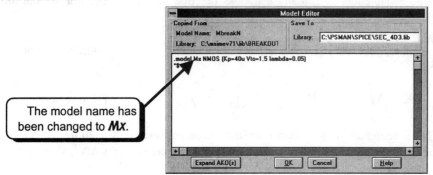

We are calling the new model **Mx**. Click the **OK** button to return to the schematic. Notice that the model for the driver MOSFET has changed from **MbreakN** to **Mx**:

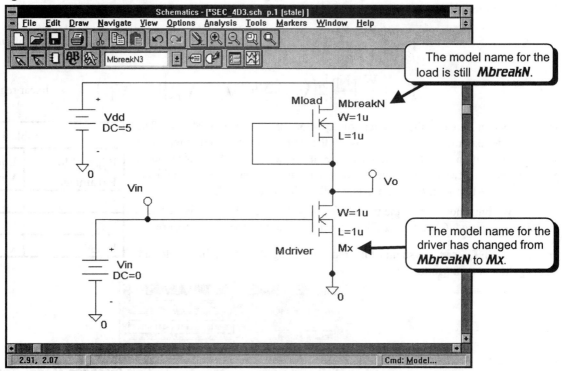

Next, we must change the model for the load MOSFET. Click the **LEFT** mouse button on the graphic symbol for the load MOSFET, ⊥⊏⊏. It should turn red, indicating that it has been selected. Select **Edit** and then **Model**:

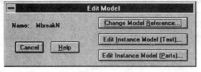

We would like to create a new model, so click the **LEFT** mouse button on the **Edit Instance Model (Text)** button:

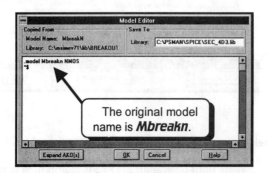

Change the model as shown:

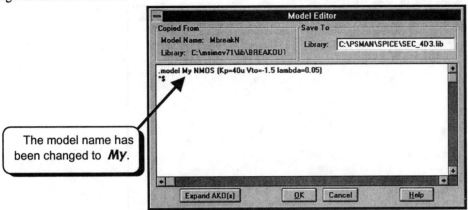

We are calling the new model **My**. Note that the symbols for enhancement and depletion mode MOSFETs are the same. PSpice knows that a MOSFET is either an enhancement or a depletion mode MOSFET by the value of the threshold voltage parameter, **Vto**. Click the **OK** button to return to the schematic. Notice that the model for the driver MOSFET has changed from MbreakN to **My**:

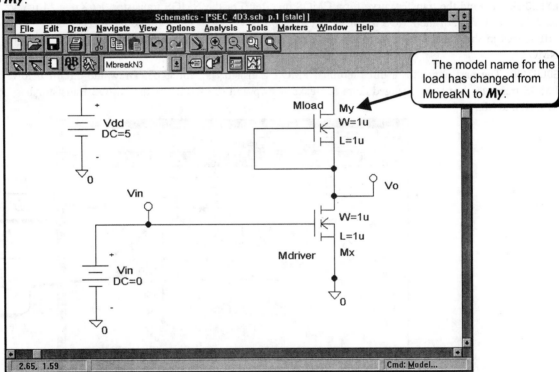

    We are now ready to set up the DC Sweep. Select **Analysis** and then **Setup**, and then click the **DC Sweep** button. We would like to sweep Vin from 0 volts to 5 volts. Fill in the dialog box as shown:

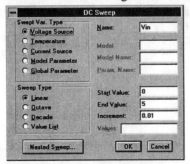

The dialog box is set to sweep **Vin** from 0 volts to 5 volts in 0.01 volt steps. A linear sweep means that points will be equally spaced. Click the **OK** button and then click the **Close** button. Run PSpice (**Analysis** and then **Simulate**). When the simulation is complete, Probe should run automatically. Add the trace **V(Vo)** (select **Trace** and then **Add**). The transfer curve is:

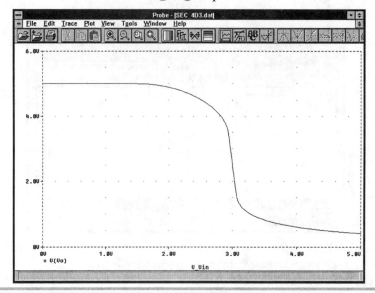

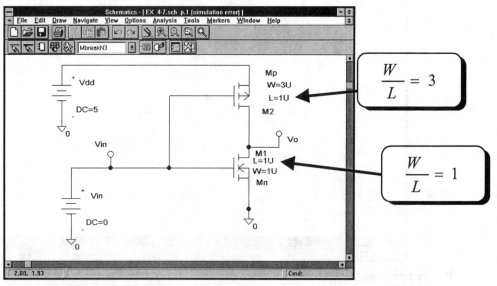

EXERCISE 4-7: Find the transfer curve for a CMOS inverter. For the NMOS transistor, let Kp = 24 μA/V², Vto = 1.5 V, and lambda = 0.01 V⁻¹. For the PMOS transistor, let Kp = 8 μA/V², Vto = −1.5 V, and lambda = 0.01 V⁻¹. To compensate for the differences in the transistors' transconductance (Kp), let the W/L ratio of the PMOS be three times the W/L ratio of the NMOS.

SOLUTION: The part name for a three-terminal PMOS transistor is Mbreakp3. Draw the circuit below. The model of the PMOS transistor has been changed from Mbreakp to Mp and the model of the NMOS transistor has been changed from Mbreakn to Mn:

Create the following models for the two transistors :

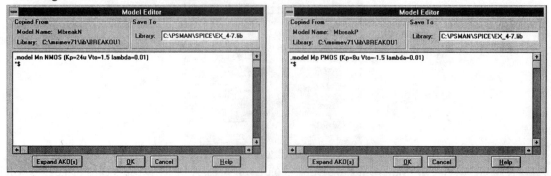

Run a DC Sweep to sweep **Vin** from 0 to 5 V in 0.001 V increments. A small increment is required for a CMOS inverter:

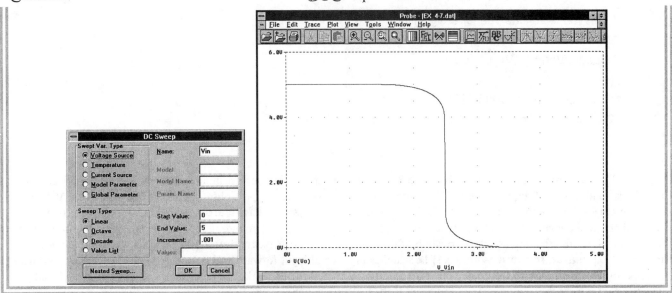

4.D.4. Goal Functions — Inverter Analysis

We will use the circuit of the previous section and find the noise margins (NM_H and NM_L) and input and output transition voltages (V_{IL}, V_{IH}, V_{OL}, V_{OH}). The transfer function of the NMOS inverter of the previous section was found on page 154, and is repeated below with approximate values of V_{IL}, V_{IH}, V_{OL}, V_{OH} shown on the plot:

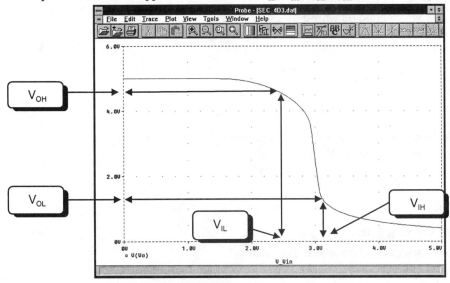

The definitions of V_{IL}, V_{IH}, V_{OL}, V_{OH}, NM_H, and NM_L are:

V_{IL} - The maximum value of the input voltage that is considered a logic zero (low input). For $0 \leq V_{IN} \leq V_{IL}$, the input is considered a logic zero.

V_{IH} - The minimum value of the input voltage that is considered a logic one (high input). For $V_{IH} \leq V_{IN} \leq V_{max}$, the input is considered a logic one.

V_{OL} - The maximum output voltage of the gate that is considered to be a logic zero (low output).

V_{OH} - The minimum output voltage of the gate that is considered to be a logic one (high output).

NM_H - High input noise margin. $NM_H = V_{OH} - V_{IH}$.

NM_L - Low input noise margin. $NM_L = V_{IL} - V_{OL}$.

How to locate V_{IL}, V_{IH}, V_{OL}, and V_{OH} for a particular gate is not always consistent and may vary for different types of logic. One definition of the location of the points V_{IL}, V_{IH}, V_{OL}, and V_{OH} is where the slope of the transfer curve is one, $dV_O/dV_{IN}=1$. We will use this definition in this example. The question is, how can we find these points on the graph? We can do this by using the goal functions available with Probe. The available goal functions are located in file c:\msimev71\msim.prb. If you edit this file using the Windows Notepad, you will see the text below near the bottom of the file:

    Voh(1,2)=y2

* This function finds Voh for a digital logic gate.

* The transition points are defined as where dVo/dVin = 1.

* Usage: Voh(d(V(Vo)), V(Vo))

{

1| search forward level (-1) !1;

2| search forward xvalue(x1) !2;

}

The name of the function is **Voh**. It has 2 input arguments **(1,2)**. The first input will be the derivative of the trace, and the second input will be the trace itself. To use this function we would type Voh(d(V(Vo)), V(Vo)), where d(V(Vo)) is the derivative of the output trace and V(Vo) is the output trace. **1|search forward level** means search the first input forward and find a level. The level we are looking for is **-1**. When the point is found, the text **!1** designates its coordinates as x1 and y1. Since the first input is the derivative of the trace, **1| search forward level (-1) !1** finds where the slope of the trace is –1. **2| search forward xvalue(x1) !2** searches the second input forward and finds the point when the x-coordinate is equal to **x1**. This point is where the slope of the transfer curve is –1. **!2** marks the coordinates of this point as x2 and y2. The function returns the y-coordinate of this point **Voh(1,2)=y2**, which is equal to V_{OH}.

A second function is:

Vil(1)=x1

* This function finds Vil for a digital logic inverter.

* The transition points are defined as where dVo/dVin = 1.

* Usage: Vil(d(V(Vo)))

{

1| search forward level (-1) !1;

}

This function is similar to Voh except that it returns the x-coordinate of the point where the slope is –1. The x-coordinate of the point where the slope is –1 is our definition of V_{IL}.

A third function is:

Vol(1,2)=y2

* This function finds Vol for a digital logic gate.

* The transition points are defined as where dVo/dVin = 1.

* Usage: Vol(d(V(Vo),V(Vo)))

{

1| search backward /End/ level (-1) !1;

2| search backward /End/ xvalue(x1) !2;

}

This function is used to find V_{OL} and is similar to the V_{OH} function. Note that V_{OL} is also defined where the slope of the transfer curve is –1. To distinguish V_{OH} and V_{OL}, this function finds the point starting from the end of the trace and searching backwards. There are two points on the trace where the slope is –1. This function finds the point closest to the end of the trace.

The goal functions are used in Probe. Follow the procedure of Section 4.D.3. When you obtain the Probe plot of the transfer curve below, continue with this section:

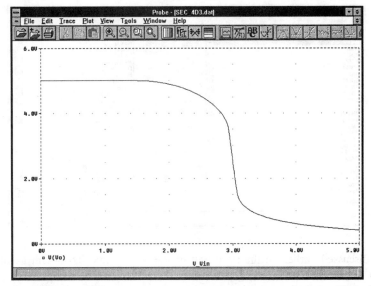

Next, we would like to find the values of V_{IL}, V_{IH}, V_{OL}, V_{OH}, NM_H, and NM_L. Click the *LEFT* mouse button on **Trace** from the Probe menu bar:

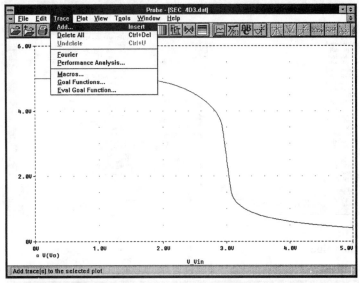

Click the *LEFT* mouse button on **Eval Goal Function**:

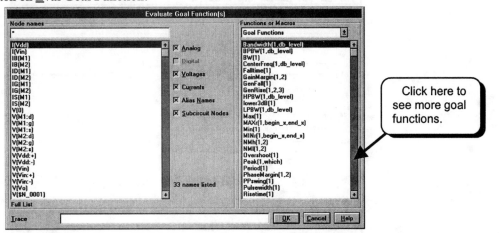

The left pane lists the voltage and current traces that we have seen previously in the Add Traces dialog box. The right pane lists the available goal functions. Click the *LEFT* mouse button on the down arrow a few times to view more goal functions:

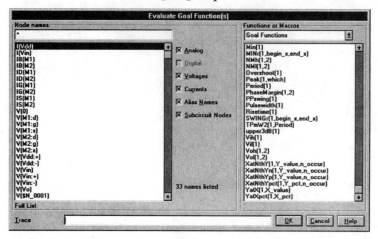

We see the goal functions discussed earlier on the right half of the dialog box. To obtain a value for V_{OH}, enter the text **Voh(d(V(Vo)),V(Vo))** in the text field next to *Trace*:

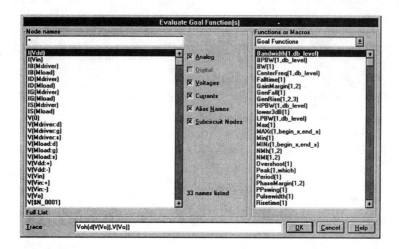

Click the *OK* button to view the value:

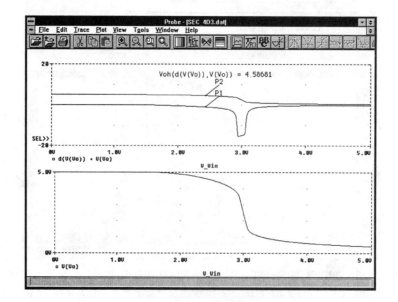

Probe draws a plot of the traces used in the goal function (**dV(Vo)**) and **V(Vo)** in this example), labels the locations of the points used in the goal function (**P1** and **P2** in this example), and then displays the value of the goal function. The value of V_{OH} is **4.58681** volts. Select **Window** and then **New** to obtain a new Probe window:

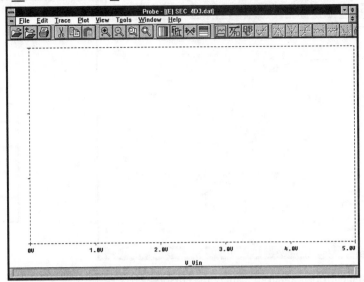

To find V_{IL}, select **Trace** and then **Eval Goal Function**, and enter the trace `Vil(d(V(Vo)))`:

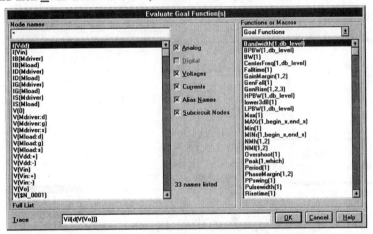

Click the **OK** button to view the value of V_{IL}:

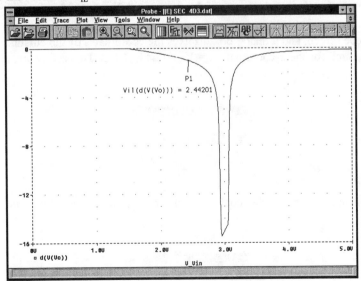

The value of V_{IL} is **2.44201** volts. Repeat the procedure above to find values for V_{IH}, V_{OL}, NM_H, and NM_L. The values are summarized in Table 4-2.

Suppose we wanted to show the values of V_{IL}, V_{IH}, V_{OL}, and V_{OH} on the Probe screen. This can be done using the cursors. First, display a trace of V(Vo) by itself on the Probe window:

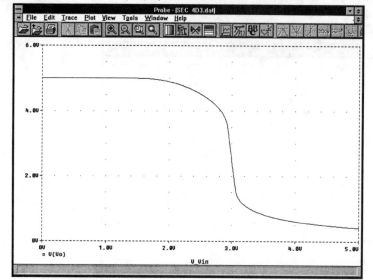

Table 4-2	
	Value
V_{IH}	3.32 V
V_{OL}	0.966 V
NM_H	1.27 V
NM_L	1.48 V

Click the cursor button in the Probe toolbar to display the cursors:

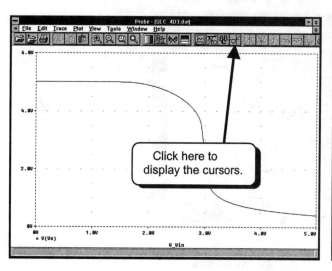 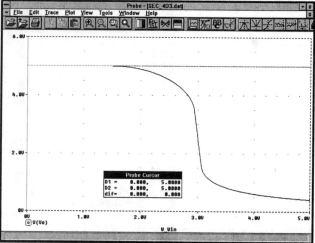

Cross hairs appear on the Probe screen, and a dialog box displays the coordinates of the cursors. We would like to mark the point V_{OH} and V_{IL}. We will search for the x-coordinate x = V_{IL} = 2.44. Select **Tools**, **Cursor**, and then **Search Command** from the Probe menus:

We will use a search command similar to the goal functions discussed earlier. Enter the text **search forward xvalue(2.44201)**:

Click the **OK** button. The cursor will move to the coordinate x = 2.44201:

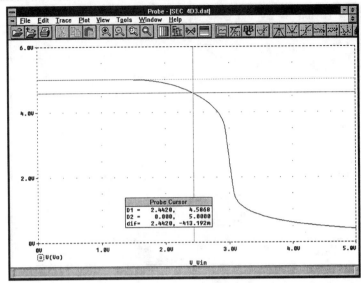

We would like to mark this point on the screen, so select **Tools**, **Label**, and then **Mark** from the Probe menus:

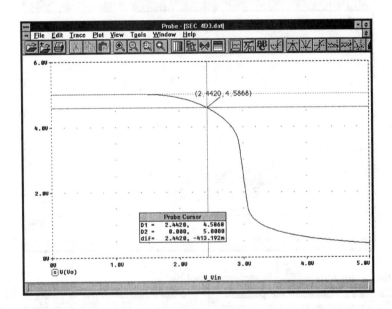

       Next, we will mark the point V_{OL} and V_{IH}. We will search for the x-coordinate x = V_{IH} = 3.32. Select **Tools**, **Cursor**, and then **Search Command** from the Probe menus:

The dialog box contains our last search command. Modify the text to `search forward xvalue(3.32)`:

Click the **OK** button. The cursor will move to the coordinates of x = 3.32:

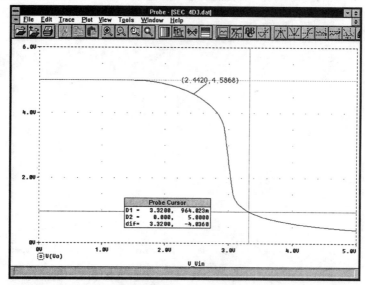

Mark the point on the screen by selecting **Tools**, **Label**, and then **Mark** from the Probe menus:

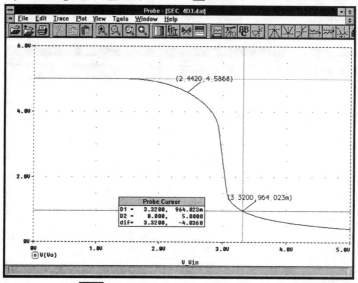

To hide the cursors, click the cursor button 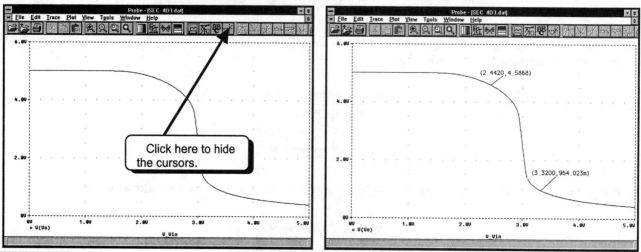 in the Probe tool bar:

EXERCISE 4-8: Find values of V_{IL}, V_{IH}, V_{OL}, V_{OH}, NM_H, and NM_L for the CMOS inverter of **EXERCISE 4-7**.

SOLUTION: V_{IL} = 2.237 V, V_{IH} = 2.763 V, V_{OL} = 0.249 V, V_{OH} = 4.751 V, NM_H = 1.988 V, and NM_L = 1.988 V.

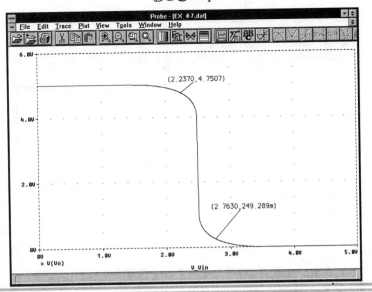

4.D.5. Parametric Sweep — Family of Transfer Curves

A question we would like to ask is, how does the transfer curve of the circuit from the previous section change as we change the driver MOSFET width-to-length ratio? We would like a family of curves which show the effect of changing the driver width. Families of transfer curves can be generated using the Parametric Sweep in conjunction with the DC Sweep or by using the DC Nested Sweep. The DC Nested Sweep was demonstrated in Section 4.D.2, so we will demonstrate the Parametric Sweep here.

A Parametric Sweep allows us to sweep a parameter. We would like to sweep the width of the driver MOSFET in the circuit of the previous section. We are assuming that you have followed that procedure and have set up the MOSFET models and DC Sweep. We must first set up a parameter for the width of the driver. We will start with the completed circuit from the previous section. It is repeated here for convenience.

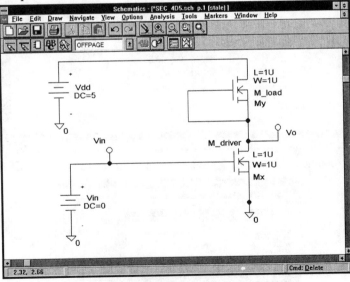

Double click the *LEFT* mouse button on the text **W=1U** next to the driver MOSFET:

We wish to enter a parameter for the value instead of *1U*. Type the text {**W_val**}:

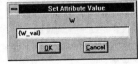

Click the **OK** button to accept the change:

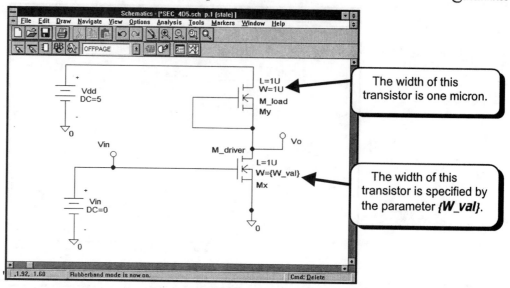

The width of this transistor is one micron.

The width of this transistor is specified by the parameter *{W_val}*.

The driver width is now the value *{W_val}*. In PSpice, any text enclosed in curly brackets {} is referred to as a parameter. A parameter can be assigned different values and can be changed during a simulation.

Next, we must declare *{W_val}* as a parameter. Get a part called PARAM and place it on your schematic.

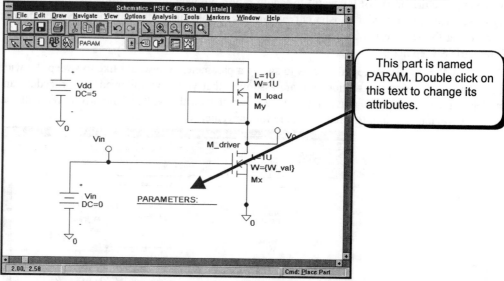

This part is named PARAM. Double click on this text to change its attributes.

When you have placed the part, double click the **LEFT** mouse button on the text **PARAMETERS**:

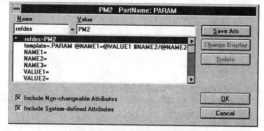

This attribute box is used to define parameters. We can define up to three parameters with this part. We will define **W_val** as a parameter and give it a default value of 1U (1 micron). Fill in the attributes as shown:

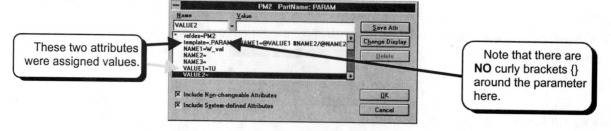

These two attributes were assigned values.

Note that there are **NO** curly brackets {} around the parameter here.

For every parameter defined (**NAME1**, **NAME2**, **NAME3**), a value must also be assigned (**VALUE1**, **VALUE2**, **VALUE3**). Click the **OK** button to accept the changes:

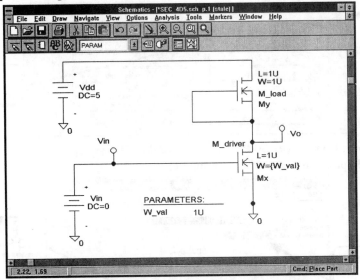

We must now set up the Parametric Sweep. Select **Analysis** and then **Setup** from the Schematics menus:

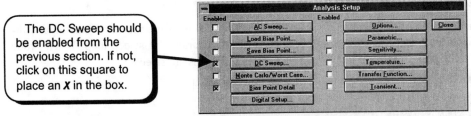

The **DC Sweep** should be enabled from the simulation of the previous section. We must now set up the Parametric Sweep. Click the **LEFT** mouse button on the **Parametric** button to obtain the Parametric sweep dialog box:

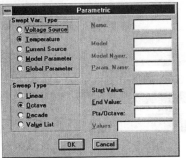

We would like to sweep the parameter W_val from 1 micron to 32 microns in factors of 2. We will use an **Octave Sweep Type**. An octave is a factor of 2. An octave would be from 1 to 2, 2 to 4, 4 to 8, 8 to 16, and so on. Fill in the dialog box as shown:

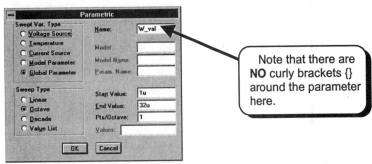

The dialog box is set to sweep the global parameter **W_val**. The sweep is from **1U** (1×10^{-6} meters = 1 micron) to **32U** (32×10^{-6} meters = 32 microns). The sweep is in octaves, 1U to 2U, 2U to 4U, 4U to 8U, 8U to 16U, and 16U to 32U. **1** point per octave is specified so the values of **W_val** will be 1U, 2U, 4U, 8U, 16U, and 32U. Click the **OK** button to accept the setup:

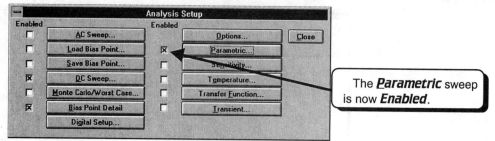

The **Parametric** sweep is now **Enabled**.

Notice in the above dialog box, both the **DC Sweep** and the **Parametric** sweep are enabled. Click the **Close** button to return to the schematic. Run PSpice (**Analysis** and then **Simulate**):

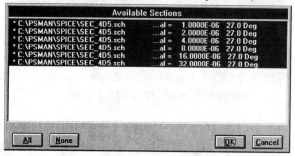

The current value of the parameter in the Parametric Sweep.

The current value of the DC Sweep variable.

The PSpice window indicates the current value of the parameter. Notice that for each value of the parameter, the input voltage is swept from 0 to 5 volts. Logically, the DC Sweep loop is executed inside the Parametric sweep loop. When Probe runs, you will be asked which values of **W_val** you would like to view. By default, all runs are selected:

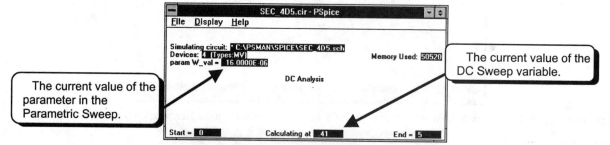

We would like to look at all of the curves, so click the **OK** button:

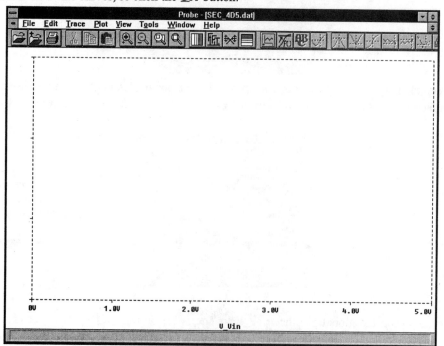

We would like to plot the output voltage. Select **Trace** and then **Add** from the Probe menus, and then add the trace **V(Vo)** :

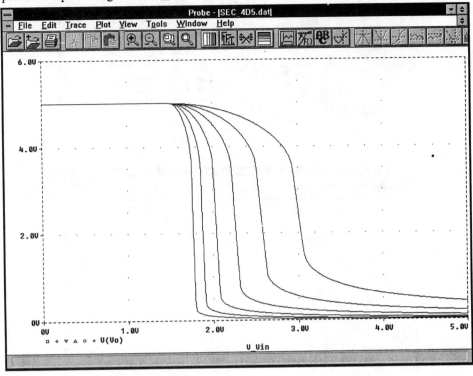

EXERCISE 4-9: In **EXERCISE 4-7** we looked at the operation of a CMOS inverter. We will now investigate how the (W/L) ratio of the PMOS transistor affects the transfer curve of the inverter. Let the (W/L) ratio of the PMOS transistor have values 1, 3, 6, 9, 12, and 15.

SOLUTION: Starting with the circuit of **EXERCISE 4-7**, let the W value of the PMOS transistor be set by the parameter Wp_val. Also, add the "PARAM" part:

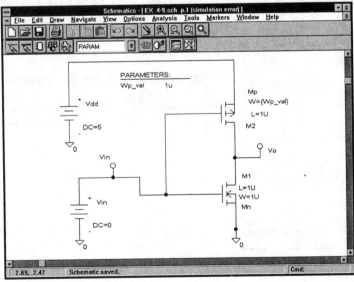

Set up the DC Sweep to sweep *Vin* from 0 V to 5 V in 0.001 V steps. Set up the Parametric Sweep to use the value list:

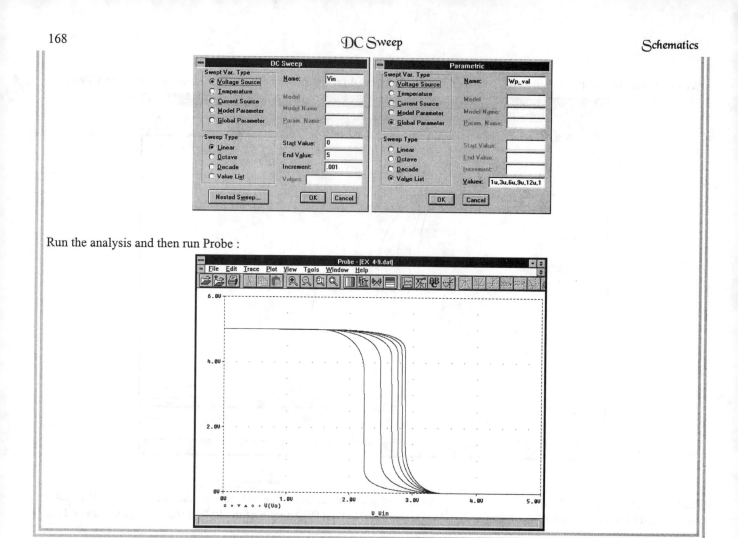

Run the analysis and then run Probe :

4.E. DC Nested Sweep — BJT Characteristic Curves

The nested DC Sweep can be used to generate characteristic curves for transistors. We will illustrate generating these curves using a BJT. Wire the circuit below:

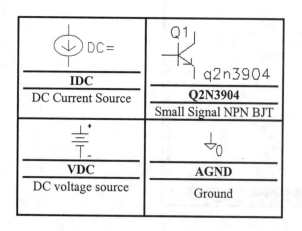

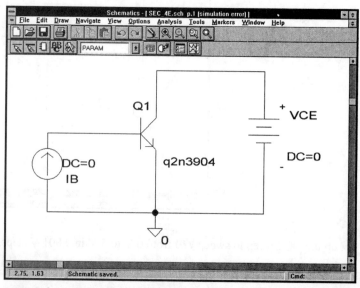

We need to set up the DC Sweep and the DC Nested Sweep. Select **Analysis** and then **Setup** from the Schematics menus, and then click the *LEFT* mouse button on the *DC Sweep* button. We would like the main *DC Sweep* to sweep *VCE* from 0 to 15 volts. Fill in the dialog box as shown:

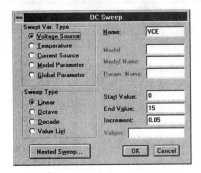

The dialog box is set to sweep **VCE** from **0** volts to **15** volts in **0.05** volt increments. The **Sweep Type** is linear, which means that the voltage points are equally spaced.

Next, we need to set up the **Nested Sweep**. Click the **LEFT** mouse button on the **Nested Sweep** button. We would like to sweep the base current from 0 mA to 1 mA in 100-μA increments. Fill in the dialog box as shown:

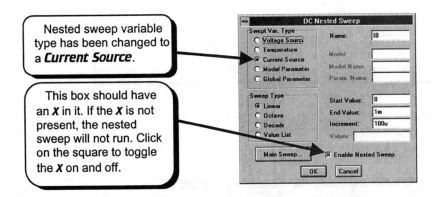

Nested sweep variable type has been changed to a **Current Source**.

This box should have an **X** in it. If the **X** is not present, the nested sweep will not run. Click on the square to toggle the **X** on and off.

Notice that the **Swept Var. Type** is set as a **Current Source**, and the **Sweep Type** is set to **Linear**. Make sure that the square next to the text **Enable Nested Sweep** has an **X** in it, ⊠.

Logically, the Main Sweep loop is executed inside the Nested Sweep loop; that is, for each value of the base current, VCE is swept from 0 to 15 volts. Return to the schematic by clicking the **OK** button and then clicking the **Close** button. Run PSpice (select **Analysis** and then **Simulate**):

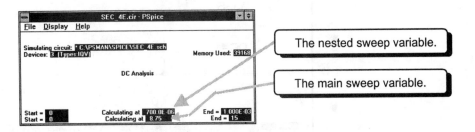

The nested sweep variable.

The main sweep variable.

The PSpice window indicates the progress of each loop. Notice that for each value of the nested variable, the main variable is swept from 0 to 15 volts. When the simulation is finished, Probe will run automatically. Add the trace **IC(Q1)**:

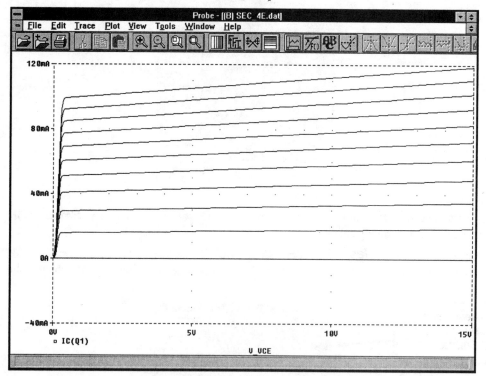

This is the characteristic family of curves for a 2N3904 NPN bipolar junction transistor.

EXERCISE 4-10: Use the DC Nested Sweep to find the characteristic curves of a 2N5951 jFET.

SOLUTION: Draw the circuit below:

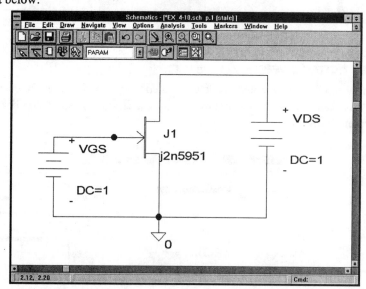

For each value of **VGS**, we want to sweep **VDS** from 0 to 15 volts. Thus, we want **VGS** to be our nested sweep variable and **VDS** to be our main sweep variable. Note also that for jFETs we must sweep **VGS** from zero to a negative value. Fill in the DC Sweep and DC Nested Sweep dialog boxes as shown:

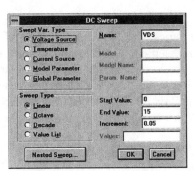
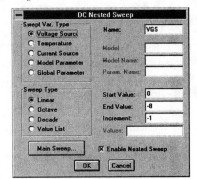

Run PSpice and then Probe. The x-axis will automatically be VDS. To plot the characteristic curves, plot the drain current ID(J1):

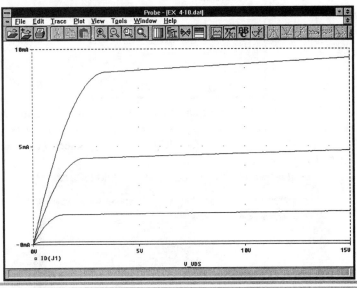

4.F. DC Current Gain of a BJT

In this section we will investigate how the DC current gain (H_{FE}) of a bipolar junction transistor varies with DC bias collector current I_{CQ}, DC bias collector-emitter voltage V_{CEQ}, and temperature. We will use the basic circuit shown below for all simulations:

⊕ DC=	Q1 q2n3904
IDC	**Q2N3904**
DC current source	Small signal NPN BJT
⊥⁺ ⊤₋	▽₀
VDC	**AGND**
DC voltage source	Ground

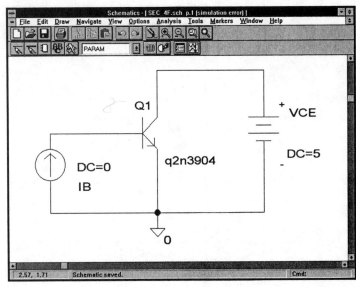

4.F.1. H_{FE} Versus Collector Current

Our first analysis will display how H_{FE} of a transistor varies with DC collector current.* In many applications we want to use a transistor where it has its maximum current gain. This plot will tell us how to choose the collector current for maximum current gain. We can then bias the transistor at this collector current. This plot is easily generated with a DC Sweep. We will generate this curve with V_{CE} constant at 5 V. Note in the circuit above that V_{CE} is a DC voltage source of 5 volts and we will not change it during the simulation. Select **Analysis** and then **Setup** from the Schematics menus, and then click the **DC Sweep** button. We will sweep IB from 1µA to 1 mA in decades with 20 points per decade. Fill in the dialog box as shown:

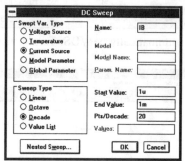

Click the **OK** button and then the **Close** button to return to the Schematic. Run the simulation and then run Probe. We would like to plot the current gain, H_{FE}. When we go to add a trace, we do not see H_{FE} listed. However, H_{FE} can be calculated as the ratio of IC/IB for the transistor. Thus, we will plot the trace IC(Q1)/IB(Q1). Select **Trace** and then **Add** from the Probe menus and enter the expression `IC(Q1)/IB(Q1)`:

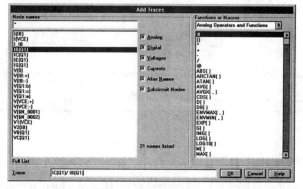

Click the **OK** button to plot the trace:

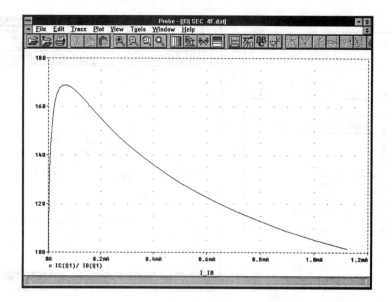

---

*Thanks to John D. Welkes of Arizona State University for this example.

Note that the x-axis is the base current. This is a plot of H_{FE} versus IB. We need to change the x-axis to collector current. We also note that the x-axis is linear. However, our sweep covered three decades, 1 μA to 1mA. We will also change the x-axis to a log scale. From the Probe menus, select **Plot** and then **X Axis Settings**:

Click on the circle ○ next to the text **Log** to specify a log scale. The circle should fill with a dot ◉, indicating that the option is selected:

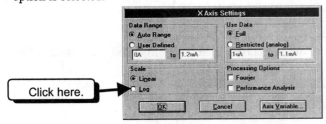

Next we must change the x-axis variable to collector current. Click the **Axis Variable** button:

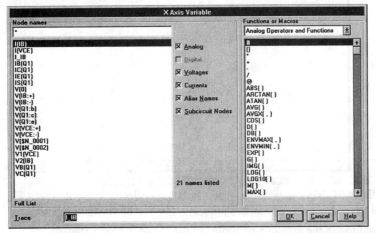

Select trace IC(Q1):

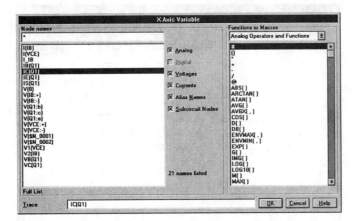

Click the **OK** button to accept the trace and then click the **OK** button again to plot the trace with the new settings:

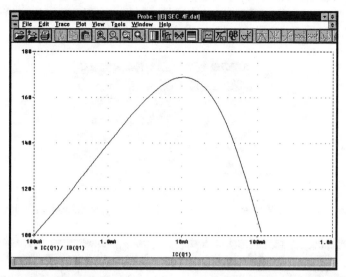

We see that the 2N3904 BJT has a maximum DC current gain at a collector current of about 10 mA. Use the cursors to find and label the maximum value of current gain:

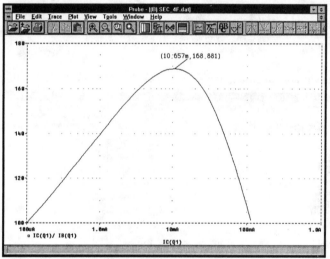

EXERCISE 4-11: For the MJE3055T NPN power transistor, find the collector current where H_{FE} is maximum. Specify the maximum value of H_{FE} and the collector current where it occurs. Let V_{CE} be constant at 5 V.

SOLUTION: Use the same circuit as in the previous example but use the MJE3055T transistor:

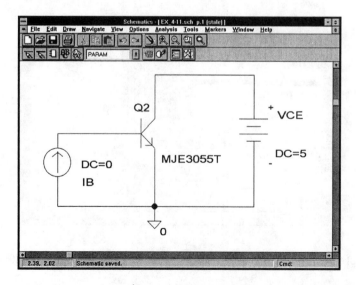

Since this is a power transistor, it can handle much higher collector currents. Sweep **IB** from 100μA to 100 mA:

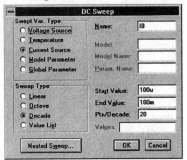

Use Probe to plot H_{FE} versus I_C and use the cursors to display the maximum value:

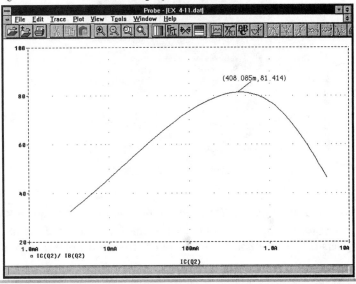

4.F.2. H_{FE} Versus I_C for Different Values of V_{CE}

The next thing we would like to do is to see how the H_{FE} versus I_C curve is affected for different values of DC collector-emitter voltage. The curve in the previous example was generated at $V_{CE} = 5$ V. We would now like to generate four curves at different values of V_{CE} and plot them all on the same graph. We will generate curves at $V_{CE} = 2$ V, 5 V, 10 V, and 15 V. We will use the same circuit as in the previous example:

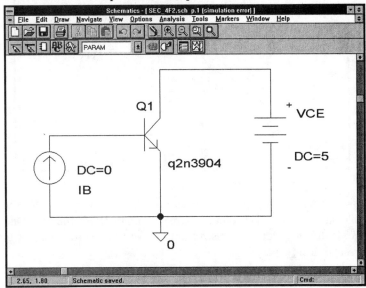

The DC Sweep is set up the same as in the previous example. We will sweep IB from 1 μA to 1mA:

To change V_{CE}, we will use a Nested DC Sweep. Click the **Nested Sweep** button and fill in the dialog box as shown:

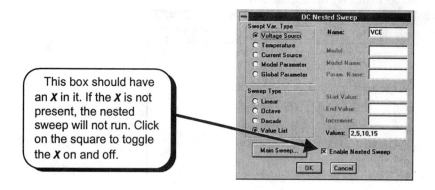

Note that the sweep variable type is a **Voltage Source**, and that there is an x in the square ⊠ to enable the nested sweep. Click the **OK** button and then the **Close** button to return to the schematic. Simulate the circuit (**Analysis** and then **Simulate**) and then run Probe. Plot H_{FE} versus I_C. Use the same procedure to generate the plot that we used on pages 172–174. Four curves will be displayed:

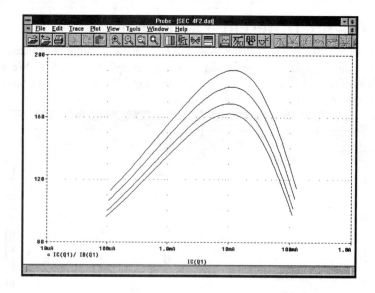

A question may arise as to how to determine which curve represents V_{CE} when it equals 2 V, 5 V, 10 V, and 15 V. In Section 4.F.1. we generated a single plot at $V_{CE} = 5$ V. This plot had a maximum of $H_{FE} = 168.881$. One of the curves on this plot is at V_{CE} equals 5 V. If we can identify the plot with this maximum, we can determine all of the other curves. Display the cursors (select **Tools**, **Cursors**, and then **Display**):

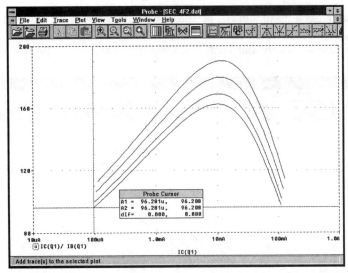

The cursor is now on the bottom curve. To find the maximum value of this curve, select **Tools**, **Cursors**, and then **Peak**. Do not select **Max**. Peak will move the cursor to the next local maximum while Max will find the absolute maximum for the entire plot. After selecting **Tools**, **Cursors**, and then **Peak**, the cursor moves to the peak of the lowest curve:

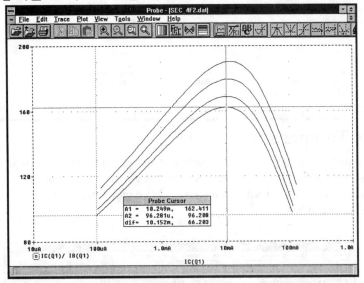

The cursor coordinates show that this curve has a maximum H_{FE} of 162.411. This is not the curve at $V_{CE} = 5$ V. Select **Tools**, **Cursors**, and then **Peak**. This will find the next local maximum, or the peak of the second curve:

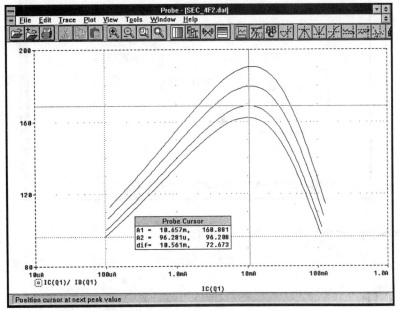

The cursors show that this curve has a maximum H_{FE} of 168.881 and thus this is the curve at $V_{CE} = 5$ V. We can conclude that the bottom curve is at $V_{CE} = 2$ V, the second curve is at $V_{CE} = 5$ V, and the top curve is at $V_{CE} = 15$ V. To be absolutely certain, you should run each case separately and verify the identity of each curve. Place text on the plot to identify each curve:

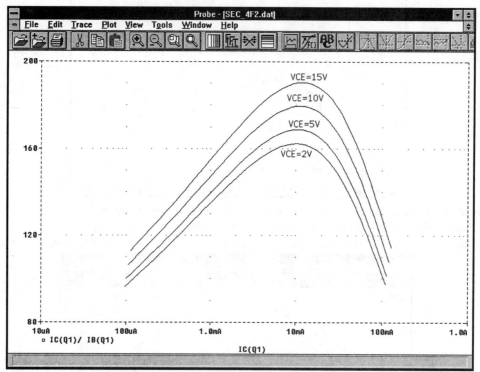

4.F.3. H_{FE} Versus Temperature

We will now generate H_{FE} versus I_C curves at different temperatures. This is a typical curve found in most data sheets for BJTs. We will use almost the same DC Nested Sweep that we used in Section 4.F.2, except for the Nested Sweep we will vary temperature rather than V_{CE}. We will use the same circuit and setup as in Section 4.F.2, but we will modify the Nested Sweep. We will use the circuit below:

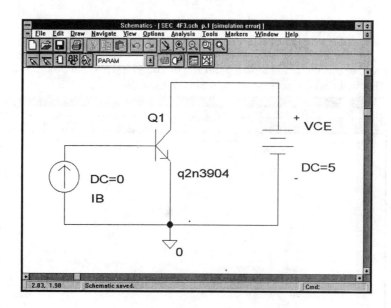

Select **Analysis** and then **Setup** and then click on the DC Sweep button:

These are the settings from Section 4.F.2. We will leave this dialog box unchanged. We will simulate the circuit at –25 °C, +25 °C, and +125 °C. Click on the **Nested Sweep** button and fill in the dialog box as shown:

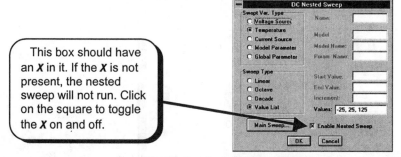

This box should have an **X** in it. If the **X** is not present, the nested sweep will not run. Click on the square to toggle the **X** on and off.

Click the **OK** button to accept the settings and click the **Close** button to return to the schematic. Run the simulation and then run Probe. Generate a plot of H_{FE} versus I_C. Three curves will be shown:

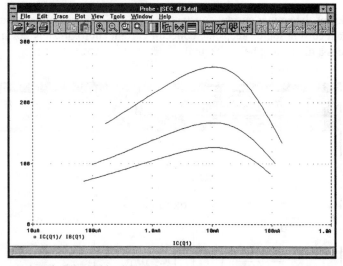

H_{FE} increases with temperature so the plot should be labeled as:

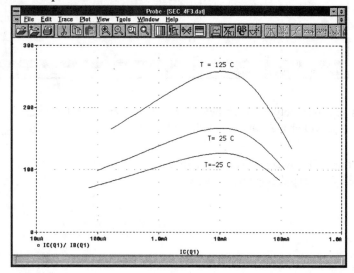

If you do not know which trace is at what temperature, you would need to run the H_{FE} versus I_C simulation three times, once at each temperature, to identify which curve is which.

The above plot shows the interesting result that at any particular temperature, the maximum value of H_{FE} always occurs at approximately the same value of I_C (10 mA in the screen capture above). An interesting question is, how does H_{FE} vary with temperature when V_{CE} is constant at 5 V and I_C is constant at 10 mA? We will need a new circuit to perform this simulation:

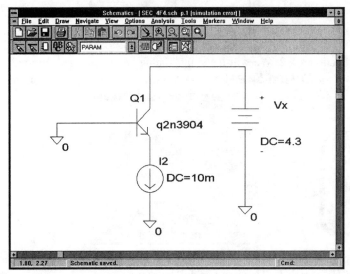

The constant current source keeps the emitter current constant at 10 mA. Since the collector current is approximately equal to the emitter current, the constant current source keeps the collector current approximately constant at 10 mA. Since the base is grounded, the emitter voltage is a diode drop below ground:

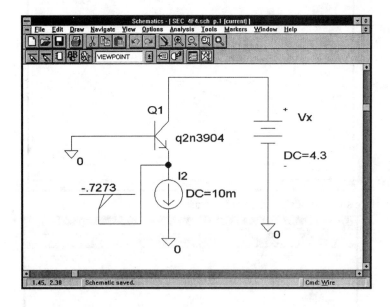

Since the emitter is approximately 0.7 volts below ground, we set Vx to 4.3 volts so that V_{CE} is approximately 5 V. As the temperature changes, V_{BE} will change, and V_{CE} will drift away from 5 V. However, V_{CE} is approximately 5 V for the simulation.

We now need to set up a DC Sweep to vary the temperature from –25 to 125 °C. We need only a DC Sweep, not a nested DC Sweep. Set up the DC Sweep dialog box as shown:

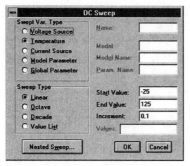

Run the simulation and then run Probe. To show that the circuit keeps I_C and V_{CE} relatively constant versus temperature, we will plot I_C and V_{CE} versus temperature:

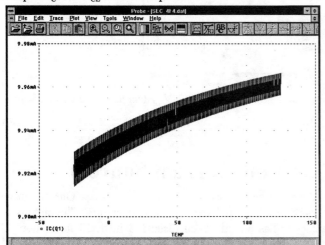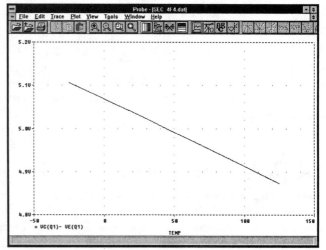

We see that I_C varies by about 40 μA and V_{CE} varies by about 200 mV over the range of the simulation. For the purposes of this simulation, these quantities will be assumed to be constant. Notice that the x-axis is automatically set to temperature. To plot H_{FE} versus temperature all we need to do is add the trace IC(Q1)/IB(Q1). The x-axis will automatically be set to temperature:

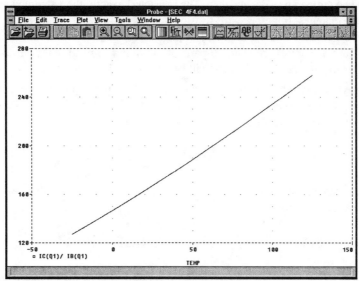

The plot is almost a straight line, indicating that H_{FE} is approximately proportional to temperature.

EXERCISE 4-12: For a 2N3904 transistor, if I_C is held constant at 10 mA, how does the base-emitter voltage vary with temperature?

SOLUTION: Use the simulation of the previous example and plot the base-emitter voltage:

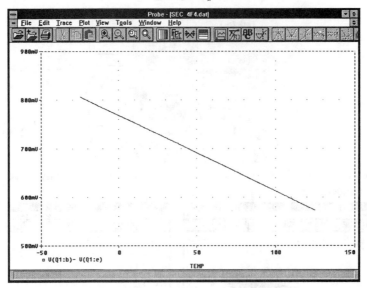

We see that V_{BE} decreases with temperature. Over the temperature range of –25 to 125 °C, V_{BE} decreases from 806 mV down to about 573 mV.

4.G. Temperature Analysis — Constant Current Sources

To demonstrate temperature effects, we will look at two circuits that can be used as constant current sources. One circuit will be greatly affected by temperature and the other is designed to be relatively independent of temperature. The circuits use transistors and resistors. The temperature dependence of transistors has already been discussed in detail in Section 4.F. Before we look at the circuits, we will look at how PSpice handles temperature characteristics of resistors.

4.G.1. Temperature Characteristics of a Resistor

When you place a resistor part called R (⊢‒\/\/\‒) in your circuit, you are using an ideal resistor; that is, the resistor has no tolerance and no temperature dependence. The resistor is exactly the value you specify and the resistance does not change as the temperature changes. In practice, all resistors have tolerance and temperature variations. If you use a 1-kΩ resistor with ±5% tolerance, the resistor will not be exactly equal to 1000 Ω. Its value could be anywhere between 950 Ω and 1050 Ω. Temperature variations are anywhere from 50 parts per million (ppm) to 400 ppm. In PSpice we can specify both temperature variations and resistor tolerance. Tolerance will be covered in Part 9.

PSpice has three ways to specify temperature dependence. Three model parameters named TC1, TC2, and TCE are available. TC1 and TC2 may be used together. If you use TCE, then TC1 and TC2 are ignored. Suppose we have a resistor with a resistance equal to Rval. If TCE is specified, then the resistance is:

$$ resistance = Rval \left[1.01^{TCE(T - Tnom)} \right] $$

TCE is called the exponential temperature coefficient. If TCE is specified, then the values of TC1 and TC2 are ignored. If TCE is not specified, then the resistance is

$$ resistance = Rval \left[1 + TC1(T - Tnom) + TC2(T - Tnom)^2 \right] $$

TC1 is referred to as the linear temperature coefficient and TC2 is referred to as the quadratic temperature coefficient. Tnom is the nominal temperature. Its default value is 27 °C.

The default values of TCE, TC1, and TC2 are zero. Thus, if you do not specify any of the coefficients, there will be no temperature dependence. In all of our previous simulations the resistors had no temperature dependence. Thus, if we varied temperature in any of the simulations, the resistors would not be affected. To add temperature dependence we must specify the temperature coefficients.

There are two ways to specify the temperature coefficients. One way is to create a resistor model and specify TC1, TC2, or TCE. The second way is to use the resistor graphic in the schematic which has an attribute for specifying TC1 and TC2. We will look first at the resistor attributes. Place a resistor part (R) in a schematic:

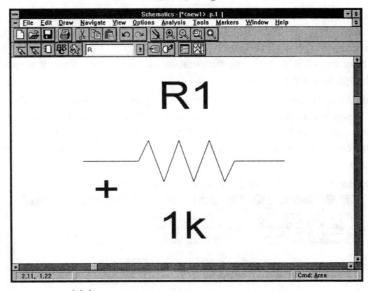

Double click on the resistor graphic 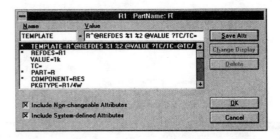 to edit its attributes:

The attribute **TC** is used to specify TC1 and TC2. TC2 is optional. To specify TC1=.0005 and TC2=0 we would set TC=**0.0005** as shown below:

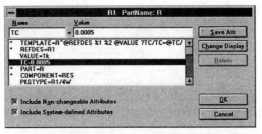

To specify TC1=0 and TC2=0.00003 we would specify TC=**0,0.00003**

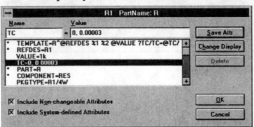

To specify TC1=0.0005 and TC2=0.00003 we would use the line TC=**0.0005,0.00003**

The advantage to using this method is that each resistor in the circuit can have different values for TC1 and TC2. Thus some resistors can have no temperature dependence, some resistors can have the same temperature dependence, and some can have a different temperature dependence. The disadvantage to using this method is that if you have a circuit with many resistors, all from the same manufacturer and all with the same temperature characteristics, you must specify the coefficients for each resistor individually even though they are the same.

The second method is a resistor model that specifies temperature dependence. Every resistor that uses the model will have the same temperature dependence. Thus, if you have a circuit with many resistors, all from the same manufacturer and all with the same temperature characteristics, the same model can be used to specify the temperature characteristics of all of the resistors. Models will be covered in more detail in Part 7. Here we will just discuss making a model for resistors that includes temperature dependence.

To create a resistor model, place a part called Rbreak in your circuit:

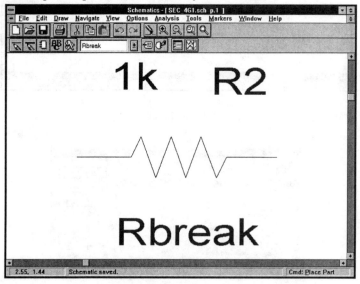

Before we continue we must save the schematic. Save the schematic with the file name sec_4g1.sch. When the schematic is saved, we can continue. Click the **LEFT** mouse button on the resistor graphic 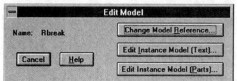 to select it. The graphic should turn red, indicating that it has been selected. Select **Edit** and then **Model** from the Schematics menus[*]:

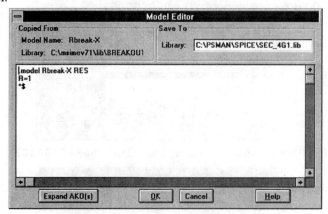

Click the **LEFT** mouse button on the **Edit Instance Model (Text)** button. This will allow us to edit the model for part Rbreak and create a new model:

---

[*]If the menu selection **Model** appears grayed out (Model), you will not be able to choose the **Model** menu selection.

Attempt to select the diode graphic again until its graphic, $+\!\!-\!\!\bigwedge\!\!\bigwedge\!\!-$, is highlighted in red.

This window is a text editor. The name of the model has been changed from Rbreak to **Rbreak-X** so that we will create a new model rather than change the existing model for Rbreak. The only parameter specified in the model is **R=1**. R is a model parameter used to specify the value of the resistance. If the value of the resistor specified in the schematic is x, then the actual value of the resistor used for calculations is x•R. Since R is set to 1, the value of resistance specified in the schematic is also the value used in calculations. We can delete the model parameter R=1 or we can leave it in the model. It will have no effect on the simulations. We will now add the temperature coefficients.

To specify a resistor with an exponential temperature coefficient of 0.0007 we would modify the model as follows:

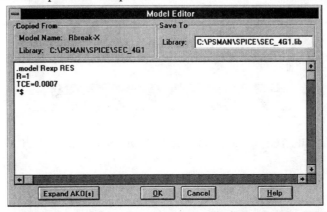

We have renamed the model to **Rexp** and added the model parameter TCE. Click the **OK** button to accept the model. You will return to the schematic:

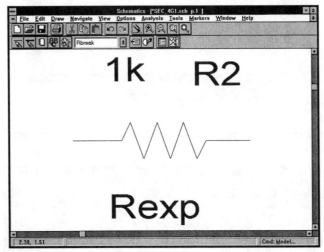

Notice that the model name **Rexp** is now displayed next to the resistor. Rbreak is referred to as a breakout part. Since breakout parts are specifically used to create models, the name of the model used by the part is displayed on the schematic.

Place three more Rbreak parts in your schematic:

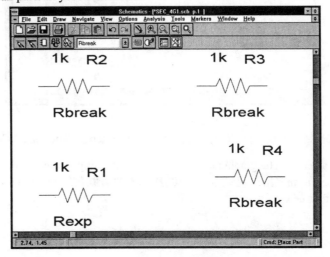

We will now create two more models. Click the *LEFT* mouse button on the resistor graphic 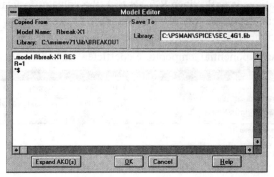 of *R2* to select it. The graphic should turn red, indicating that it has been selected. Select **Edit** and then **Model** from the Schematics menus and then select the ***Edit Instance Model*** button. This will allow us to edit the model for part Rbreak and create another new model:

We will now create a new model with TC1=0.0001. TC2 and TCE will be zero by default. We will call this model Rlinear:

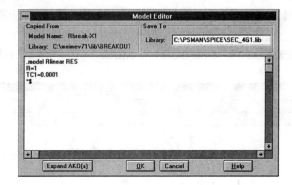

Click the ***OK*** button to save the changes to the model. R2 should be displayed with the new model name next to it:

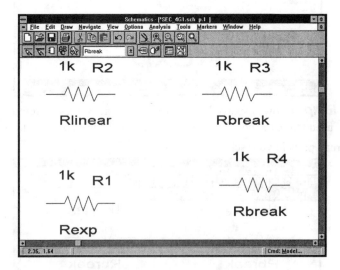

Next we will create a model for R4 that has quadratic temperature dependence. Click the *LEFT* mouse button on the resistor graphic for R3 to select it. The graphic should turn red, indicating that it has been selected. Select **Edit** and then **Model** from the Schematics menus and then select the ***Edit Instance Model*** button. This will allow us to edit the model for part Rbreak and create another new model:

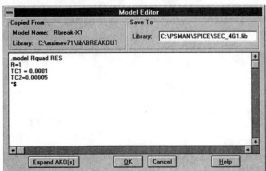

We will create a model called Rquad with TC1 = 0.0001 and TC2=0.00005:

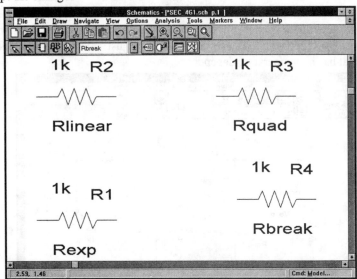

Click the **OK** button to accept the changes:

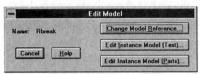

Last, we will show how to use one of the new models for **R4**. Click the *LEFT* mouse button on the resistor graphic for **R4** to select it. The graphic should turn red, indicating that it has been selected. Select **Edit** and then **Model** from the Schematics menus:

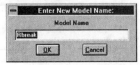

Click the *LEFT* mouse button on **Change Model Reference**:

The present model used by R4 is **Rbreak**. We would like **R4** to use model **Rexp**. Type the text `Rexp`

Click the **OK** button to accept the model name:

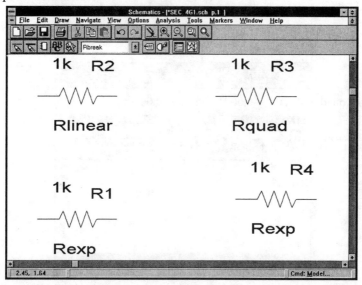

Both R4 and R1 use the same model. If we had 100 resistors, we would use this last method to make them all use the same model. Thus we would have to create only a single model, and then change the model reference of each resistor.

   To get an idea of how these resistors change with temperature, we will plot their resistance from –25 °C to 125 °C. Create the circuit below. Resistors R1, R2, and R3 use the models we just created. R4 has been deleted. Rx is a regular resistor (get a part named R) that has no temperature dependence.

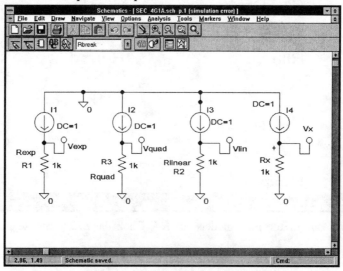

Since each resistor has 1 A of current flowing through it, the voltage across the resistor is numerically equal to the resistance.

We will set up a DC Sweep to vary the temperature. Select **Analysis** and then **Setup** from the Schematics menus, and then select the **DC Sweep** button. Fill in the dialog box as shown:

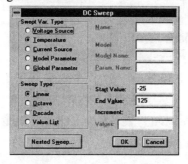

The sweep variable is set to **Temperature**. We have chosen a **Linear** sweep with temperature going from –25 °C to 125 °C in 1°C increments. Click the **OK** button and then the **Close** button to return to the schematic. Select **Analysis** and then **Simulate** to simulate the circuit. When the simulation is complete, Probe will run. Display the traces V(Vexp), V(Vquad), V(Vlin), and V(Vx) to display the value of each resistor:

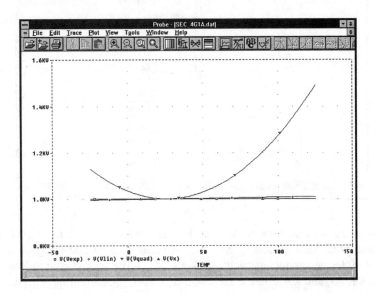

The resistor with the quadratic temperature coefficient appears to have a drastic resistance change. This is unrealistic because the coefficient we chose was too large. Remove trace V(Vquad) so that we can see the other traces more clearly:

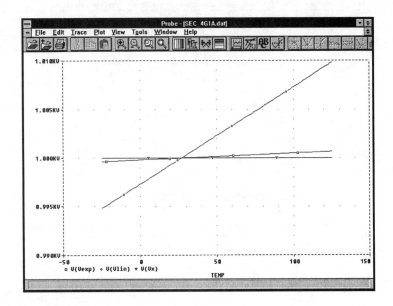

We can now see the temperature variations more clearly. Note that V(Vx) was constant. V(Vx) was the voltage across the ideal resistor and has no temperature variation.

The curves generated here are arbitrary because we just randomly picked the temperature coefficients. To accurately model your resistors, you would need to get a data sheet on the resistors you are using and find out if the temperature dependence is linear, quadratic, or exponential, and also find the correct coefficients. The coefficients used here were just for illustration.

4.G.2. BJT Constant Current Source

For an example of a temperature dependence, we will look at the circuit below:

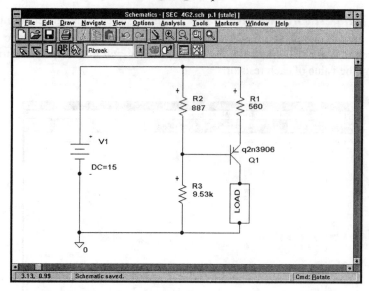

There are two ways to look at this circuit. One is that this is a self-biasing circuit for a BJT amplifier. If this were an amplifier, the load would most likely be a resistor. (The term "self-bias" means that the goal of the circuit is to make the collector current independent of device parameters such as H_{FE} and V_{BE}.) The second way to look at the circuit is that as far as the load is concerned, Q1, R1, R2, and R3 form a current source; that is, the current through the load is determined by Q1, R1, R2, and R3. If this circuit were designed as a current source for the load, we could view the circuit as:

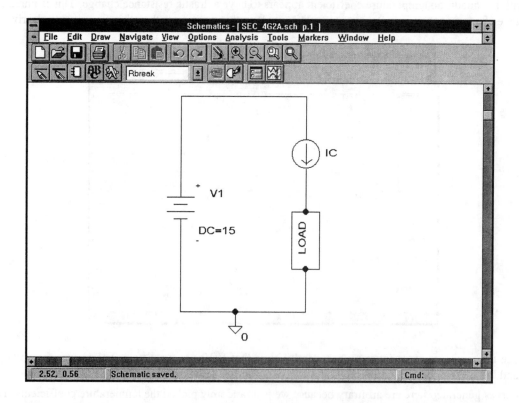

No matter what the application, the circuit is designed to make the collector current independent of V_{BE} and H_{FE}. We will assume that the resistors used have no tolerance and have their exact value specified in the circuit. To see how the tolerance of resistors affects the collector current, see Section 9.C. From previous sections we know that the resistance of a resistor, and H_{FE} and V_{BE} of a BJT, are affected by temperature. The question is, if the temperature changes, how much does the collector current change?

We will perform two analyses on the circuit. The first will be a plot of collector current versus temperature with ideal resistors. The resistors will have no temperature dependence. The only temperature-dependent device will be the BJT.

The second simulation will use temperature dependence for resistors and the BJT. For simulation, we will use the circuit below:

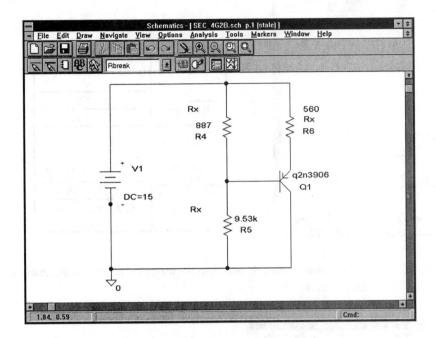

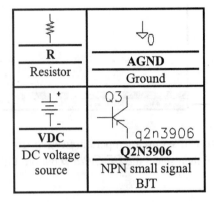

First, we will use PSpice to find the collector current without temperature dependence. We will add a current probe to display the current on the screen:

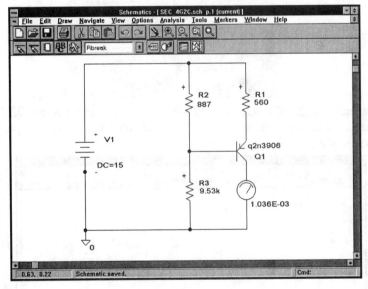

The resistors were tweaked slightly to achieve a current close to 1 mA. We will now see how this circuit performs with temperature changes.

4.G.2.a. Current Versus Temperature — No Resistor Temperature Dependence

Our first simulation will not specify temperature coefficients for any of the resistors. This will make the resistors independent of temperature. The model for the 2N3906 includes temperature dependence and will be the only device in the circuit that varies with temperature. Both H_{FE} and V_{BE} are functions of temperature, and these parameters will affect I_C.

Create the circuit shown below:

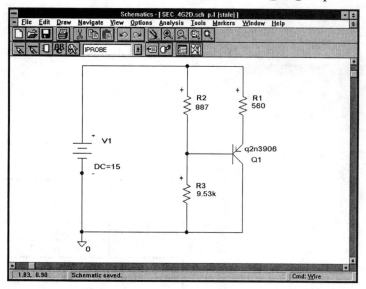
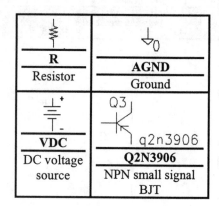

We will set up a temperature sweep from –25 °C to +125 °C. Select **Analysis** and then **Setup** from the Schematics menus, and then select the **DC Sweep** button. Fill in the dialog box as shown:

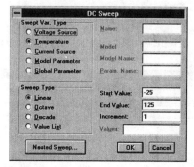

We have specified a linear sweep from –25 °C to 125 °C with 1-degree increments. Click the **OK** button and then the **Close** button to return to the schematic. Simulate the circuit and run Probe. Add the trace **IC(Q1)** to plot the collector current. Use the cursors to label the end points of the range.

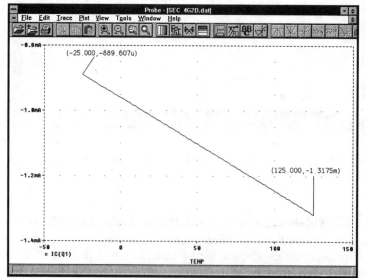

We see that, over the temperature range, the collector current varies from –889 µA to –1.317 mA. It is not very independent of temperature. Remember that this variation is without resistor temperature dependence. We could make the circuit more temperature independent by choosing smaller values for R2 and R3. However, this will consume more power.

4.G.2.b. Current Versus Temperature — With Resistor Temperature Dependence

We will now add temperature dependence to the previous circuit. A typical 1% resistor has a linear temperature coefficient of 100 parts per million, or 100×10^{-6}. We will create a resistor model with this temperature coefficient. First, we will replace all resistors in the above circuit with the Rbreak part:

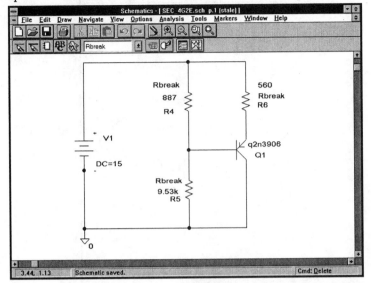

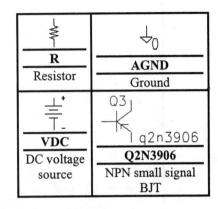

We will now edit the Rbreak model and create a model with the specified temperature coefficient:

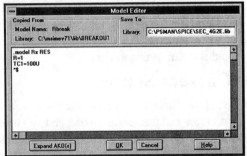

We have named the model **Rx** and specified the value of TC1 to be 100U, or 100×10^{-6}. We will use this model for all of the resistors:

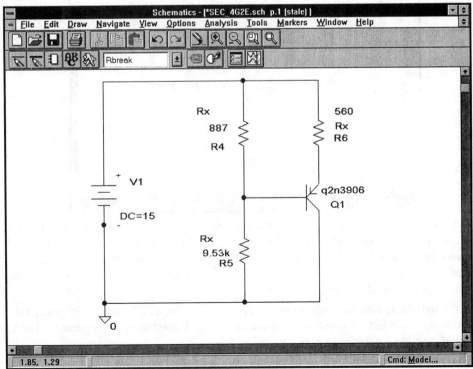

We will now run the same temperature sweep that we did in the previous section:

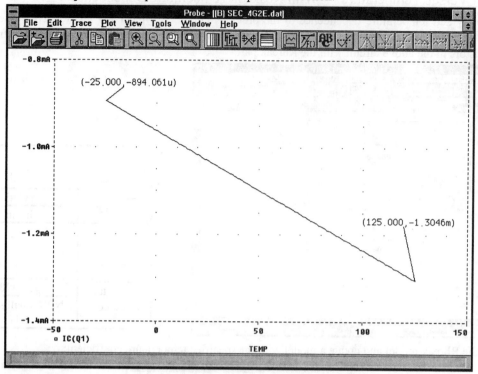

We see that with both resistor and BJT temperature dependence, the collector varies from −894 μA to −1.304 mA. In our simulation with only BJT temperature dependence, the collector current ranged from −889 μA to −1.317 mA. Our conclusion is that most of the temperature dependence in this circuit is due to the BJT.

4.G.3. Op-Amp Constant Current Source

The op-amp constant current source below is designed to eliminate the effects of temperature on the BJT used in the current source. This current source is a very accurate and temperature-independent current source:

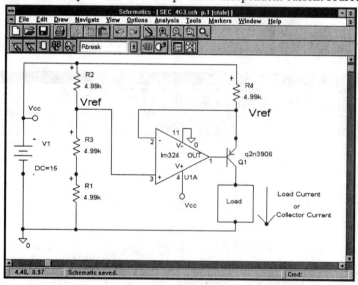

Resistors R1, R2, and R3 form a voltage divider. The voltage at the negative terminal of R2 is Vref = 10 V. Due to the negative feedback of the circuit, the voltage at the negative terminal of R4 is also Vref. The positive terminal of R4 is hooked to the supply so the voltage across R4 is held constant at 5 V. If the voltage across R4 is constant at 5 V, then the current through R4 is constant at 1 mA. This circuit is designed to keep the voltage across R4 constant. By keeping the voltage constant, the current through R4 is held constant.

The current through R4 is also the emitter current of Q1. Since for a BJT, the collector current is approximately equal to the emitter current, the collector current and load current are held constant at approximately 1 mA. We will place a current probe (IProbe) in the collector to measure the current:

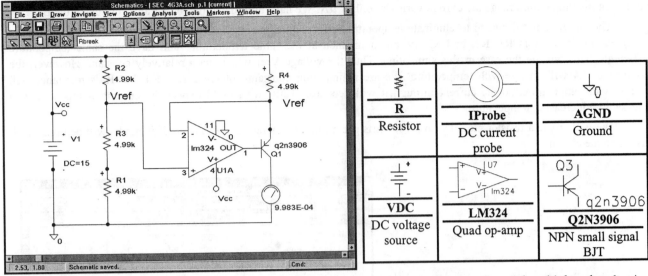

We see that the collector current is 998 µA. Since $I_C=[\beta/(\beta+1)]I_E$, we expect the collector current to be a bit less than 1 mA.

We will now run a linear temperature sweep from –25 to 125 °C to see how this circuit is affected by temperature. For the first simulation only the BJT will have temperature dependence. We will use normal resistors (part name R), which are independent of temperature. Note that the Q2n3906 BJT model includes temperature dependence. Set up the DC Sweep as shown:

Run the simulation and plot the collector current :

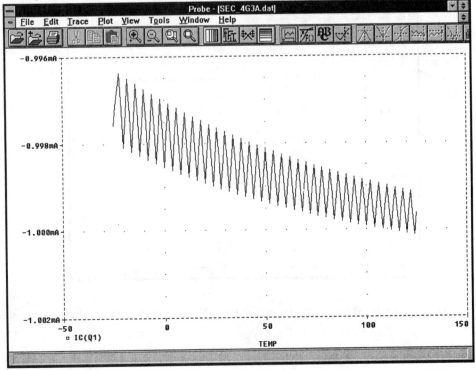

Over the entire temperature range the current varies by only 4 μA. We have eliminated the temperature effects of the BJT.

The circuit is not designed to eliminate temperature effects of resistors. The voltage Vref will be very independent of temperature if resistors R1, R2, and R3 are all maintained at the same temperature and have the same temperature coefficients. This will be the case in this simulation. Thus the voltage Vref will remain relatively constant. However, the resistance of R4 will change with temperature. We are maintaining constant voltage across R4, but R4's resistance will increase with temperature, so the current through it will decrease. Thus, we should expect the current of this source to decrease with increasing temperature.

We will create a resistor model with a 100 parts per million temperature coefficient. This model will be used by all resistors in the circuit:

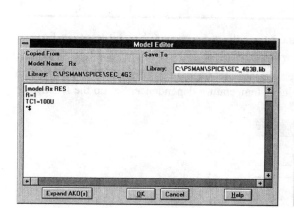

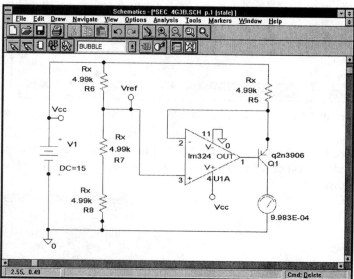

Notice that we have added a bubble at the voltage reference node so that we can easily plot the voltage. We will run the same temperature analysis as we did before. Sweep the temperature from –25 to +125 °C. We will first plot the reference voltage to see if it is independent of temperature:

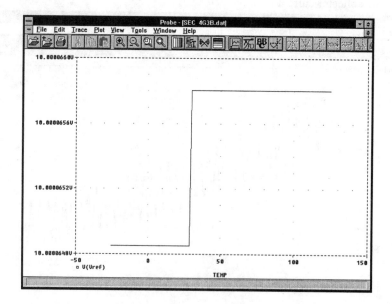

We see that the voltage is, for all practical purposes, constant at 10 V. Next we will plot the collector current of the BJT:

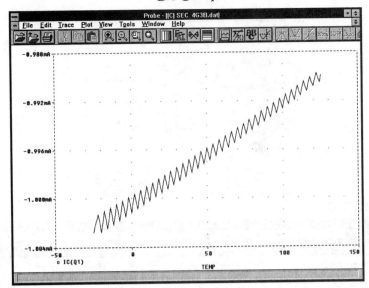

Even with resistor temperature dependence, the collector current changes by only about 13 µA over the entire temperature range. This circuit is a very temperature-independent constant current source.

EXERCISE 4-13: The circuit below is a current mirror and is a constant current source of about 50 µA. Display how the current varies with temperature. Let resistors have a linear temperature coefficient of 200 ppm.

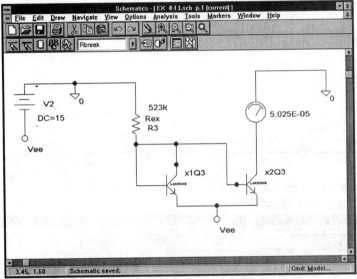

SOLUTION: Edit the resistor model to create a model with temperature dependence:

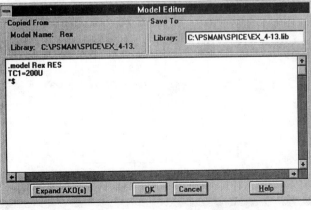

Set up a DC Sweep from –25 °C to +125 °C:

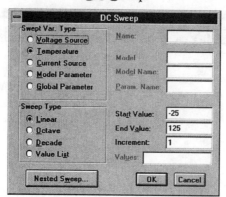

Simulate the circuit and plot IC(X2Q3):

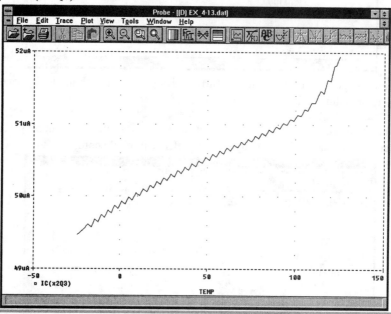

EXERCISE 4-14: The circuit below is a Widlar current source of about 50 μA. Display how the current varies with temperature. Let resistors have a linear temperature coefficient of 200 PPM.

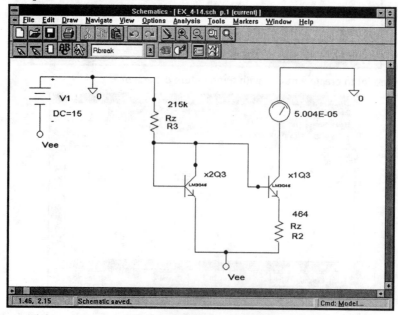

SOLUTION: Edit the resistor model to create a model with temperature dependence:

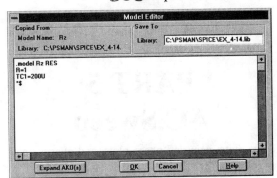

Set up a DC Sweep from –25 °C to +125 °C:

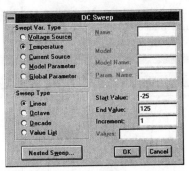

Simulate the circuit and plot IC(X1Q3):

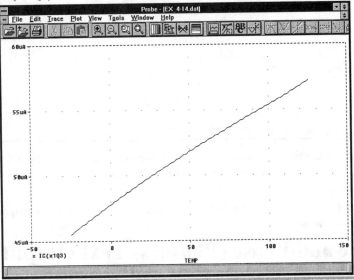

4.H. Summary

- The DC Sweep is used to find DC voltages and currents in a circuit.
- The output can be viewed graphically using Probe.
- The DC Sweep can be used to find DC voltages and currents for multiple values of DC sources. It can answer the following question in one simulation: "What is the voltage at node 1 for V1=10 VDC, V1=11 VDC, and V1=12 VDC?" This question could be answered using the DC Nodal Analysis if the DC Nodal Analysis were run three times.
- All capacitors are replaced by open circuits.
- All inductors are replaced by short circuits.
- All AC and time-varying sources are set to zero (IAC, VAC, Vsin, Isin, Vpulse, etc.).
- The DC Sweep may be used to view I-V characteristics of a device.
- The Nested DC Sweep and the Parametric Sweep can be used to see how values of devices affect the performance of a circuit.
- The Nested DC Sweep and the Parametric Sweep can be used to generate families of curves.
- Goal Functions can be used to obtain numerical data from Probe graphs. The result of evaluating a Goal Function is a single numerical value.

PART 5
AC Sweep

The AC Sweep is used for Bode plots, gain and phase plots, and phasor analysis. The circuit can be analyzed at a single frequency or at multiple frequencies. In this part we will illustrate its use at a single frequency for magnitude and phase results (phasors), and at multiple frequencies for Bode plots.

It is important to realize the difference between the AC Sweep and the Transient Analysis discussed in Part 6. The AC Sweep is used to find the magnitude and phase of voltages and currents. The Transient Analysis is used to look at waveforms versus time. An example of a waveform versus time is:

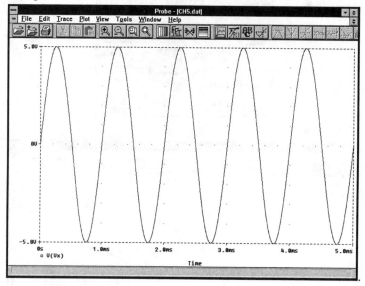

This graph shows us a voltage versus time. If you want to look at the magnitude and phase of voltages and currents, use the AC Sweep. The magnitude of the above waveform is 5 V and the phase is zero degrees — in phasor notation, 5|0. The AC Sweep will give a result such as 5|0.

The AC Sweep uses the sources VAC and IAC. These sources are functions of magnitude and frequency. **Do not use the source VSIN for the AC Sweep**.

5.A. Magnitude and Phase (Phasors) Text Output

Wire the circuit shown below:

R Resistor	**C** Capacitor	**L** Inductor	**AGND** Ground
V1 AC=1 Phase=0 **VAC** AC source	**BUBBLE** Bubble	**PRINT** Print results in output file	

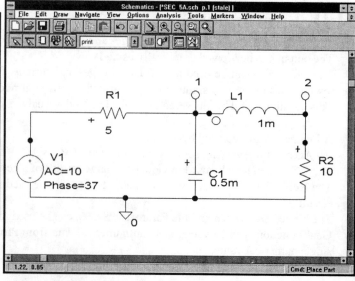

The source is an AC source with a magnitude of 10 volts and a phase of 37 degrees, 10|37 in phasor notation. This magnitude can be interpreted as either peak or RMS. If you specify 10|37 as the magnitude of the source, and the number 10 is an RMS value, then the magnitudes of all of your results will be RMS values. If you specify 10|37 as the magnitude of the source, and the number 10 is a peak value, then the magnitudes of all of your results will be peak values.

We must now add a part to the circuit that prints the results we would like. Get the part called **PRINT**. When you get the **PRINT** part, the cursor will be replaced by a rectangle, [□ *PRINT=*]. Move the rectangle to where you want to put the **PRINT** statement and click the *LEFT* mouse button. When you click the button, a second rectangle appears. Move the rectangle away from the first **PRINT** part and click the *LEFT* mouse button. Since we need only two **PRINT** parts, click the *RIGHT* mouse button to stop placing parts. You should have a schematic similar to the one below:

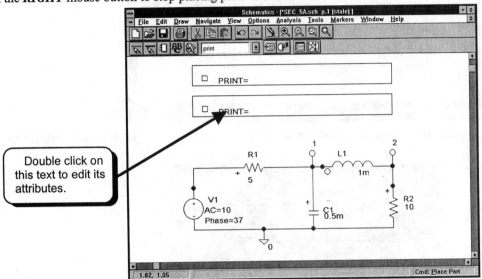

To print voltages and currents we must edit the attributes of the **PRINT** part. Double click the *LEFT* mouse button on one of the **PRINT=** parts. The dialog box below will appear:

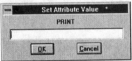

Type in the text `AC V(1,0) I(R_R1) I(C_C1)`:

Click the **OK** button to accept the attribute. This **PRINT** statement prints the magnitudes of the specified voltages and currents. For example, **V(1,0)** prints the magnitude of the voltage at node **1** relative to node **0** (ground), and **I(C_C1)** prints the magnitude of the current through capacitor **C1**. The **AC** in the attribute tells the **PRINT** statement to print the results from the AC Sweep. If we run an analysis other than the AC Sweep, this print part will generate no output because it specifies output for the AC Sweep.

Double click the *LEFT* mouse button on the other **PRINT=** text. Enter the text `AC VP(2,0) IP(C_C1) IP(L_L1)`:

This **PRINT** statement prints the phases of the specified voltages and currents. For example, **VP(2,0)** prints the phase of the voltage at node **2** relative to node **0** (ground), and **IP(L_L1)** prints the phase of the current through inductor **L1**. The **AC** in the attribute tells the **PRINT** statement to print the results of the AC Sweep. When you change the attributes you should have the following schematic:

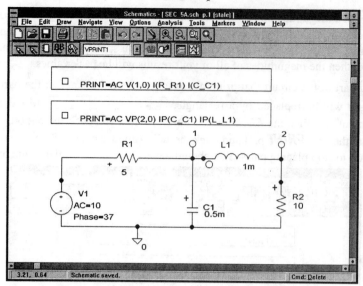

We are now ready to set up the simulation. From the Schematics menu bar select **Analysis** and then **Setup**. The screen below will appear:

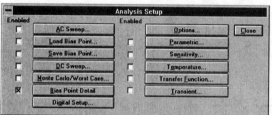

Click the *LEFT* mouse button on **AC Sweep**. We would like to find the desired voltages and currents at frequencies of 100, 200, and 300 Hz. The dialog box below has been set up for these frequencies:

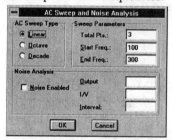

A linear sweep means that points are equally spaced between the starting and ending frequencies. If you were only interested in a single frequency, you would set the total points to 1 and set the starting and ending frequencies to the same value. Click the **OK** button to accept the settings:

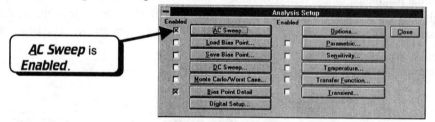

Click the **Close** button to return to the schematic.

Since we are using print statements, the results are saved* in the output file and there is no need to run the Probe program. If you wish, you may disable the Probe Auto-run feature.* To run the simulation select **Analysis** and then **Simulate** from the Schematics menu bar. At this point one of two things will happen:

1. The PSpice simulation window will pop up:

---

*See Section 4.B, page 130, to alter the Auto-run feature.

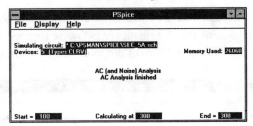

This screen tells you the progress of the simulation.

2. If a simulation has already been run once and you reduced the PSpice window to an icon, the PSpice window will not open. Instead, the Schematics window will remain in front. This screen tells you the progress of the simulation:

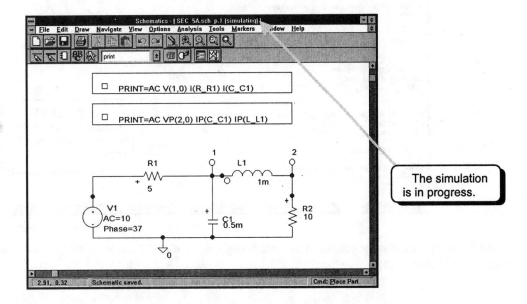

Notice that at the top of the screen Schematics says that it is *(simulating)*. This indicates that a simulation is in progress. When *(simulating)* changes to *(current)*, the simulation is complete. If you wish to see the PSpice simulation window, you can find the window by holding down the ALT key and pressing the TAB key. Remember to keep holding down the ALT key. Each time you type the TAB key, an icon will show on the screen:

Continue pressing the TAB key until you see the PSpice icon:

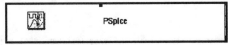

Release the ALT key. The PSpice window will open:

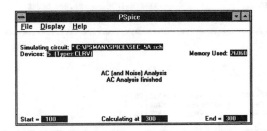

Since we used a print part, the results of the simulation are printed in the output file. When the simulation is finished, press the ALT-ESC key sequence until the Schematics window is on top. From the Schematics menu bar select **Analysis** and then **Examine Output.** The Windows Notepad program will now run and display the output file. You should see the window below:

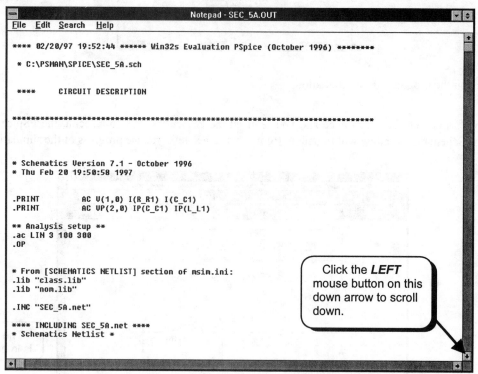

The results are printed near the bottom of the output file. Click the *LEFT* mouse button on the down arrow 🔽 until you see the text shown below:

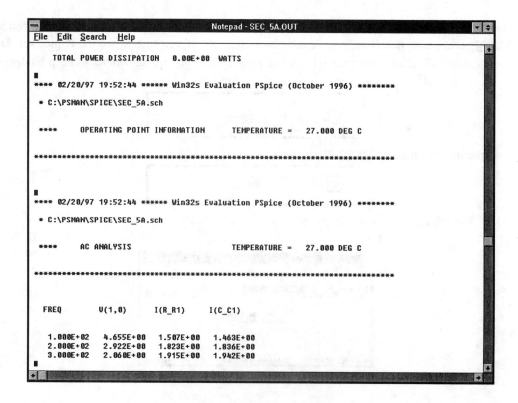

The text shows the results of one of the print statements: at 100 Hz the voltage at node 1 is **4.655** V. At 200 Hz the current through R1 is **1.823** A. At 300 Hz the current through C1 is **1.942** A.

If you click the **LEFT** mouse button on the down arrow several more times, you will reach the text of the second print statement:

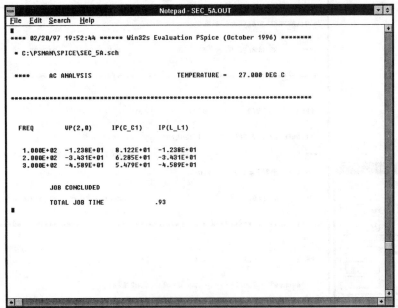

The text shows the results of the other print statement: at a frequency of 100 Hz the phase of the voltage at node 2 is –12.38 degrees; at 200 Hz the phase of the current through C1 is 62.85 degrees; at 300 Hz the phase of the current through L1 is –45.89 degrees.

The order in which the results of the print statement are written to the output file may be different for your simulation. The order of the displayed results depends on the order in which you placed the print statements in your schematic. To exit the Notepad program select **File** and then **Exit** from the Notepad menu bar.

EXERCISE 5-1: Find the magnitude and phase of the voltages at nodes 1, 2, and 3 at a frequency of 1 kHz:

SOLUTION: Use the print part to print the node voltages. The AC Sweep setup is shown below. The results are stored in the output file:

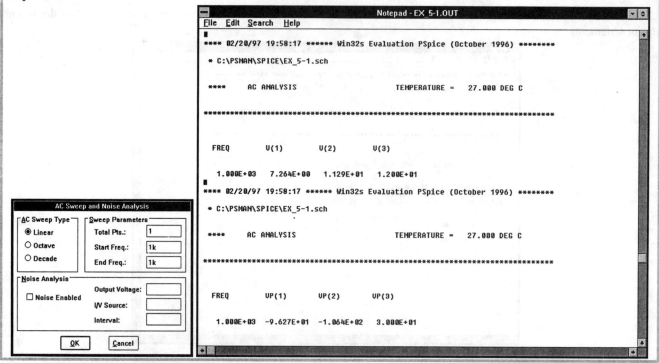

EXERCISE 5-2: Find the magnitude and phase of the voltage at node 1 and the current through **R1** at a frequency of 1 Hz:

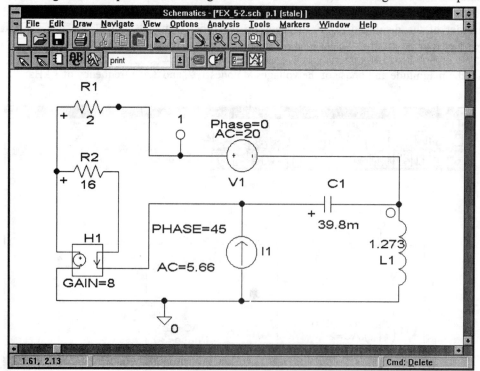

HINT: **H1** is a current-controlled voltage source. The voltage of **H1** is **8** times the current through **R2**.

SOLUTION: Use the print part to print voltages and currents. The results are stored in the output file:

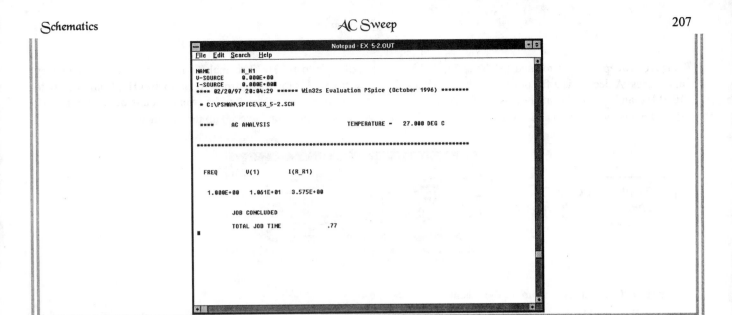

5.B. Magnitude and Phase (Phasors) Graphical Output

Wire the circuit shown below:

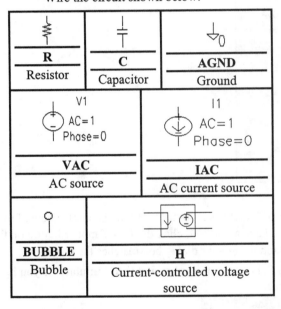

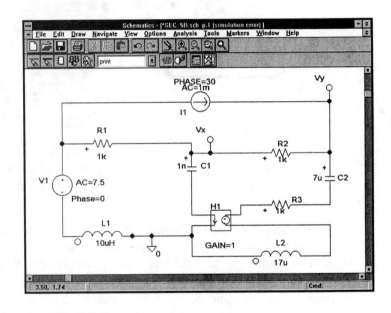

In PSpice, the suffixes **U** and **u** stand for micro, and the suffix **n** stands for nano. Thus, **L1** is 10^{-5} H and **C1** is 10^{-9} F. **I1** is an independent current source. Its magnitude is 1 mA and its phase is 30 degrees.

We will now set up the AC Sweep for this circuit. Remember that the AC Sweep gives us the phase and magnitude of sinusoidal waveforms at specified frequencies. For example, if **Vy** = 5sin(1000t + 30), the result of the AC Sweep will be 5 for the magnitude and 30 for the phase. If you want to see **Vy** displayed as a function of time, you must run a Transient Analysis.

To set up the analysis select **Analysis**, **Setup**, and then **AC Sweep** from the Schematics menu bar. Fill in the dialog box as shown below.

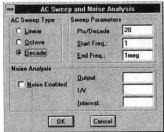

This AC Sweep is set up to simulate the circuit for frequencies of 1 Hz to 10^6 Hz. The analysis is set to sweep the frequency in decades. A decade is a factor of 10 in frequency. A decade is from 1 Hz to 10 Hz, from 10 Hz to 100 Hz, from 100 Hz to 1000 Hz, and so on. The analysis is set for 20 frequency points per decade, so there will be twenty points between 1 and 10 Hz, 20 points between 10 and 100 Hz, and so on. Click the **OK** button when you have set up the parameters:

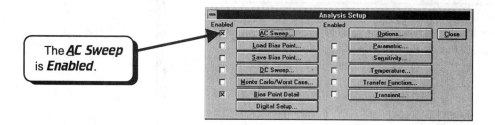

Click the **Close** button to return to the schematic.

We would like to view the results graphically with Probe. If you disabled the Probe Auto-run feature while following the example of the previous section, you will need to enable the Probe Auto-run feature. Select **Analysis** from the Schematics menu bar and then select **Probe Setup**:

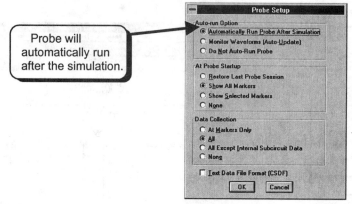

If the circle next to the text **Automatically Run Probe After Simulation** has a dot in it ◉ like the circle above, Probe will automatically run after the PSpice simulation. If the circle does not have a dot in it, click the **LEFT** mouse button on the circle to fill it with a dot. Click the **LEFT** mouse button on the **OK** button when your dialog box matches the one above.

We are now set to run the analysis. Select **Analysis** from the Schematics menu bar and then **Simulate**. If you have no errors in your schematic, the PSpice simulation window should appear:

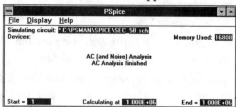

If you do not see the PSpice window, press the **ALT-ESC** key sequence to toggle the active window.

When PSpice has finished the simulation, the Probe program will automatically run and the Probe window will appear:

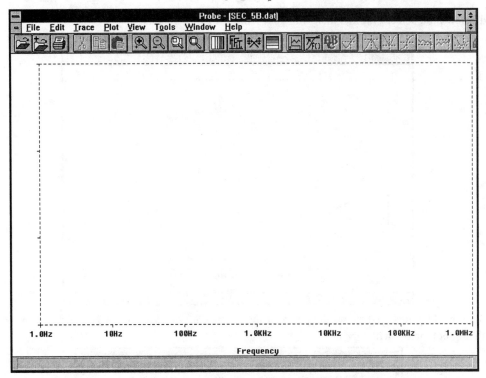

If you do not see this window, press the **ALT-ESC** key sequence to toggle the active window.

Add the trace **V(Vx)** :

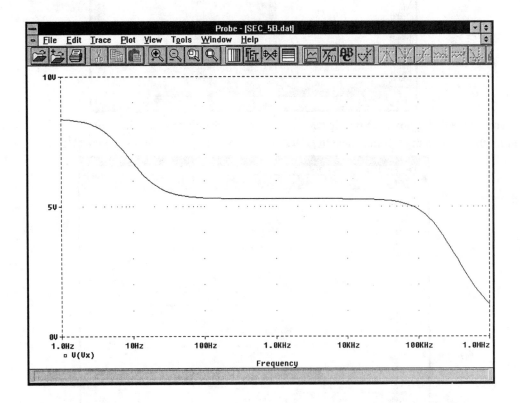

The trace above shows the magnitude of the voltage at node Vx as a function of frequency. If we ask for a voltage or current trace like V(Vx) or I(L2), the plot will be the magnitude of the voltage or current. To plot the phase of a voltage or current we would plot the traces Vp(Vx) or Ip(L2). Vp means "voltage phase" and Ip means "current phase." We will now plot the phase of the voltage at node Vx. Since the phase of any voltage or current will be between 0 and 360, we will add a new plot because the y-axis scale needed to plot the magnitude of a voltage is very different from the scale needed to plot the phase.

To add a new plot, select **Plot** and then **Add Plot** from the Probe menus:

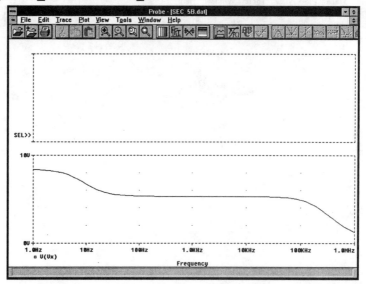

We will now add a trace that shows the phase of the voltage at node **Vx**. Select **Trace** and then **Add** from the Probe menu bar. The dialog box below will appear:

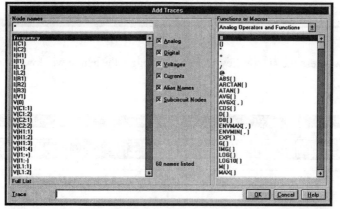

None of the traces in the dialog box are Vp or Ip commands. To plot the phase of a voltage or current, we will need to type in the command for the trace. In the text field next to ***Trace:*** enter the text **Vp (Vx)** and press the **ENTER** key:

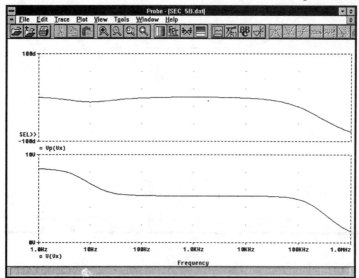

To exit Probe, select **File** and then **Exit** from the Probe menu bar.

EXERCISE 5-3: Find the magnitude and phase of the voltage at node 1 for frequencies from 1 Hz to 100 Hz. Set your AC Sweep to decades with 10 points per decade.

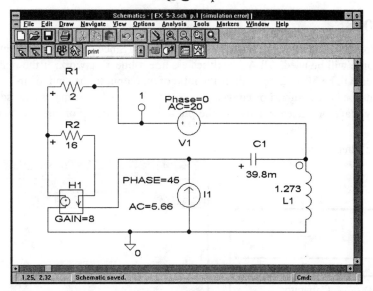

HINT: **H1** is a current-controlled voltage source. The voltage of **H1** is **8** times the current through **R2**.

SOLUTION: The results are displayed using Probe. Obtain two separate plots, since the scale of the magnitude of V(1) is much smaller than the scale of the phase of V(1):

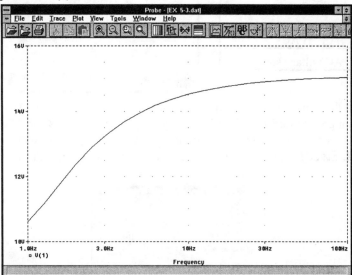

To add a second window in Probe, select **Window** and then **New**. Add the trace **VP(1)**:

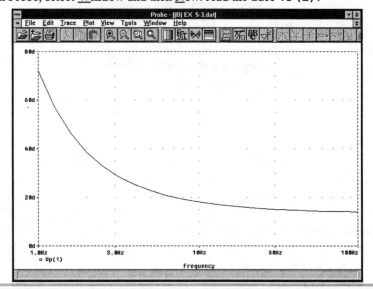

5.C. Bode Plots

Bode plots are plots of magnitude and phase versus frequency. Since we are usually interested in the magnitude of the gain, an AC 1-volt source will be used. All AC analyses assume a linear network: if the output for a 1 V source is 3 V, the output for a 10 V source will be 30 V. Since gain is the ratio of an output to the source, and since the networks are linear, the magnitude of the input does not matter. For convenience we will use a 1-volt magnitude source. Bode phase plots will give us the phase of any voltage or current relative to the phase of the source. For simplicity we will set the phase of the source to zero.

Wire the low-pass filter shown:

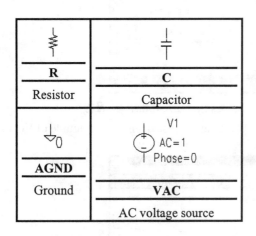

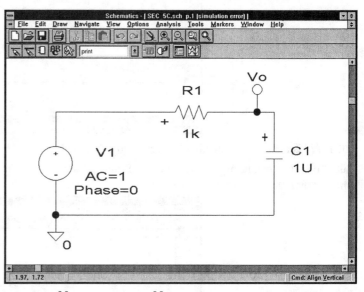

At low frequencies the capacitor is an open circuit and **Vo** should equal **V1**. At high frequencies the capacitor becomes a short, and the gain goes toward zero. The 3 dB frequency of the circuit is $\omega = 1/RC = 1000$ rad/s $\cong 159$ Hz. We will set up an AC Sweep to sweep the frequency from 1 Hz to 10 kHz. Select **Analysis** and then **Setup** from the Schematics menus, and then click the **AC Sweep** button. The AC Sweep dialog box will appear. Fill in the parameters as shown in the AC Sweep dialog box below:

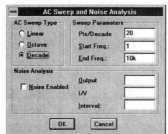

Click the **OK** button and then click the **Close** button to return to the schematic. Run the analysis by selecting **Analysis** and then **Simulate** from the Schematics menus. The PSpice simulation window will appear:

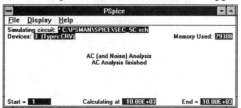

If you cannot see the window, press the ALT-ESC key sequence to toggle the active window. You may have to press the ALT-ESC sequence several times if you have several open windows. When PSpice has successfully completed the simulation, the Probe window should appear. If the Probe window does not appear,[*] press the ALT-ESC key sequence to toggle the active window or select **Analysis** and then **Run Probe** from the Schematics menu bar. The Probe window will appear:

---

[*]Check the status of the Probe Auto-run feature. See Section 4.B, page 130.

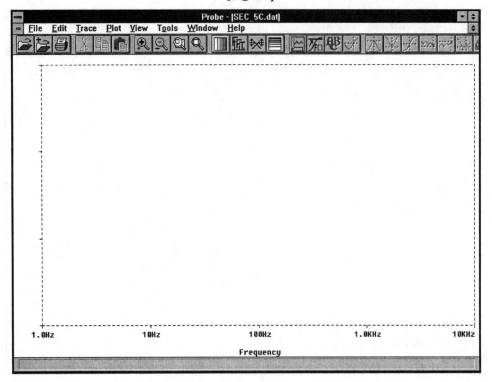

The phase and magnitude can be displayed as shown in previous sections. In this example we would like to show a Bode plot that displays the magnitude in decibels. To display Vo in decibels, we need to display the trace dB(V(Vo)). Select **Trace** and then **A**dd from the Probe menu bar. In the text field next to **Trace** enter the text **dB (V (Vo))**:

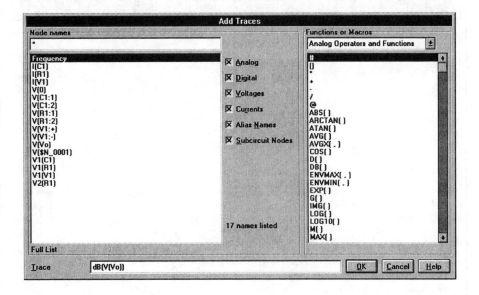

The **dB** command takes $20\log_{10}$ of the specified voltage. Thus, **dB(V(Vo))** is equivalent to $20\log(V(Vo))$. Since our source voltage (**V1**) had a magnitude of 1 V, **dB(V(Vo))** is equivalent to dB(V(Vo)/V1),[*] which is the Bode plot of the gain of this circuit. Click the **OK** button to plot the trace:

---

[*]Note that "V1" is not a valid expression for the **Trace** command. It is used only for clarity. Instead of "V1" you would have to enter **V(R1:1)**. Note that **V(R1:1)** is listed as one of the choices in the **Add Traces** dialog box.

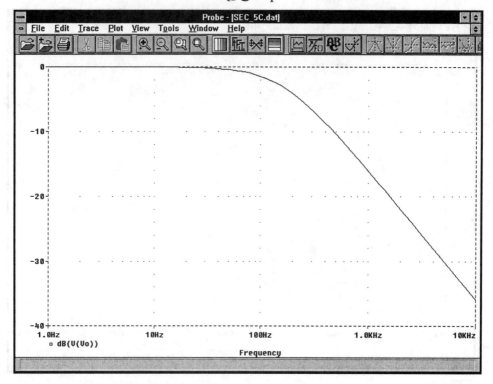

Note that the y-axis is in dB. A gain of 1 is equal to 0 dB.

We can now use the cursors to find the 3 dB frequency. (See Section 2.K for a full explanation of cursors.) To display the cursors, select **Tools**, **Cursors**, and then **Display** from the Probe menus. The cursors will appear:

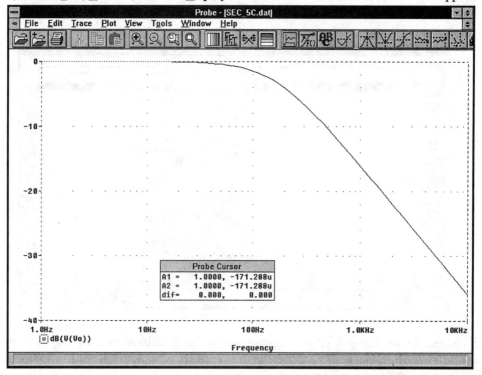

Notice that the values displayed by the cursors are in hertz and decibels. The first cursor is controlled by the *LEFT* mouse button or the arrow keys (◁▷). The second cursor is controlled by the *RIGHT* mouse button or the shift key plus the arrow keys. The difference, ***dif***, is the difference between the values of the first cursor and the values of the second cursor. To find the 3 dB point, we want to leave the second cursor in its original place and move the first cursor until the difference in magnitude is –3. To move the first cursor press and hold the RIGHT ARROW key. Move the first cursor until the difference is approximately –3:

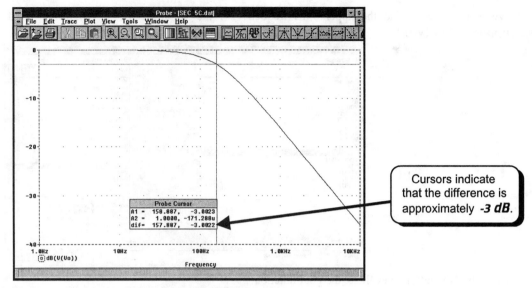

The cursors show that the 3 dB frequency is approximately at 158.887 Hz.

5.C.1. Using Goal Functions to Find the Upper 3 dB Frequency

A second method can be used to find the upper 3 dB frequency found in the previous section. We will continue from the end of the previous section. First, hide the cursors by selecting **Tools**, **Cursors**, and then **Display** from the Probe menus. Next, select **Window** and then **New** to open a new Probe window:

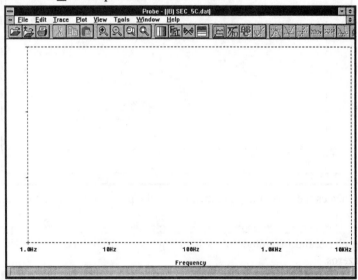

We will now evaluate the goal function. In Probe, select **Trace** and then **Eval Goal Function**:

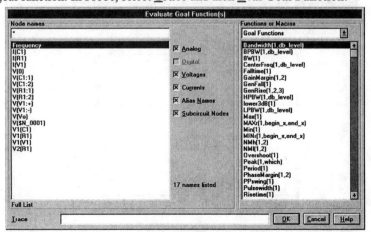

We would like to find the upper 3 dB frequency of the voltage trace V(Vo). Enter the following **Trace** command, `upper3dB(V(Vo))`:

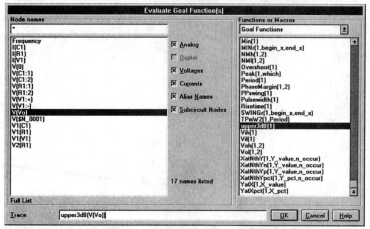

Click the **OK** button to evaluate the goal function:

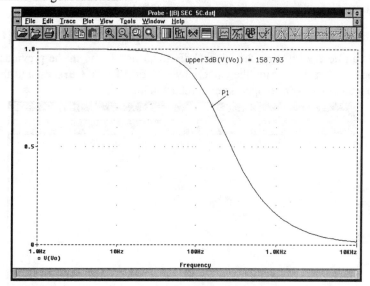

Probe plots the trace V(Vo), locates the 3dB point, and then marks the point on the plot. Note that the plot of V(Vo) is not in decibels.

EXERCISE 5-4: Plot the Bode phase and magnitude plots for frequencies from 1 Hz to 1 MHz. Use the cursors to find the frequencies of the poles and zeros.

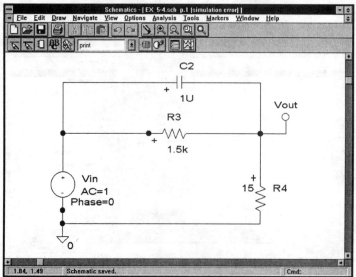

ANALYTIC SOLUTION: At low frequencies, the capacitor is an open circuit and the voltage at **Vout** can be obtained by the voltage divider of **R3** and **R4**:

$$\frac{V_{out}}{V_{in}} = \frac{R_4}{R_3 + R_4} = 0.0099 = -40 \text{ dB}$$

At high frequencies, the capacitor is a short circuit and **Vout = Vin** so

$$\frac{V_{out}}{V_{in}} = 1 = 0 \text{ dB}$$

Thus, we should expect the magnitude plot to start at –40 dB and finish at 0 dB. The next question is, where are the poles and zeros? Let

$$Z = Z_c \| R_3 = \frac{(1/j\omega C_2)R_3}{(1/j\omega C_2) + R_3} = \frac{R_3}{1 + j\omega R_3 C_2}$$

Substituting Z for $Z_c \| R_3$ yields the equivalent circuit:

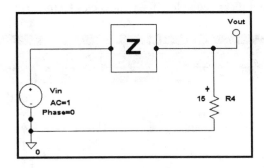

An expression for the gain can now be obtained from the voltage divider of **R4** and **Z**:

$$\frac{V_{out}}{V_{in}} = \frac{R_4}{Z + R_4} = \frac{R_4}{\left(\dfrac{R_3}{1 + j\omega R_3 C_2}\right) + R_4}$$

Multiplying the numerator and denominator by $1+j\omega R_3 C_2$ gives

$$\frac{V_{out}}{V_{in}} = \frac{R_4(1 + j\omega R_3 C_2)}{R_3 + R_4 + j\omega C_2 R_3 R_4}$$

We see that the zero is at $1 = \omega R_3 C_2$, or

$$\omega_z = \frac{1}{R_3 C_2} = \frac{1}{(1500\ \Omega)(1\ \mu F)} = 666 \text{ rad}/\text{s} = 106 \text{ Hz}$$

The pole is at $R_3 + R_4 = \omega C_2 R_3 R_4$, or

$$\omega_p = \frac{R_3 + R_4}{C_2 R_3 R_4} = \frac{1}{C_2\left(\dfrac{R_3 R_4}{R_3 + R_4}\right)} = \frac{1}{C_2(R_3 \| R_4)}$$

$$= \frac{1}{(1\mu F)(1500\Omega \| 15\Omega)} = 66.6 \text{ krad/sec} = 10.6 \text{ kHz}$$

SOLUTION: Set up an AC Sweep from 1 Hz to 1 MHz. Simulate the circuit and then run Probe :

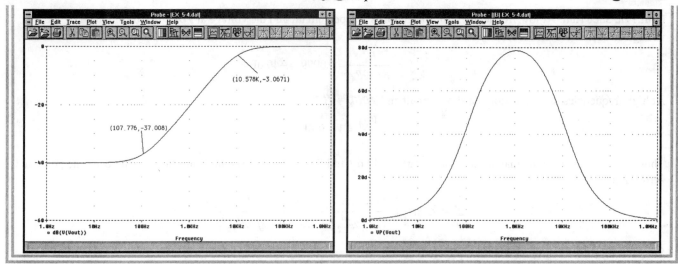

EXERCISE 5-5: Plot the Bode phase and magnitude plots for frequencies from 1 Hz to 1 MHz. Use the cursors to find the frequencies of the poles and zeros.

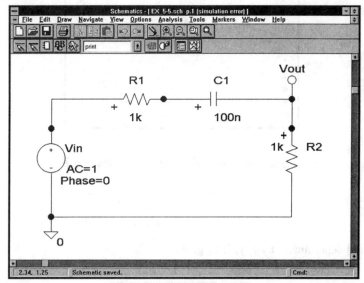

ANALYTIC SOLUTION: At low frequencies, the capacitor is an open circuit and the voltage at **Vout** is zero. As the frequency is increased, the impedance of the capacitor will decrease and the voltage at **Vout** will increase. Thus, at low frequencies we should expect the gain in decibels to be a large negative number and to increase at a rate of 20 dB/decade. At high frequencies, the capacitor is a short circuit and **Vout** can be obtained from the voltage divider of R1 and R2:

$$\frac{V_{out}}{V_{in}} = \frac{R_2}{R_2 + R_1} = 0.5 = -6 \text{ dB}$$

Thus, we should expect the magnitude plot to start at a negative value in decibels and finish at –6 dB. The next question is, where are the poles and zeros? The gain of the circuit can be obtained from the voltage divider of **R1**, **R2**, and **C1**:

$$\frac{V_{out}}{V_{in}} = \frac{R_2}{R_1 + R_2 + Z_c} = \frac{R_2}{R_1 + R_2 + \left(\dfrac{1}{j\omega C_1}\right)} = \frac{j\omega C_1 R_2}{1 + j\omega C_1 \left(R_1 + R_2\right)}$$

We see that there is a zero at ω=0 rad/sec and a pole at $1=\omega C_1(R_1+R_2)$, or

$$\omega_p = \frac{1}{C_1(R_1 + R_2)} = \frac{1}{(100 \text{ nF})(1000 \ \Omega + 1000 \ \Omega)} = 5000 \text{ rad / s} = 795.8 \text{ Hz}$$

SOLUTION: The results are displayed using Probe :

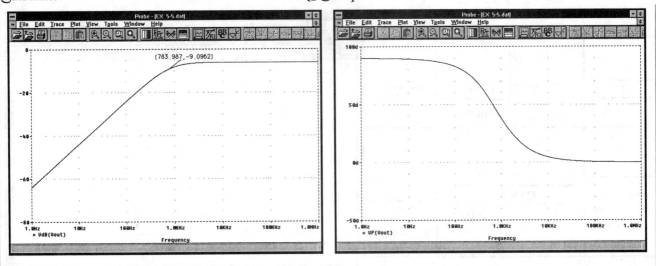

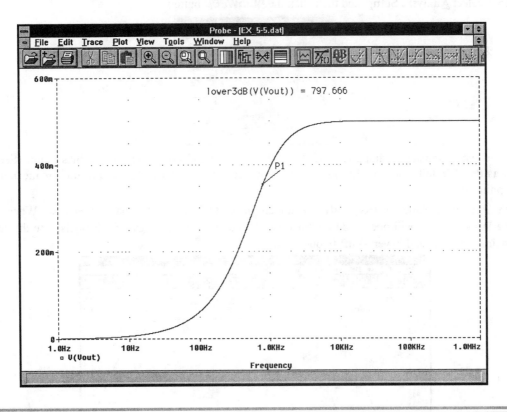

The goal functions can also be used to find the lower 3 dB frequency. Select **Trace** and then **Eval Goal Function**. Evaluate the expression `lower3dB(V(Vout))`:

5.D. Amplifier Gain Analysis

One of the most important applications of the AC Sweep is to see the frequency response of an amplifier. If an AC Sweep is performed on a circuit with a transistor, the DC bias point is calculated and the transistor is replaced by a small-signal model around the bias point. The AC Sweep is then performed on the linearized model of the transistor. The AC Sweep can only be used to find the small-signal gain and frequency response. Voltage swing, clipping, and saturation information must be obtained from the transient simulation, or by using the operating point information.

Wire the amplifier circuit as shown:

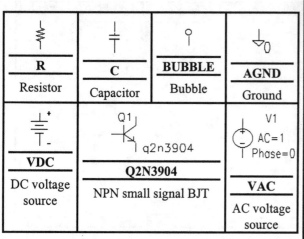

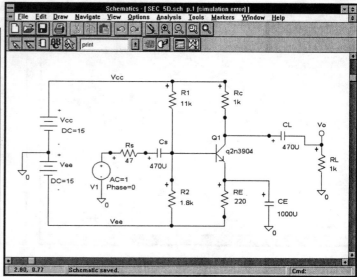

We would like to see the gain of this amplifier from 1 Hz to 100 MHz. Set up the AC Sweep as shown below. To obtain the dialog box below, select **Analysis, Setup,** and then click the **AC Sweep** button:

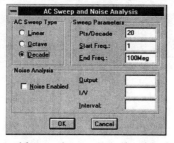

Whenever you simulate an amplifier with transistors you should always check the bias point. (See Section 3.E, "**Transistor Bias Point Detail,**" on page 121.) If the bias is not correct, then the AC Sweep results are meaningless. Always check the bias point first.

Run the PSpice simulation: select **Analysis** and then **Simulate** from the Schematics menu bar. When the simulation is complete, the Probe window will open. Add the trace **dB(V(Vo))** [*] to plot the gain in decibels. Use the cursors to label the mid-band gain, and upper and lower –3 dB frequencies.

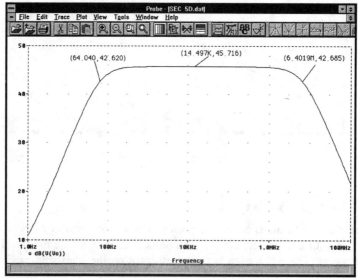

We can see that the mid-band gain is 45.7 dB, the upper 3 dB frequency is 6.4 MHz, and the lower 3 dB frequency is 64 Hz.

EXERCISE 5-6: Find the mid-band gain and upper and lower 3 dB frequencies of the common-base amplifier shown below.

---

[*]See page 213 for an explanation of the "dB" command.

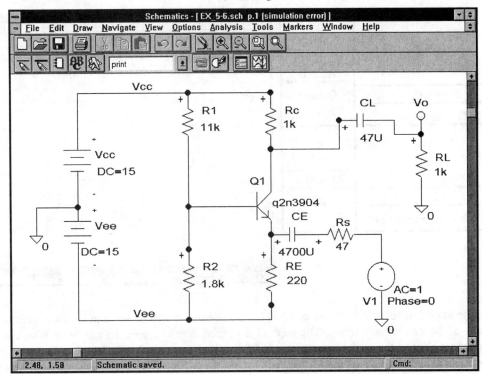

SOLUTION: The results are displayed using Probe :

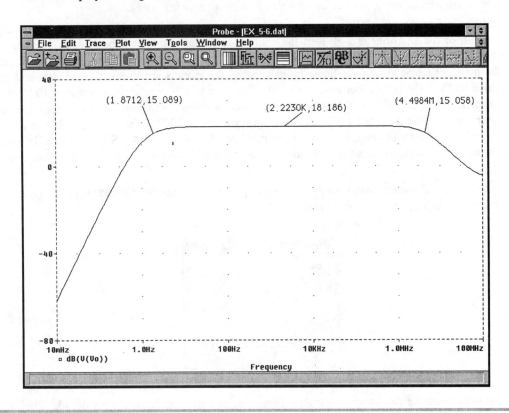

5.E. Operational Amplifier Gain

In this section we will use PSpice to determine the gain and bandwidth of an operational amplifier with negative feedback. Wire the circuit shown below:

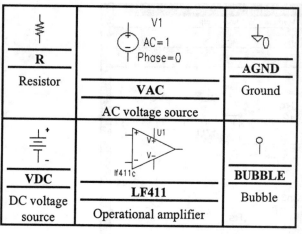

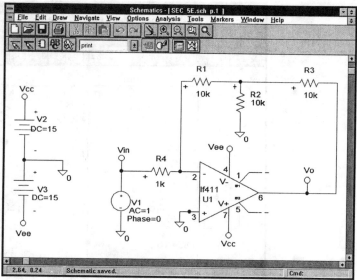

This example uses the non-ideal op-amp model of an LF411. Since there is only one op-amp in this circuit, we will not reach the component limit of the evaluation version. The method described here can be used for circuits with several op-amps. However, the component limit of the evaluation version of PSpice limits us to only two non-ideal op-amp models. If more than two op-amps are needed, the Ideal_OPAMP model can be used. Since the ideal op-amp model has no frequency limitations, it cannot be used to find the bandwidth, but it can be used to find the gain.

Notice that the **lf411** has power supply terminals **V+** and **V-**. They must be connected to the appropriate DC supplies. In the circuit, BUBBLEs are used to make connections without drawing wires.

We would like to find the gain of this amplifier, V_o/V_{in}. Since the magnitude of V_{in} is one volt, the magnitude of V_o is the gain. If the gain of the amplifier is 200, then the magnitude of the output will be 200 V. This may seem a little unreasonable since the DC supplies are ±15 V. When an AC Sweep is performed, the operating points of all non-linear parts are found, and the parts are replaced by their linear models. Since the models are linear, the voltages and currents in the circuit are not limited by the DC supplies. For a linear circuit, if an input of 1 V produces an output of 10 V, then an input of 1,000,000 V will produce an output of 10,000,000 V. Thus, for gain purposes the magnitude of the input does not matter and a 1 V input is chosen for convenience. For an AC simulation a magnitude of 200 V is not unreasonable. A magnitude of 200 V is not physically possible for our op-amp, but it is a valid number when using the AC Sweep to calculate the gain. If you wish to observe the maximum voltage output of a circuit, you must run a Transient Analysis. This is done in Section 6.G on page 273.

We must set up an AC Sweep. Select **Analysis**, **Setup**, and then click the **AC Sweep** button. The AC Sweep dialog box will appear. Fill in the dialog box as shown below:

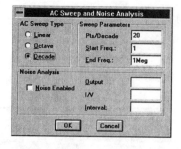

The sweep is set up to simulate the circuit for frequencies from 1 Hz to 1 MHz at 20 points per decade. Click the **OK** button to accept the settings, and then click the **Close** button to return to the schematic. Run PSpice by selecting **Analysis** and then **Simulate** from the Schematics menu bar. When Probe runs, add the trace **V(Vo)**. To add a trace, select **Trace** and then **Add** from the Probe menu bar. You will see the amplifier gain as a function of frequency:

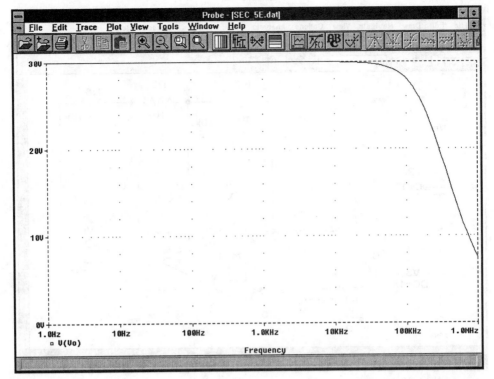

This plot shows us that the gain of the amplifier is 30. If we wish to find the 3 dB frequency for this amplifier, we must display the trace in dB.* We will add a second window to display the new trace. Select **Window** and then **New** to open a new window and then add the trace dB(V(Vo)). Use the cursors to locate the −3 dB frequency:

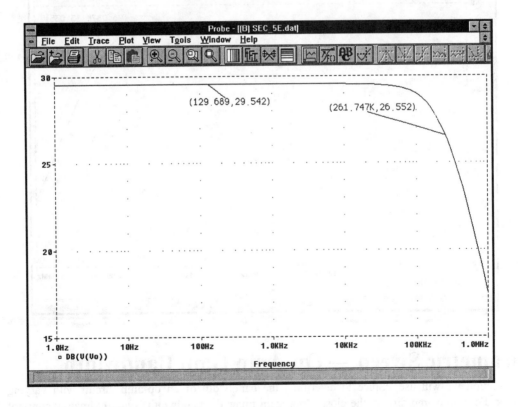

EXERCISE 5-7: Find the mid-band gain and upper 3 dB frequency for the amplifier shown below:

---

*See page 213 for an explanation of the "dB" command.

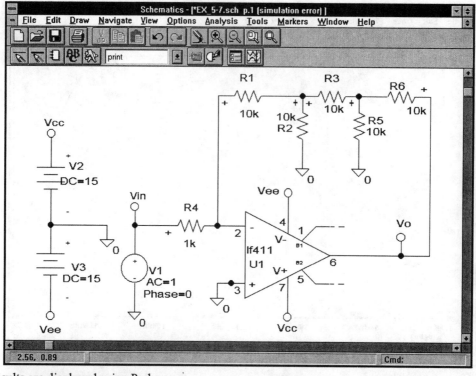

SOLUTION: The results are displayed using Probe :

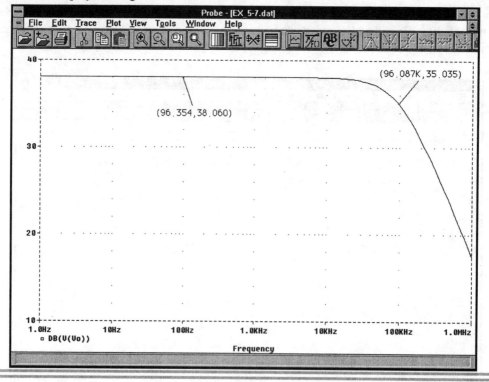

5.F. Parametric Sweep — Op-Amp Gain Bandwidth

In this section we will use PSpice to determine the bandwidth of an op-amp circuit with varying amounts of negative feedback. For an op-amp circuit, the closed-loop gain times the bandwidth is approximately constant. To observe this property, we will run a simulation that creates a Bode plot for several different closed-loop gains. We will use the circuit below:

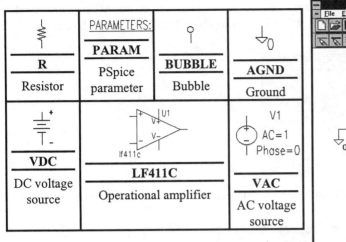

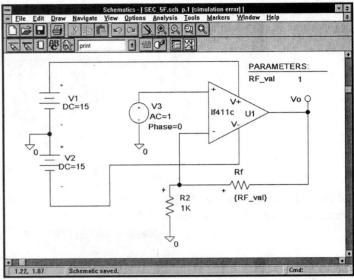

This circuit uses a **lf411c** op-amp macro model. All op-amp models, except the ideal op-amp model, include bias currents, offset voltages, slew rate limitations, and frequency limitations. Also note that the op-amp model requires DC supplies. The model will not work without the supplies. In the evaluation version, only two op-amp models can be used before reaching the component limitation. If you need more than two op-amps, use the ideal op-amp model.

Looking at the schematic, we see that resistor **Rf** looks a little different. **Rf** does not have a numerical value but has the text **{RF_val}**. **{RF_val}** is called a parameter and is defined by the **PARAMETERS:** part. The default value of **{RF_val}** is set by the **PARAMETERS:** part, in this case 1 Ω. Parameters are used to easily change values in a circuit, pass values to a subcircuit, or change the values of components during a simulation. We will use the parameter **{RF_val}** to change the value of the resistor **Rf** during the simulation. This will allow us to change the gain of the op-amp circuit.

First, we will look at the attributes of **Rf** to see how the parameter is entered. Double click on the resistor graphic, ⊣◦◦◦⊢, for **Rf**. The dialog box below will appear:

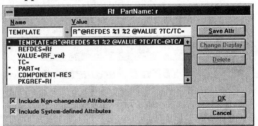

Where it says **VALUE=**, you would normally enter the value of the resistor, 1k, for example. In this case, the value of the resistor is the value of the parameter **{RF_val}**. This parameter must be defined somewhere in the circuit. To define a parameter you must place a part called PARAM in your circuit. This part has been placed in the circuit above and is shown on the schematic as the text **PARAMETERS:**. To edit the attributes of the **PARAMETERS:** part, double click the **LEFT** mouse button on the text **PARAMETERS:**. The dialog box below will appear.

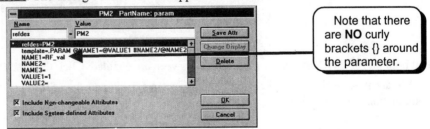

Note that there are **NO** curly brackets {} around the parameter.

This part has several attributes. The **NAME1=** attribute defines the name of the first parameter. The **VALUE1=** attribute defines the default value of the first parameter. For each parameter part, three parameters can be defined. In the dialog box above only one parameter has been defined. The line **NAME1=RF_val** declares **{RF_val}** as a parameter. The line **VALUE1=1** specifies the default value of **{RF_val}** to be **1**.

We would like to see the frequency response of the op-amp circuit, so we must set up an AC Sweep. Select **Analysis, Setup**, and then click the **AC Sweep** button to obtain the AC Sweep dialog box. Fill in the dialog box as shown below:

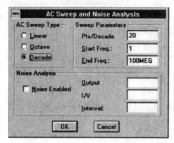

The dialog box is set to sweep frequency from 1 Hz to 100 MHz at 20 points per decade. When you are done setting up the parameters for the AC Sweep, click the **OK** button:

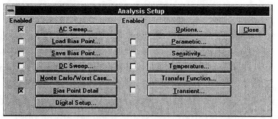

We would now like to set up values for the parameter **RF_val**. Parameters can be changed using the Parametric setup dialog box. From the dialog box above, click the **Parametric** button. Fill in the dialog box as shown:

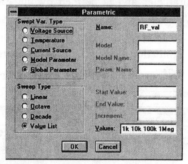

The parameter we have defined (**RF_val**) is a **Global Parameter**. The sweep types are similar to the DC and AC Sweeps discussed previously. We would like specific values for the parameter, so we will use the value list. Logically, the parametric sweep is executed outside the AC Sweep. First, **{RF_val}** will be set to **1k** and then the AC Sweep will be performed. Next, **{RF_val}** will be set to **10k** and then the AC Sweep will be performed. Then, **{RF_val}** will be set to **100k**, and so on. Click the **OK** button to accept the settings:

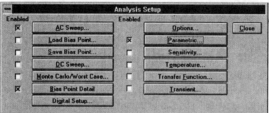

Note that both the **AC Sweep** and the **Parametric** sweep are enabled. Click the **Close** button to return to the schematic.

We are now ready to run the simulation. Select **Analysis** and then **Simulate** from the Schematics menu bar. The PSpice simulation window will appear:

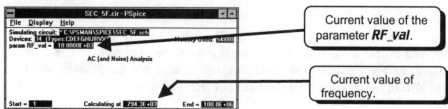

As you watch the simulation's progress you will see that the AC Sweep runs four times, once for each value of the parameter *{RF_val}*. In the window above, the current AC Sweep is for *{RF_val}* = 10,000.

When the simulation is complete, Probe will run. You will be presented with the following screen:

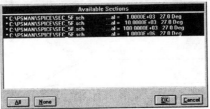

Since there were four simulations, one for each value of parameter RF_val. Probe is asking which of the simulations we would like to view. You can look at any combination of the runs. By default, all of the simulations are selected. We would like to look at all of the simulations at the same time, so click the **OK** button to accept the runs. You will be presented with the standard blank Probe screen. Add the trace **dB(V(Vo))**.* To add the trace, select **Trace** and then **Add** from the Probe menu bar. This trace shows the Bode magnitude plot of the gain. You will see the screen below.

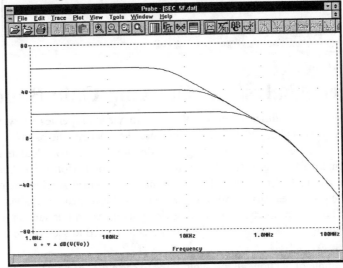

The plot shows four Bode plots, one for each value of *{RF_val}*. We see that for larger gains the bandwidth is reduced.

EXERCISE 5-8: For the inverting op-amp below, find the upper 3 dB frequency for mid-band gains of −1, −10, −100, and −1000. Show the Bode magnitude plots for all of the gains on the same graph.

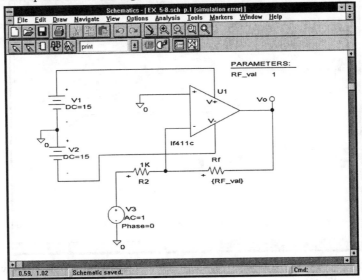

---

*See page 213 for an explanation of the "dB" command.

SOLUTION: The results are displayed using Probe :

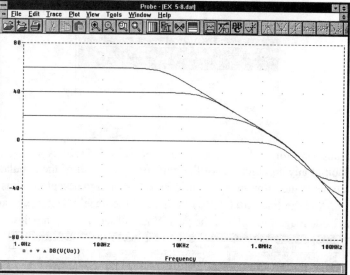

5.G. Performance Analysis — Op-Amp Gain Bandwidth

The Performance Analysis capabilities of Probe are used to view properties of waveforms that are not easily described. Examples are amplifier bandwidth, rise time, and overshoot. To calculate the bandwidth of a circuit, you must find the maximum gain, and then find the frequency where the gain is down by 3 dB. To calculate rise time, you must find the 10% and 90% points, and then find the time difference between the points. The Performance Analysis gives us the capability to plot these properties versus a parameter or device tolerances. The Performance Analysis is used in conjunction with the Parametric Sweep to see how the properties vary versus a parameter. The Performance Analysis is used in conjunction with the Monte Carlo Analysis to see how the properties vary with device tolerances. In this section we will plot the bandwidth of an amplifier versus the value of the feedback resistor. See Sections 9.B.3 and 9.E to see how to use the Performance Analysis in conjunction with the Monte Carlo Analysis.

Suppose that for the example of Section 5.F, we would like to see a plot of how the upper 3 dB frequency is affected by the value of the feedback resistor, **Rf**. This plot can be accomplished using the Performance Analysis capabilities of Probe. Repeat the procedure of Section 5.F. When you obtain the plot on page 227, you may continue with this section. The plot on page 227 is repeated as follows:

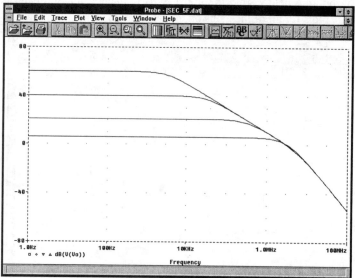

Before we look at the Performance Analysis we would like to create a new plot window. Select **Window** and then **New**. A second empty window will appear:

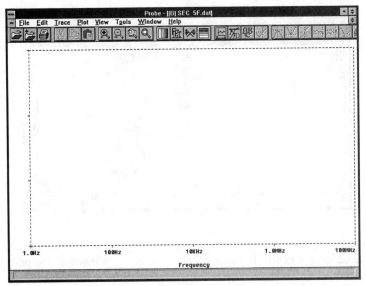

To plot the upper 3 dB bandwidth versus the parameter **RF_val** we must enable the Performance Analysis. From the Probe menus select **Plot** and then **X Axis Settings**:

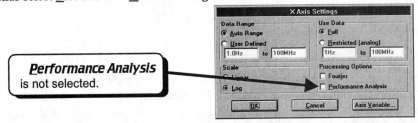

Under **Processing Options** we see that the square □ next to **Performance Analysis** does not have an **X** in it, indicating that it is not enabled. To enable the **Performance Analysis** click the *LEFT* mouse button on the text **Performance Analysis**. The square should fill with an **X**, indicating that the **Performance Analysis** is enabled:

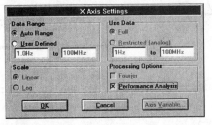

Click the **OK** button to return to the Probe screen:

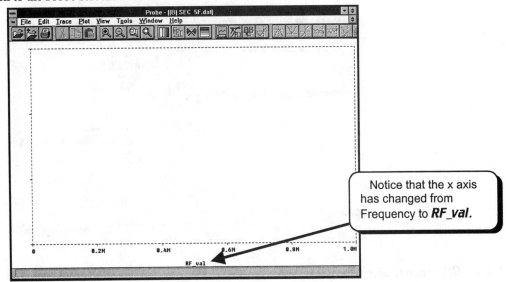

We now wish to plot the upper 3 dB frequency versus **RF_val**. Select **Trace** and then **Add**:

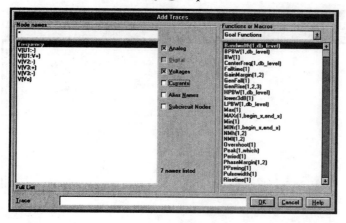

The left pane shows the normal voltage and current traces available to us. The right pane contains goal functions. Goal functions are defined in a file called msim.prb. If you view this file using Windows Notepad, you will see the following function near the middle of the file:

```
upper3dB(1) = x1
* Find the upper 3 dB frequency
{
1|sfle(max-3dB,n) !1;
}
```

The name of the function is **upper3dB**. It has **1** input argument. **1|sfle** means search the first input forward and find a level. The level we are looking for is 3 dB less than the maximum (**max-3**). The **n** means find the specified level when the trace has a negative-going slope. When the point is found, the text **!1** designates its coordinates as x1 and y1. The function returns the x-coordinate of the point (**upper3dB(1) = x1**). The x-axis of a frequency trace is frequency, so this function returns the frequency of the upper 3 dB point. A second function is:

```
lower3dB(1) = x1
* Find the lower 3 dB frequency
{
1|sfle(max-3dB,p) !1;
}
```

Indicates a positive-going slope.

This function is similar to the **upper3dB** function, except that it finds the 3 dB point when the trace has a positive-going slope. This will mark the coordinates of the lower 3 dB point.

Type in the trace **upper3dB(V(Vo))**:

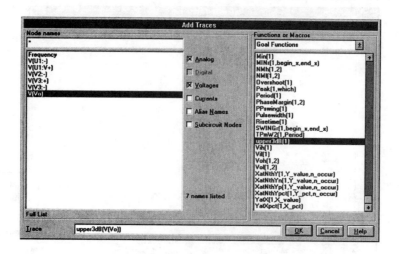

Click the **OK** button to accept the trace:

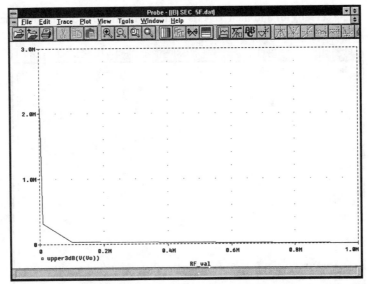

This plot is not too informative when plotted on a linear scale. A log-log plot is much more useful. To change the x-axis to a log scale select **Plot** and then **X Axis Settings** from the Probe menu:

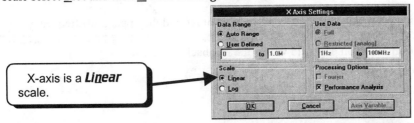

We see that the x-axis is plotted on a linear scale. To change the x-axis to a log scale, click the *LEFT* mouse button on the text *Log*. The circle next to the text *Log* should fill with a dot ⊙, indicating that a log scale is selected:

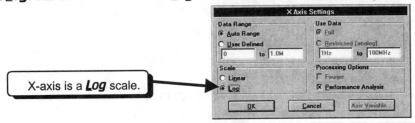

Click the *OK* button to return to the plot:

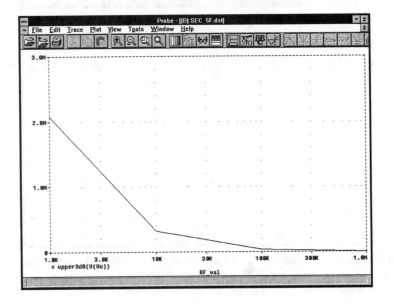

To change the y-axis to a log scale select **Plot** and then **Y Axis Settings** from the Probe menu. Change the axis to a log scale:

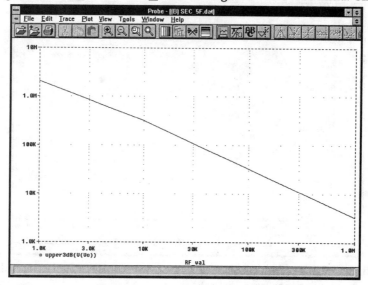

When plotted on a log scale, we can easily see the relationship between the value of **Rf** and the upper 3 dB frequency.

EXERCISE 5-9: Plot the gain bandwidth of the amplifier on page 225 versus the feedback resistor R_F.

SOLUTION: Rerun the simulation with more detail in the Parametric Sweep:

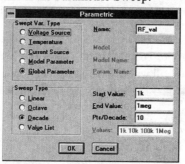

In Probe use the Performance Analysis and add the trace **Max(V(Vo))*upper3dB(V(Vo))**. Change the x-axis to a log scale:

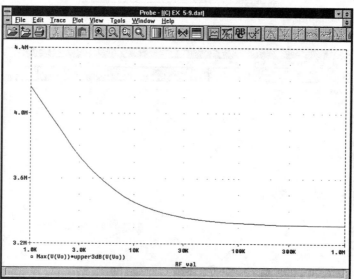

The function Max(V(Vo)) gives us the maximum value of Vo for each trace. Since the input was one volt, the Max(V(Vo)) is also the maximum gain for each trace. Thus Max(V(Vo))*upper3dB(V(Vo)) is the maximum gain for each trace times the upper 3 dB frequency for each trace, or the gain times bandwidth for each trace.

We see from the plot that as **Rf_val** changes by a factor of 1000 (the gain changes by a factor of 1000), the gain-bandwidth product is relatively constant, ranging between 3.3 MHz and 4.2 MHz.

5.H. Mutual Inductance

Mutual inductance requires two parts, the inductors (L) and the coupling between inductors (K). We will illustrate the use of the coupling part K with two circuits. The first circuit will have three inductors with unequal coupling. The second circuit will have four inductors with equal coupling. Wire the circuit shown below. The dots on the inductors are critical since they indicate the polarity of the mutual coupling. Make sure the dots on your schematic agree with the ones on the schematic shown.

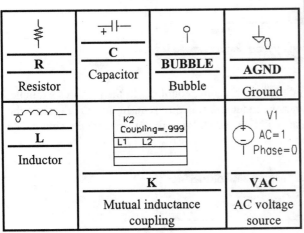

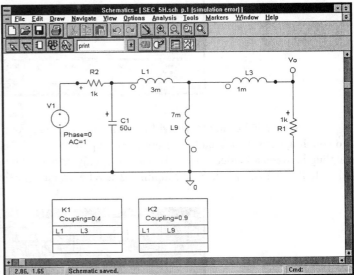

Note that there is coupling only between **L1** and **L3** and between **L1** and **L9**. If you want coupling between **L3** and **L9**, you will have to add another mutual inductance coupling part (K). To change the value of the coupling, double click the **LEFT** mouse button on the text **Coupling=**. To change the inductor names in the coupling part, double click the **LEFT** mouse button on the inductor names inside the coupling part box. For example, to change **L1** in coupling part **K1**, double click the **LEFT** mouse button on the text **L1**. To change **L9** in coupling part **K2**, double click the **LEFT** mouse button on the text **L9**. This circuit required two coupling parts (K) because the coupling between the pairs of inductors is different.

We would like to run an AC Sweep to see how the magnitude and phase of **Vo** change with frequency. Set up an AC Sweep for frequencies from 1 Hz to 1 MHz at 20 points per decade. To set up the AC Sweep, select **Analysis**, **Setup**, and then click the **AC Sweep** button. Fill in the AC Sweep dialog box as shown below:

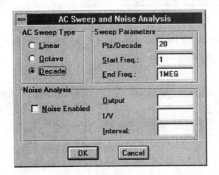

Run PSpice: select **Analysis** and then **Simulate** from the Schematics menu bar. When the simulation is complete, run Probe. Add the trace **V(Vo)** (select **Trace** and then **Add** from the Probe menus). The trace below will appear:

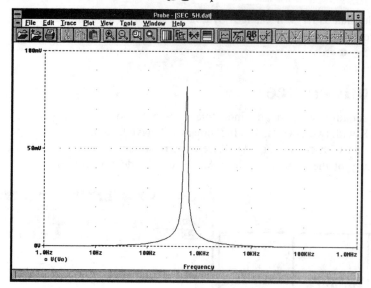

As a second example, we will illustrate a circuit with four inductors with equal coupling. In the previous example, two coupling parts were required because the coupling between the various inductors was different. In this circuit, since the coupling between all inductors is to be the same, only one coupling part will be needed. We will not simulate the circuit. We will only illustrate how to use the coupling part. Wire the circuit below:

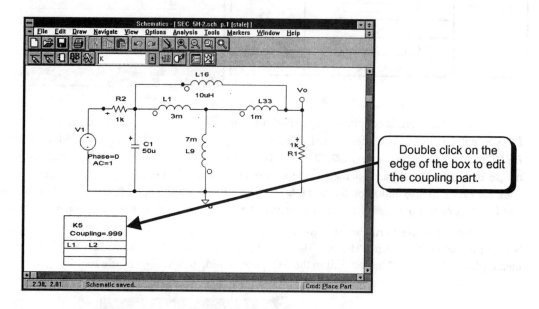

When you get the coupling part it will show only two inductors, L1 and L2, as being coupled. Double click the *LEFT* mouse button on the box around the coupling part to edit its attributes:

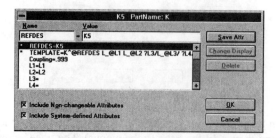

If you scroll down through the dialog box, you will notice that you can name up to nine inductors. This part can be used to couple nine different inductors, all with the same coefficient of coupling. We would like to specify the coefficient of coupling to be 0.85 between all four inductors. Fill in the dialog box as shown:

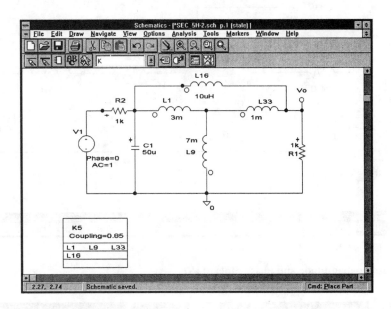

The dialog box must be filled out sequentially; that is, L3 must be specified before L4, L4 must be specified before L5, and so on. When you click the **OK** button, the new inductor names will fill the mutual inductance part:

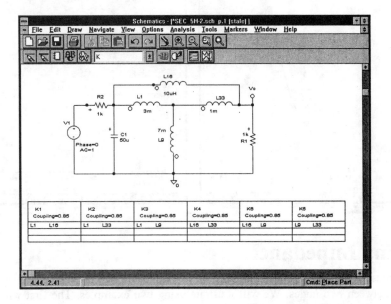

The following circuit is equivalent to the one above:

EXERCISE 5-10: Find the magnitude and phase of the voltage Vo for frequencies from 10 Hz to 1 kHz. Note that L1 is coupled to L2, but L3 is not coupled to either L1 or L2. The 100 GΩ resistor is used to isolate the two loops. In an actual circuit there would not be a connection between the two loops. However, PSpice requires all portions of a circuit to be referenced to ground. If the 100 GΩ resistor were replaced by an open circuit, the circuit would still function in the same way if tested in the lab, but the right loop would not have a ground reference. Without the 100 GΩ resistor, PSpice will generate an error message and will not simulate the circuit.

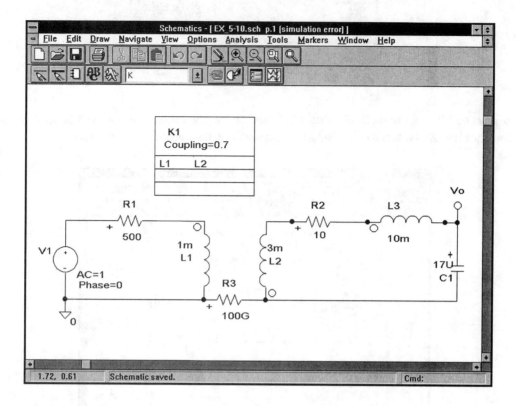

SOLUTION: Use Probe to display magnitude and phase plots of **Vo**:

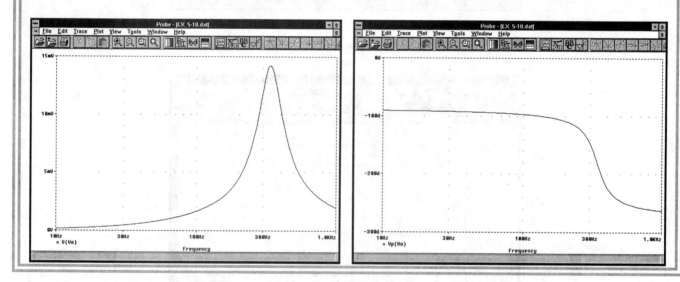

5.I. Measuring Impedance

The technique presented here can be used to measure the impedance or resistance between any two nodes. We will find the AC impedance between two nodes. We will illustrate using two examples. The first will be a passive circuit with resistors only. The second will be a jFET source follower.

5.I.1. Impedance Measurement of a Passive Circuit

We will find the impedance between nodes **1** and **0** in the circuit below:

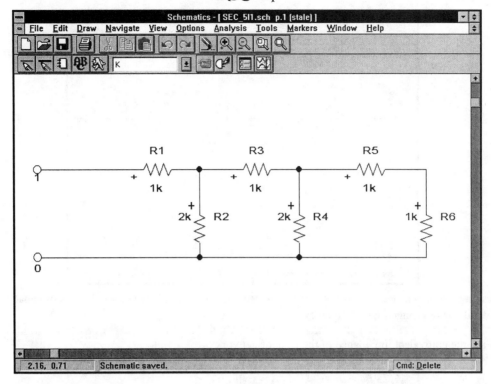

Add a test current source as shown. The test current source is added between nodes 1 and 0 because we are interested in the impedance between those two nodes. Either an AC current or an AC voltage source can be used. Wire the circuit shown:

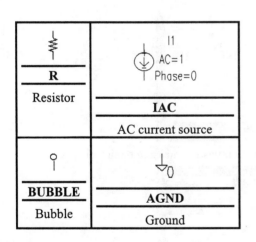

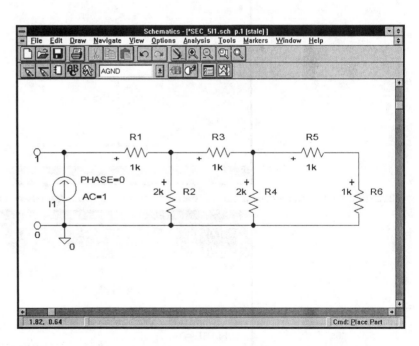

Although this circuit has no dependent sources, inductors, or capacitors, the circuit could contain any circuit element and we could use this method to find the impedance. This circuit was chosen because the impedance is easily calculated and can be compared to the PSpice result. For this circuit, the calculated impedance between nodes *1* and *0* is 2 kΩ.

Set up an AC Sweep (**Analysis, Setup, AC Sweep**) to sweep frequencies from 1 Hz to 1 MHz at 20 points per decade. Run PSpice (**Analysis, Simulate**). When Probe runs, add the trace V(1,0)/I(I1). Remember that voltage divided by current is impedance. We are dividing the voltage between nodes *1* and *0* by the current flowing into and out of those nodes. This is the impedance between those nodes. You will see this trace:

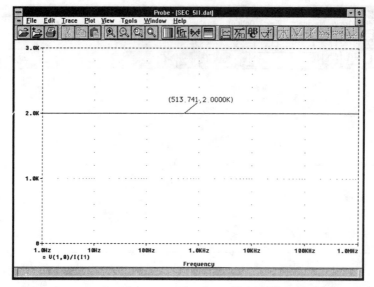

This trace shows that the impedance is constant at the expected value of 2 kΩ. If we had inductors or capacitors in the circuit, the impedance would have changed with frequency.

EXERCISE 5-11: Find the equivalent impedance of the resistor network:

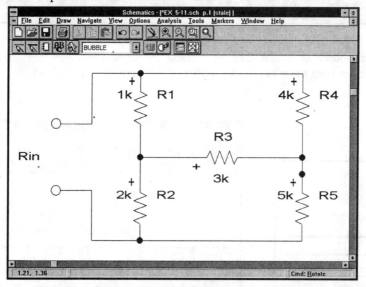

SOLUTION: Add an independent current source between the two terminals where you wish to measure the impedance:

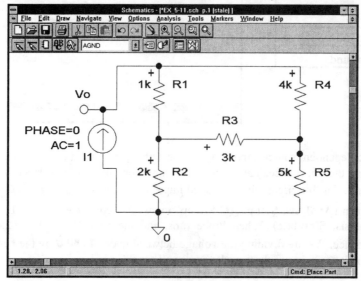

The equivalent impedance is $V(Vo)/I(I1)$:

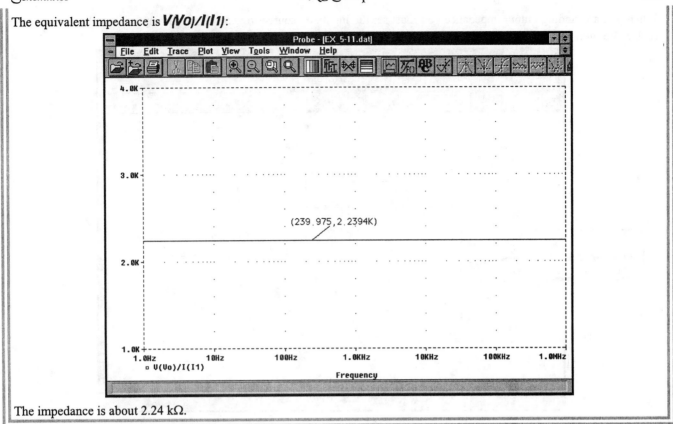

The impedance is about 2.24 kΩ.

5.I.2. Impedance Measurement of an Active Circuit

We would now like to find the output impedance of the jFET source follower circuit shown below. The output is taken at the source terminal of the jFET, so the impedance we are interested in is between nodes **Vo** and **0** (ground).

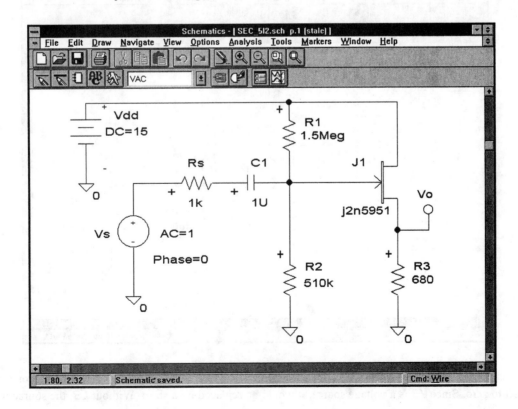

When you are finding output impedance you must set the input AC source to zero. Since the input is a voltage source, we replace **Vs** with a short:

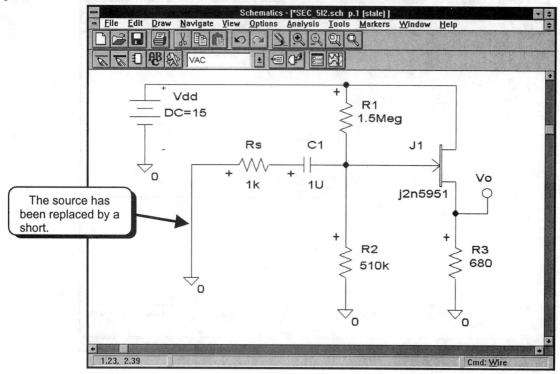

Next we need to add an AC source at the output to measure the impedance. We could use either an AC voltage source or an AC current source. If you use a voltage source, wire the circuit shown below:

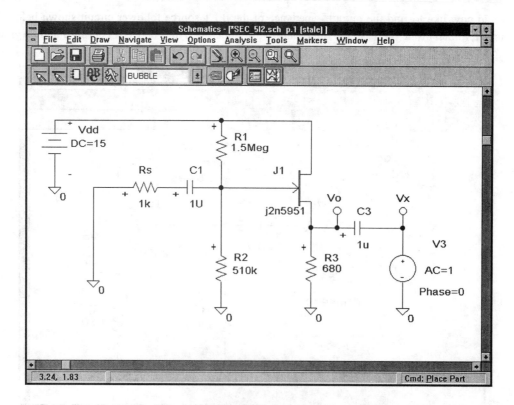

Capacitor **C3** is necessary to preserve the bias of the amplifier. Remember that **V3** is an AC source. For biasing, all AC sources are set to zero. Since **V3** is a voltage source, it would be replaced by a short. Without **C3**, the source terminal of the jFET would be grounded when calculating the bias. This would destroy the bias and render the impedance measurement invalid.

The output impedance of the source follower can be observed by plotting the ratio `V(Vo)/I(V3)` in the circuit above. It is important to note that the ratio V(Vx)/I(V3) is not the output impedance of the source follower, but the output impedance plus the impedance of the capacitor.

The second method to measure the impedance of a circuit uses an AC current source:

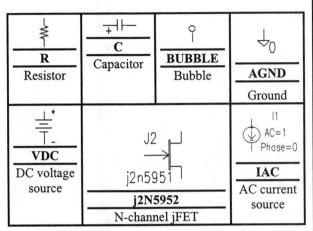

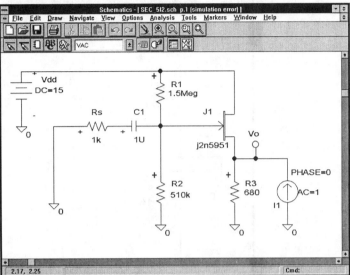

I1 is an AC current source. Note that there is no blocking capacitor between the current source and the amplifier. When the bias is calculated, all AC sources are set to zero. When a current source is zero, it is replaced by an open circuit. An open circuit at *Vo* is equivalent to the original circuit for bias calculations. The output impedance of the circuit can be calculated by the ratio `V(Vo)/I(I1)`.

Set up an AC Sweep (**Analysis**, **Setup**, **AC Sweep**) to sweep frequencies from 1 Hz to 100 MHz at 20 points per decade. Run PSpice (**Analysis**, **Simulate**). When Probe runs, add the trace (**Trace**, **Add**) `V(Vo)/I(I1)`. Remember that voltage divided by current is impedance. We are dividing the voltage between nodes *Vo* and *0* (ground) by the current flowing into and out of those nodes. This is the impedance between those nodes. You will see this trace:

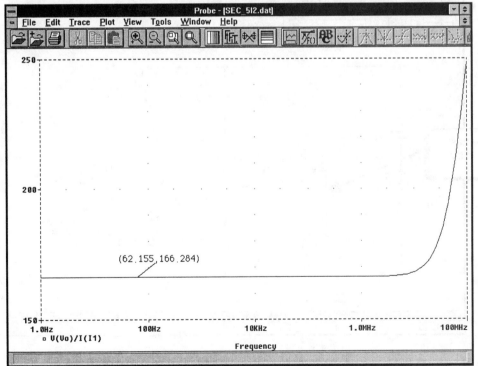

The trace shows that the impedance is 166 Ω for frequencies up to 1 MHz.

EXERCISE 5-12: Find the output impedance of the emitter-follower:

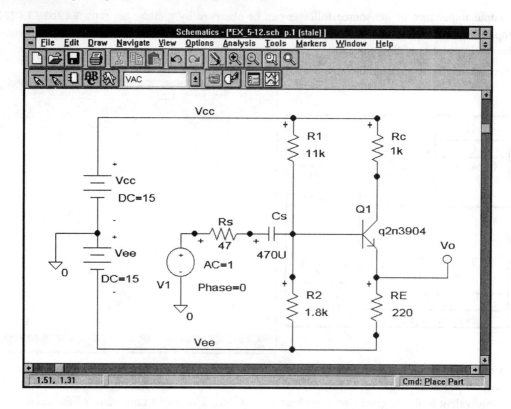

SOLUTION: Replace the independent voltage source with a short circuit and add an independent current source between the two terminals where you wish to measure the impedance:

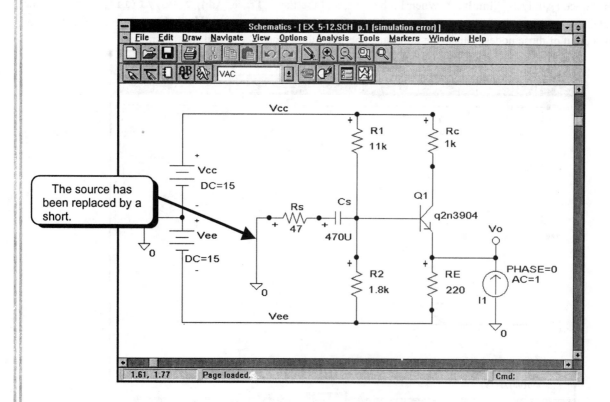

The output impedance is `V(Vo)/I(I1)`:

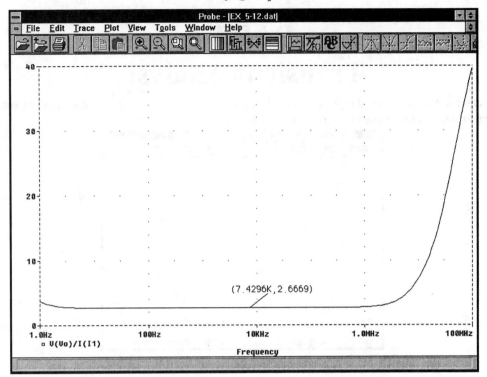

At mid-band, the impedance is about 2.7 Ω. The change in impedance at low frequencies is due to **Cs**.

5.J. Summary

- The AC Sweep is used to find the magnitude and phase of voltages and currents. It is not used to look at waveforms versus time. It is used with the sources VAC and IAC only.
- All DC sources, such as VDC and IDC, and all time-varying sources, such as Vsin, Isin, Vpulse, and Ipulse, are set to zero.
- The results can be obtained for a single frequency or multiple frequencies.
- For multiple frequencies, the results can be viewed graphically using Probe. For a single frequency, the results can be viewed as text in the output file. Use the "Print" part to generate text output.
- The AC Sweep is used to generate plots of magnitude versus frequency, or plots of phase versus frequency.
- To plot the magnitude of a voltage, specify V(*expression*). For example, the magnitude of the voltage at node 1 would be specified as V(1).
- To plot the phase of a voltage, specify Vp(*expression*). For example, the phase of the voltage at node Vx would be specified as Vp(Vx).
- To plot the magnitude of a current, specify I(*expression*). For example, the magnitude of the current through R5 would be specified as I(R5).
- To plot the phase of a current, specify Ip(*expression*). For example, the phase of the current through C6 would be specified as Ip(C6).
- Use the dB command to plot a trace in decibels. The command dB(*expression*) is equivalent to $20\log_{10}(expression)$.
- The AC Sweep and the dB command are used to generate Bode plots.
- The AC Sweep is used to find the gain and frequency response of an amplifier.
- Mutual inductance requires two parts: the inductors (L) and the coupling (K).
- The Parametric Sweep can be used to see how values of devices affect the performance of a circuit.
- Goal Functions can be used to obtain numerical data from Probe graphs. The result of evaluating a Goal Function is a single numerical value.
- The Performance Analysis is the use of a Goal Function in conjunction with a Parametric Sweep. The result of the Goal Function is plotted versus the swept parameter.

PART 6
Transient Analysis

The Transient Analysis is used to look at waveforms versus time. Waveforms are displayed as you would see them on an oscilloscope screen. An example of a waveform versus time is:

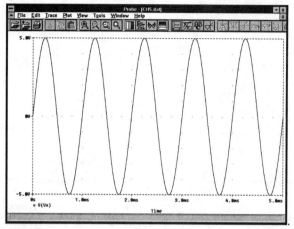

This graph shows us a voltage versus time. Use the Transient Analysis to obtain a waveform versus time. If you want to look at the magnitude and phase of voltages and currents, use the AC Sweep. The magnitude of the above waveform is 5 V and the phase is zero degrees — in phasor notation, 5|0. The AC Sweep analysis will give us the result 5|0. The Transient Analysis will give us the graph above. The Transient Analysis uses voltage and current sources that are functions of time. AC sources such as VAC and IAC are functions of magnitude and frequency. These sources are used for the AC Sweep only. The AC sources are set to zero for the Transient Analysis.

6.A. Introduction

This section covers three topics that are important when using the Transient Analysis. The topics covered apply to all examples covered in this part. The topics may not be necessary, but they are good to keep in the back of your mind if problems arise.

6.A.1. Sources Used with the Transient Analysis

The sources below are meant to be used with the Transient Analysis:

- VSIN, ISIN - Sinusoidal voltage or current source. Typical voltage waveform: v(t)=5sin(2000t+30°)
- VEXP, IEXP - Can be used to create an exponential waveform. Typical current waveform: i(t)=5[1–exp(t/τ)].
- Vpulse, IPULSE - Pulse waveform. Can be used to create a square wave.
- VPWL, IPWL - Can create any arbitrary waveform that is made up of straight lines.
- VSFFM, ISFFM - Used to create a frequency-modulated sine wave.
- Vsq - A square wave voltage source. This source uses the pulsed voltage source to make a square wave. It is a special case of Vpulse.
- Vtri - A triangle wave voltage source. This source uses the pulsed voltage source to make a triangle wave. It is a special case of Vpulse.
- Vramp - A saw-tooth voltage source. This source uses the pulsed voltage source to make a saw-tooth wave. It is a special case of Vpulse.
- V_ttl - A 0 to 5 V square wave with adjustable frequency and duty cycle. This source uses the pulsed voltage source to make a square wave. It is a special case of Vpulse.

6.A.2. Step Ceiling

One of the features of the Transient Analysis that causes much confusion is the step ceiling argument in the Transient dialog box. Suppose we wish to simulate a circuit that has a sinusoidal source. We expect the voltages and currents to look sinusoidal, as follows:

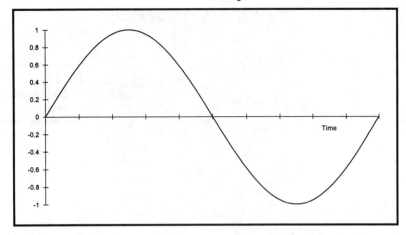

When PSpice runs a Transient Analysis, it solves differential equations to find voltages and currents versus time. The time between simulation points is chosen to be as large as possible while keeping the simulation error below a specified maximum. In some cases, where PSpice can take large time steps, you may get a graph which does not look sinusoidal:

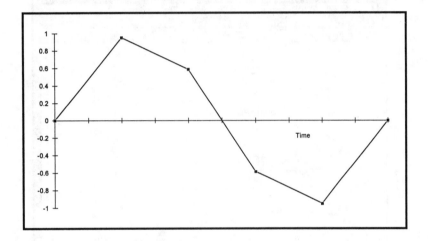

The graph does not look much like a sine wave because the time between points is so large. If we decrease the time between points, we see that the points do lie on a sinusoidal curve:

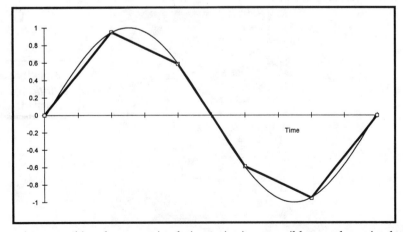

PSpice chooses as large a time step (time between simulation points) as possible to reduce simulation time, but the graphs constructed with these points may not look aesthetically pleasing. To increase the number of points, PSpice provides a Step Ceiling argument for the Transient Analysis. **The Step Ceiling is the maximum time between simulation points.** Reducing the Step Ceiling increases the number of simulation points. The following Transient dialog box shows the location of the **Step Ceiling** argument:

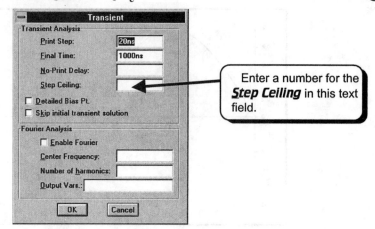

The **Step Ceiling** is initially blank so that PSpice will choose as large a time step as possible. If you need more points in your simulation, you must enter a number for the **Step Ceiling**.

　　　　As an example, we will simulate the DC power supply below. A more detailed example of a DC power supply is shown in Section 6.E.

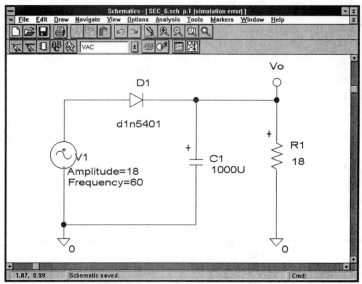

The source has a frequency of 60 Hz, corresponding to a period of 16.67 ms. We will simulate the circuit for three 60 Hz cycles, or 50 ms. The **Transient** dialog box below is set to simulate the circuit for 50 ms:

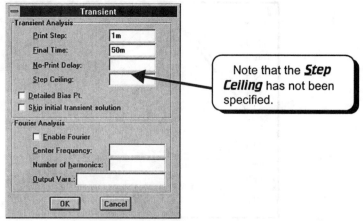

Note that the **Step Ceiling** has not been specified, allowing PSpice to take as much time between simulation points as possible. If we simulate the circuit and plot the capacitor current, we obtain the plot:

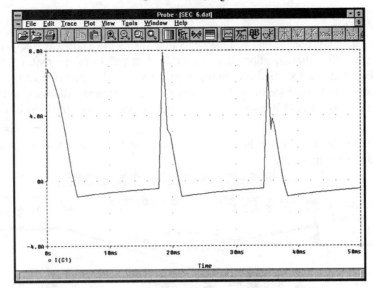

This plot is not what we expect for this circuit. The plot is jagged because of the large amount of time between points.

We will now set the **Step Ceiling** to obtain a smoother plot. The amount of time for one 60 Hz cycle is 16.67 ms. We would like to see 1000 points in each 60 Hz cycle. For 1000 points per cycle, the time between points is:

$$\text{Time Between Points} = \frac{16.67 \text{ ms}}{1000} = 0.01667 \text{ ms}$$

We need to set the **Step Ceiling** to this value. The Transient dialog box below specifies the value of **Step Ceiling**:

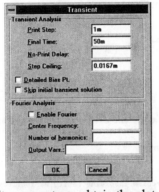

When we run the simulation and plot the capacitor current, we obtain the plot:

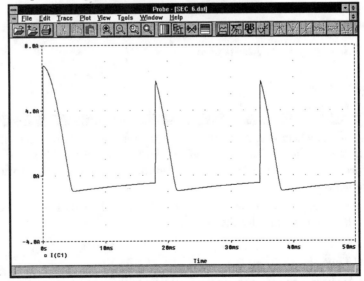

This plot agrees with what we expect the capacitor current to look like. However, the simulation does take longer to run.

6.A.3. Convergence

One of the most frustrating problems that you may encounter using the Transient Analysis is convergence problems. When PSpice is simulating a differential equation, it calculates a data point and estimates the error associated with the calculation. If the error is larger than a specified maximum, PSpice reduces the time step and recalculates the point and the error for the new point. Reducing the time step usually reduces the error. PSpice will continue reducing the time step until the error is within acceptable limits, or until PSpice reaches the limit on the number of times it is allowed to reduce the time step.

If PSpice reaches the limit on reducing the time step, PSpice will announce that the simulation failed to converge. The PSpice window will display the message:

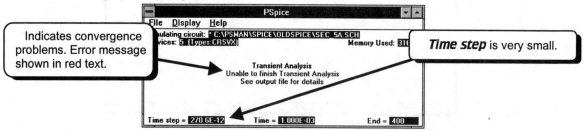

Aside from the message in the center of the screen, a second indication that the simulation did not converge is that the **Time step** is very small. It is small because PSpice went through the process of reducing the time step in order to reduce the error until it reached the limit on reducing the time step. To get more information, you can look at the output file. To view the output file, select **Analysis** and then **Examine Output** from the Schematics menu. Near the bottom of the file you will see text indicating a problem:

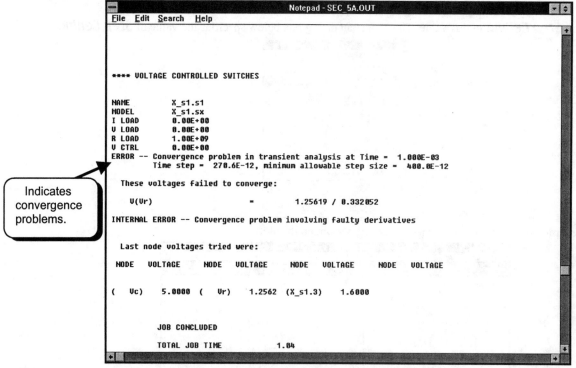

Although this message does not really help you, if you missed the message on the PSpice screen, this message reaffirms the problem. **Important note: If you are viewing the results of a Transient Analysis with Probe and the time axis does not cover the complete time specified in the Transient Analysis setup, chances are the run was terminated prematurely because of convergence problems.**

There are three ways to help prevent convergence problems in the Transient Analysis. Hopefully, one of the methods or a combination of the methods will solve the problem.

Method 1: Reduce the Final Time.

When you set up a Transient Analysis, you must specify the final time of the simulation:

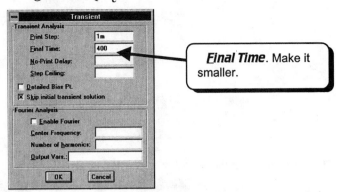

If the ***Step Ceiling*** is not specified, as in the dialog box above, PSpice must make a guess for an appropriate value for the time step. The guess it makes is the ***Final Time*** divided by some number. It starts with this time step and then reduces the time step until the error is acceptable. If the ***Final Time*** is a large number, the initial time step is large and will yield a large amount of error. After reducing the time step a fixed number of times, the time step will still be large because it started out large, and PSpice will quit because the error is large and the limit on reducing the time step was reached. To avoid the problem, reduce the ***Final Time***.

Method 2: Specify a value for the Step Ceiling and make it small.

If you do not wish to change the value of the Final Time, you can specify a value for the Step Ceiling.

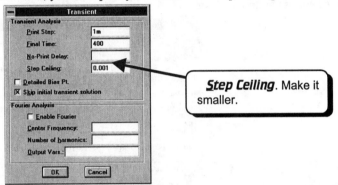

The Step Ceiling is the largest value allowed for the time step. The initial value of the time step will be the value specified by ***Step Ceiling***. If the error is too large, PSpice will reduce the time step. Since the step size can start with a small value (the value of ***Step Ceiling***), after a number of reductions, the step size will become small enough to make the error acceptable. If, after specifying a value for the ***Step Ceiling***, the simulation fails to converge, make the ***Step Ceiling*** smaller. Note that making the ***Step Ceiling*** smaller increases the simulation time.

Method 3: Increasing the number of time step reductions.

If the two methods above fail to solve the problem, you can increase the limit on the number of times the time step can be reduced. To increase the limit, select **<u>A</u>nalysis** and then **Se<u>t</u>up** from the Schematics menus:

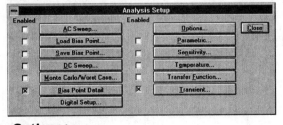

Click the *LEFT* mouse button on the ***Options*** button:

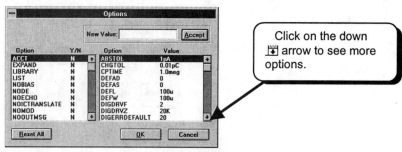

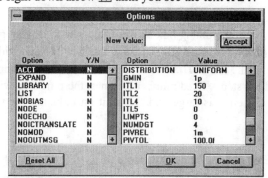

Click the *LEFT* mouse button on the right down arrow 🔽 until you see the text *ITL4*:

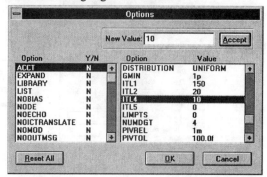

Click the *LEFT* mouse button on the text *ITL4* to highlight the text:

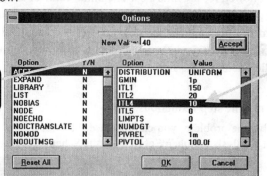

Change the value to **40** as shown below:

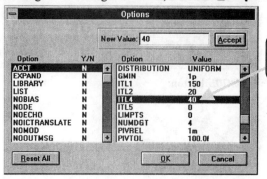

Note that the value of *ITL4* has not yet changed. To change the value, click the *Accept* button:

Click the **OK** button to return to the setup dialog box.

6.B. Capacitor Circuit with Initial Conditions

In this section we will observe the transient response of a capacitor circuit with an initial condition. We will use the circuit below:

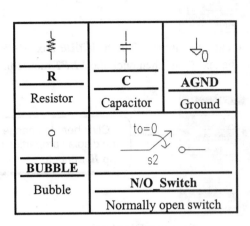

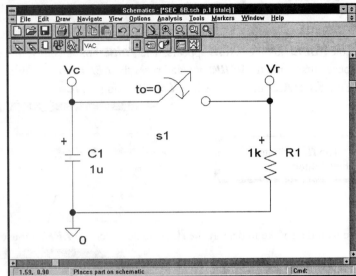

Notice the plus (**+**) sign on the capacitor. This plus sign indicates the positive voltage side of the capacitor for initial conditions. Make sure the plus sign is oriented as shown in the schematic. We would like the capacitor to have an initial condition of 5 V, and we would like the switch to close at t = 1 ms. The initial condition of the capacitor is not one of the displayed attributes, so we must edit all the attributes of the capacitor. Click the **LEFT** mouse button on the capacitor graphic, ┬┠, to select the part. When the graphic is selected it will be highlighted in red. This may take a few tries. When the capacitor is highlighted, select **Edit** and then **Attributes** from the Schematics menu bar. The attributes dialog box for **C1** will appear:

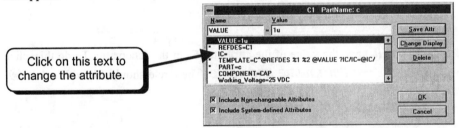

This box shows all the attributes for the capacitor. Some of the attributes are described below:

1. **VALUE** - This is the value of the capacitor in farads. The default is 1 nF.

2. **IC** - This is the initial condition of the capacitor in volts. The default is unspecified.

3. **TEMPLATE** - This is the line that Schematics uses to create the netlist for the capacitor. It cannot be changed.

4. **Working Voltage** - This is for documentation purposes only. This attribute has no effect on the simulation. It is provided to allow the user to display working voltages on schematics.

5. **REFDES** - The name of the capacitor. In this case, the name of the capacitor is **C1**.

Of all the attributes, only the value (1μ) and the name (**C1**) are displayed. All other attributes have their specified value, but are not displayed.

We would like to give the capacitor an initial voltage of 5 V. Click the **LEFT** mouse button on the text **IC=**. It will be highlighted:

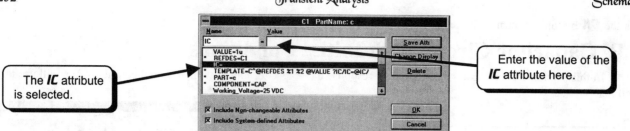

The **IC** attribute is selected.

Enter the value of the **IC** attribute here.

In the **Name** box the text **IC** appears. The **Value** box is empty. We need to enter the number 5 in the **Value** box. Move the mouse cursor into the **Value** box and press the **LEFT** mouse button. Type the text **5** and then click the **LEFT** mouse button on the **Save Attr** button. The line **IC=** will change to **IC=5**:

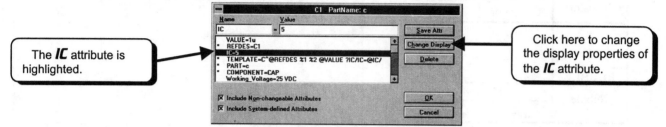

The **IC** attribute is highlighted.

Click here to change the display properties of the **IC** attribute.

We would also like to display the **IC** attribute. Click the **LEFT** mouse button on the **Change Display** button:

We would like to display the name of the attribute (**IC**) and the value of the attribute (**5**) in the schematic. To do this, click the **LEFT** mouse button on the circle ○ next to the text **Both name and value**. The circle should fill with a dot, ◉. When you have properly made the changes your dialog box will appear as shown:

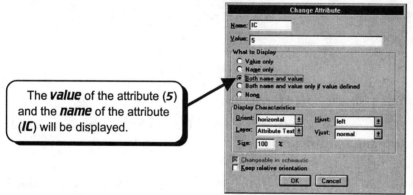

The **value** of the attribute (**5**) and the **name** of the attribute (**IC**) will be displayed.

Click the **OK** button to accept the values. The attributes dialog box will appear. Click the **OK** button to accept the attributes. You will return to the schematic as shown below.

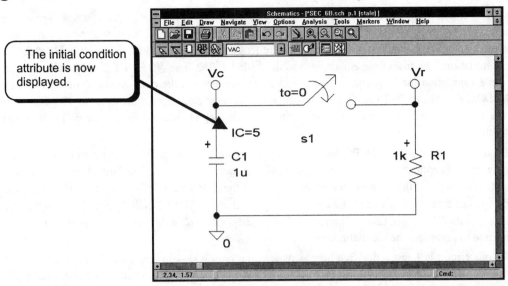

The initial condition attribute is now displayed.

We see that the initial condition attribute is displayed on the schematic. The labels may be a little crowded, so you may have to move some to make the schematic more readable.

The next thing we need to do is change the attributes of the switch so that it closes at t = 1 ms. Double click the *LEFT* mouse button on the text *t=0*. The dialog box below will appear.

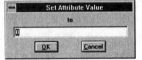

Enter the text **1m** and click the **OK** button. You should have the completed schematic as shown:

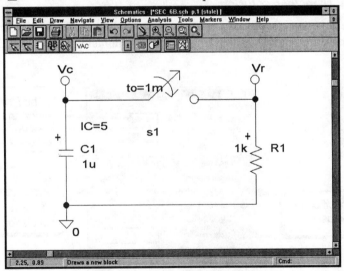

Note that the initial condition of the capacitor is **5** (5 volts) and the switch closes at time equals **1m** (1 millisecond). The next thing we need to do is set up the Transient Analysis. Select **Analysis**, **Setup**, and then click the **Transient** button to obtain the Transient Analysis dialog box:

The **Print Step** is used for printing text output with the print part. Every **Print Step** seconds, the print part will print out the specified values in the output file. We will be using Probe to view the results of the Transient Analysis, so this parameter is not very important. However, if its value is too small, it may affect the length of the simulation. The **Final Time** is the length of the simulation. The simulation runs from time equals zero to the **Final Time**. The **No-Print Delay** value is used for long simulations. If you ran a simulation for 1 second and you were interested only in the data from 990 ms to 1 s, then you would set the **No-Print Delay** value to 990 ms. PSpice would not save any data from time zero to 990 ms. This parameter is used to reduce the size of the output data file for large circuits with long simulation times. We will ignore this argument since this is a short run.

The **Step Ceiling** value is important. When PSpice simulates a circuit, it chooses the time between simulation points to be as large as possible while keeping the error below a specified maximum. The larger the time step, the less time the simulation takes to complete. To keep simulation time to a minimum, leave the **Step Ceiling** value blank. However, the step size PSpice takes may be too large for you. If this is the case, specify a value for **Step Ceiling**. **Step Ceiling** is the longest time between simulation points PSpice will take. A smaller value of **Step Ceiling** will give you more points in your simulation, but will take more time to complete the simulation.

The time constant for our circuit is 1 ms. After five time constants the circuit should reach steady state, so we will run the simulation for five time constants. Remember that the switch closes at 1 ms. To let the capacitor transient run for 5 ms we need a total simulation time of 6 ms. We would like to see at least 500 points during the capacitor transient, so set the **Step Ceiling** to 5 ms/500 or 0.01 ms.*

In the Transient dialog box there is a box labeled **Skip initial transient solution**. If this box is checked, PSpice will use the initial conditions specified in the circuit for the transient run. If the box is not checked, PSpice will calculate the initial conditions from the circuit. When PSpice calculates the initial conditions, all capacitor voltages and inductor currents are calculated assuming that all capacitors are open circuits and all inductors are short circuits. This may be different from the initial conditions specified in the circuit. For our circuit, there are no independent sources, so the initial condition of the capacitor will be zero. We wish to use the initial condition specified in the circuit. To use the specified initial conditions, the box next to **Skip initial transient solution** should have an **X** in it. To place an **X** in the box, click the *LEFT* mouse button on the text **Skip initial transient solution**.

You should have the Transient dialog box filled out as shown below. The simulation is set to run for 6 ms. The maximum time between simulation points is 0.01 ms, so the minimum number of points in the simulation is $\dfrac{6 \text{ ms}}{0.01 \text{ ms}} = 600$ points.

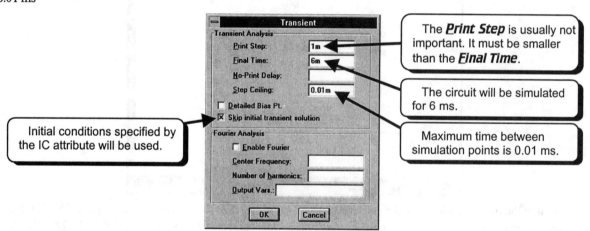

Click the **OK** box to accept the settings, and then click the **Close** button to return to the schematic. To run the simulation select **Analysis** and then **Simulate** from the Schematics menu bar. The PSpice simulation window will appear as shown below. If you do not see this window, press the ALT-ESC or ALT-TAB key sequences to toggle the active window.

---

*Convergence problems have been observed in some transient simulations of circuits with switches, such as the circuit on page 253. You will notice this error because the transient simulation terminates prematurely. When Probe runs, the time axis will not reach the **Final Time** specified by the transient simulation. If you look at the output file (**Analysis**, **Examine Output** from the Schematics menu bar), an error message near the end of the file will indicate a convergence problem. The problem can be fixed by choosing a smaller value of **Step Ceiling**, or by using one of the methods mentioned in Section 6.A.3.

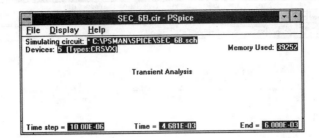

When the simulation is complete, Probe will automatically run. Add the trace **V(Vr)** (select **Trace** and then **Add** from the Probe menu bar). The trace of the resistor voltage will appear as shown below.

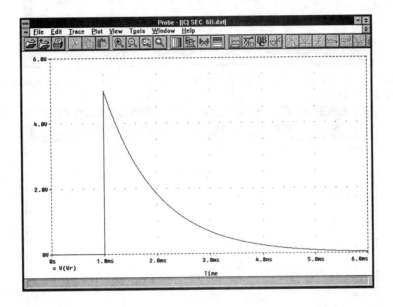

Next we would like to look at the capacitor voltage. Add a new window in Probe by selecting **Window** and then **New**. Add the capacitor voltage trace **V(Vc)** by selecting **Trace** and then **Add** from the Probe menu bar:

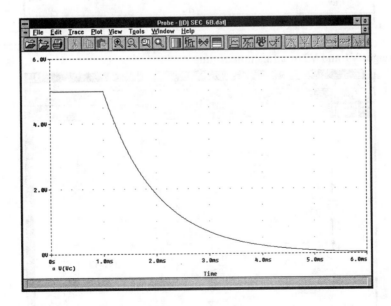

EXERCISE 6-1: Find the voltage **Vo** as a function of time:

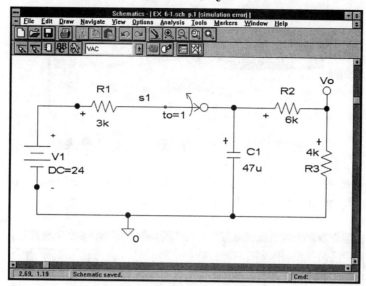

SOLUTION: The initial condition of the capacitor does not need to be specified because PSpice will determine the initial conditions from the circuit. The part name for the normally closed switch is N/C_switch.

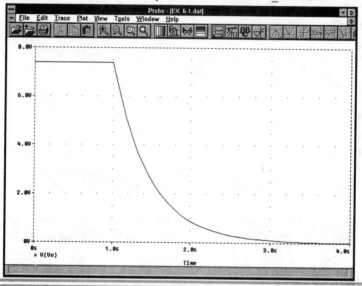

EXERCISE 6-2: Find the voltages V1 and V2 as a function of time:

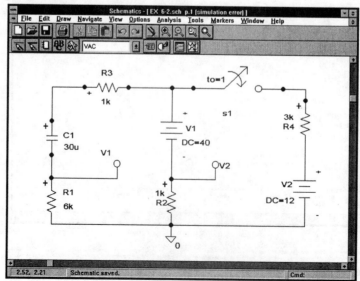

SOLUTION: The initial condition of the capacitor does not need to be specified because PSpice will determine the initial conditions from the circuit.

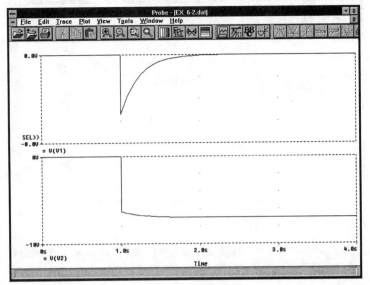

To add a second plot, as in the screen capture above, select **Plot** and then **Add Plot** from the Probe menus.

6.B.1. Note on Initial Conditions

Instead of specifying an initial condition on the capacitor, the initial voltages of nodes can be specified using the parts IC1 and IC2. For the circuit of Section 6.B, we will use the part IC1 to specify the initial condition of the capacitor. Get a part called IC1 and place it in your circuit:

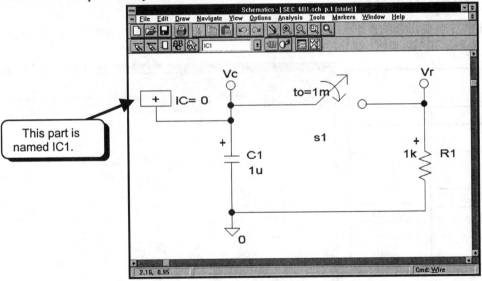

The default value of the initial condition is **0** V. To change the initial condition, double click on the text **0**. The dialog box below will open[*]:

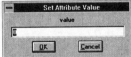

Note that the dialog box is changing the attribute **value**. Type in the text for the initial condition, **5** in this case:

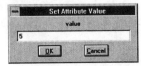

---

[*] Do not click on the text **IC=**. Doing so will change the text from **IC=** to something else.

Click the **OK** button to accept the change. The initial condition will be changed in the circuit:

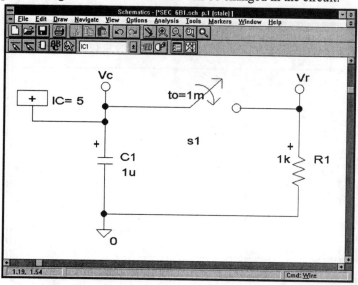

Note that the value specified is the voltage of the node relative to ground. The part IC2 allows you to specify the initial voltage between two nodes. You can use the IC2 part to specify the initial voltage of a capacitor when the capacitor does not have one of its leads grounded. When you use the IC1 or IC2 part, you do not need to specify the **Skip initial transient solution** option in the Transient Analysis dialog box.

6.C. Capacitor Step Response

In this section we will look at the response of an RC circuit to a pulse input. We will work with the circuit below:

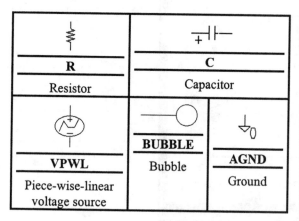
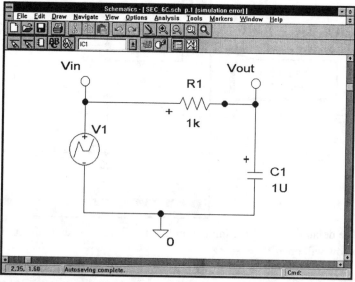

The time constant of the circuit is $\tau = RC = (1k\Omega)(1\mu F) = 1ms$. The response of an inductive or capacitive circuit will reach 99% of its final value in 3 time constants, and will reach steady state in about 5 time constants. We will run two simulations. The first will let the capacitor charge and discharge for 3 time constants (3 ms in this case) and the second will let the capacitor charge and discharge for 5 time constants (5 ms in this example). For the first simulation we will use the waveform below:

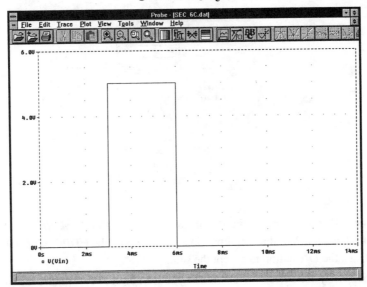

The length of the simulation is 9 ms. The pulse will start at zero volts and remain there for 3 ms, then the pulse will go to 5 V for 3 ms, and then the pulse will return to 0 for another 3 ms. To create this waveform we can use either a pulsed voltage source (VPULSE) or a piecewise linear voltage source (VPWL). For this example we will use the VPWL source. To edit the attributes of the PWL source, double click the **LEFT** mouse button on the PWL graphic, ⊸⊙⊢. The dialog box below will appear:

The DC and AC attributes are for the DC and AC analyses. We are running a Transient Analysis, so they will be ignored. The operation of the PWL source is the following: At t1 the voltage is v1; at t2 the voltage is v2; between t1 and t2 the voltage changes linearly from v1 to v2; at t3 the voltage is v3; between t2 and t3 the voltage changes linearly from v2 to v3; at t4 the voltage is v4; and so on. We would like to set up the pulsed waveform shown above. Set the PWL attributes

Table 6-1				
t1	t2	t3	t4	t5
0	2.999m	3m	5.999m	6m
v1	v2	v3	v4	v5
0	0	5	5	0

as shown in **Table 6-1**. These settings specify that from 0 to 2.999 ms the voltage will be zero. From 2.999 ms to 3 ms the voltage changes from 0 to 5 V. From 3 ms to 5.999 ms the voltage is constant at 5 V. From 5.999 ms to 6 ms the voltage changes from 5 to 0 V. For time greater than 6 ms, the voltage will remain constant at 0 V. Fill in the dialog box with these settings:

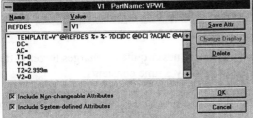
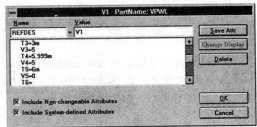

When you have made the changes to the attributes of the PWL source, click the **OK** button to accept the changes.

We must now set up a Transient Analysis to run the simulation. We would like the capacitor to discharge for 3 ms after the pulse voltage goes to 0 V. Thus, we will run the Transient Analysis for 9 ms. Select **Analysis**, **Setup**, and then click the **Transient** button. Fill in the dialog box as shown:

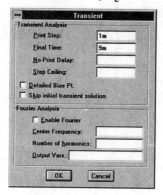

The **_Final Time_** is set to 9 ms. This will run the simulation for 9 ms. The **_Print Step_** is set to 1 ms. The value of this parameter is not important, but it must be less than the final time. If you make the **_Print Step_** too small, the simulation may take more time to complete. Click the **_OK_** button to accept the settings.

Note that we did not specify initial conditions. When you do not specify initial conditions, PSpice will determine them itself. PSpice will determine the initial conditions from the state of the circuit at t=0. In this case, the voltage of the PWL source is zero at t=0. The initial conditions will be determined with this source set to zero. Since this is the only source in the circuit, all sources are zero at t=0, so the initial condition of the capacitor must be zero at t=0.

Run the simulation (**Analysis** and then **Simulate**). When the simulation is complete, Probe will automatically run. Add the traces `V(Vin)` and `V(Vout)` :

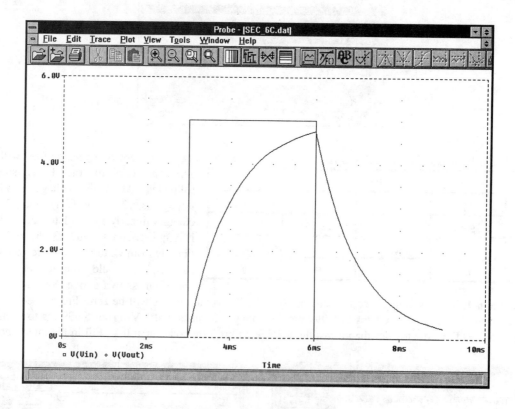

We see that the capacitor voltage never quite reaches 5 V during the pulse, and never quite discharges to zero after the pulse returns to zero. This is because we let the capacitor charge and discharge for only 3 time constants.

EXERCISE 6-3: For the circuit in the previous example, allow the capacitor to charge and discharge for 5 ms rather than 3 ms.

SOLUTION: Change the PWL source to lengthen the amount of time the pulse is at 5 V:

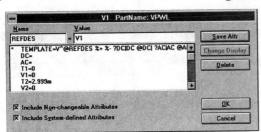

To allow the capacitor to discharge for 5 ms, we must allow the Transient Analysis to run for 5 ms after the pulse returns to zero. Fill in the **Transient** dialog box as shown:

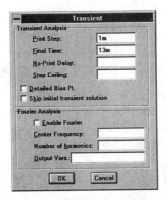

The results of the simulation are:

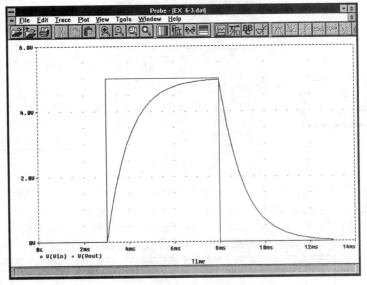

We see that the capacitor almost reaches its final value after 5 time constants.

6.D. Inductor Transient Response

      In this section we will demonstrate the transient response of an inductor circuit with a switch that is normally closed. The initial condition of the inductor will not be specified by an "IC=" line in the circuit. Instead, the initial condition will be determined by PSpice from the initial state of the circuit before the switch changes position. If you wish to specify the initial condition of the inductor, it is specified in the same way as the initial condition of a capacitor. For an inductor, the direction for positive current is into the dotted terminal, as shown in Figure 6-1. The dot is always shown on the inductor graphic. The graphic should be rotated to obtain the desired direction of positive current flow.

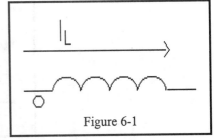

Figure 6-1

We will simulate the circuit below:

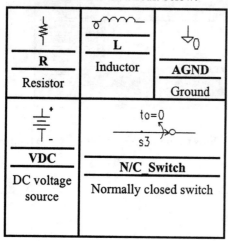

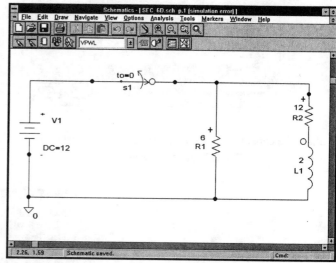

For this circuit, the initial current through the inductor is 1 A. The current will decay to zero with a time constant of 111 ms. We will simulate this circuit for nearly five time constants (550 ms). Note for this circuit that the switch is normally closed, and opens at t = 0 ms. It is important that the dot in your schematic match the dot in the schematic above. If the dots do not match, your results will be –1 times the results presented here.

We now need to set up a Transient Analysis. To obtain the Transient Analysis dialog box select **Analysis**, **Setup**, and then click the **Transient** button. We would like to run the transient simulation for 550 ms, so the **Final Time** should be set to **550m**. I would like to see at least 550 points in the simulation, so I will set the **Step Ceiling** to **1m**. The **Print Step** is irrelevant but must be less than the **Final Time**. I will leave it at the default value. Fill out the dialog box as shown:

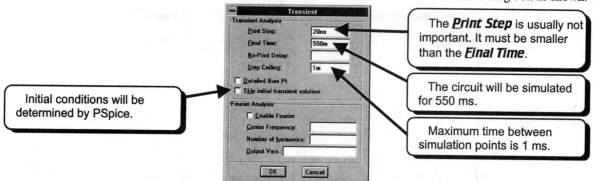

Click the **OK** button to accept the settings, and then click the **Close** button to return to the schematic.

To run the simulation, select **Analysis** and then **Simulate** from the Schematics menu bar. When Probe runs, add the trace **I(L1)**. To add a trace select **Trace** and then **Add** from the Probe menu bar. You will see the following Probe window:

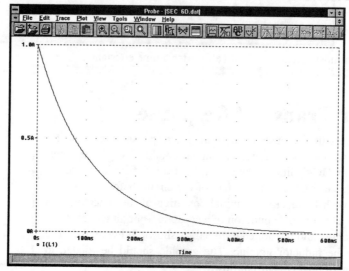

EXERCISE 6-4: Find the inductor current I_L as a function of time:

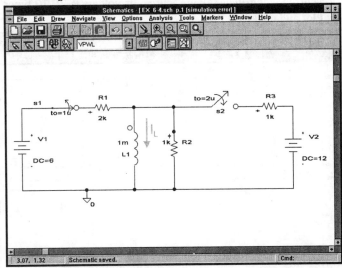

SOLUTION: The initial condition of the inductor does not need to be specified because PSpice will determine the initial conditions from the circuit:

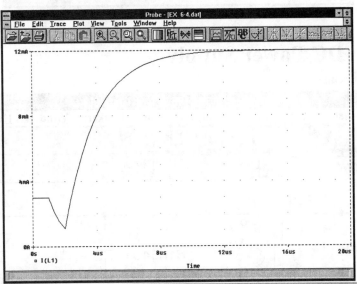

EXERCISE 6-5: Find the voltages V1 and V2 as a function of time:

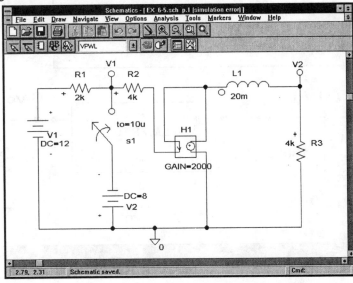

SOLUTION: The initial condition of the inductor does not need to be specified because PSpice will determine the initial conditions from the circuit. **H1** is a current-controlled voltage source. Its voltage is **2000** times the current through **R2**. To add a second plot in Probe, select **Plot** and then **Add Plot**.

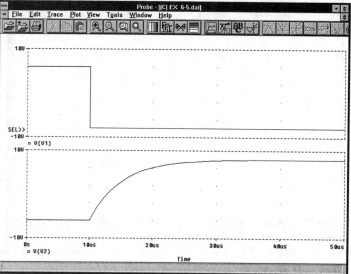

6.E. Regulated DC Power Supply

Wire the following circuit:

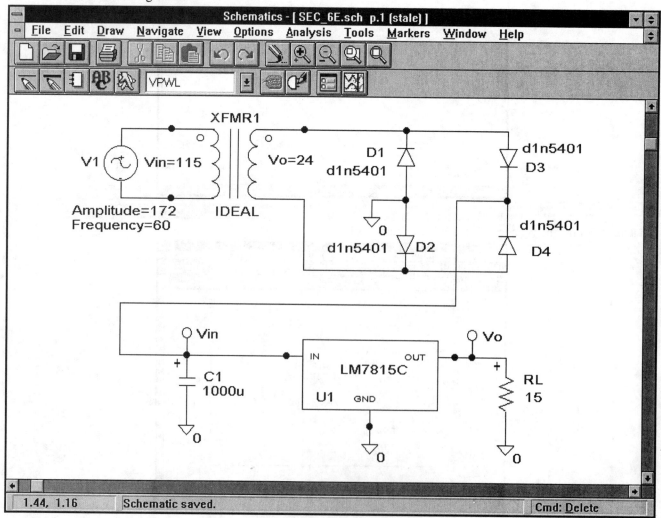

The Transient Analysis uses the sinusoidal voltage waveform when it performs the simulation. This circuit would not work if an AC Sweep were used. To set up a Transient Analysis, select **Analysis**, **Setup**, and then click the **Transient** button.

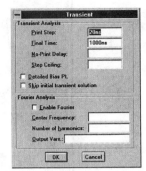

The **Print Step** is used for the print part. Since we will be using Probe, this parameter can be ignored. However, it sometimes seems to affect the run time of the simulation, so set it to a larger value than the default and less than or equal to the **Final Time**. The **Final Time** is the total time of the simulation, which runs from time equals zero to the **Final Time**. The **No-Print Delay** value is used for long simulations. If you ran a simulation for 1 second and you were only interested in the data from 990 ms to 1 s, then you would set the **No-Print Delay** value to 990 ms. PSpice would not save any data from time zero to 990 ms. This parameter is used to reduce the size of the output data file for large circuits with long simulation times. We will ignore this argument since this is a short run.

The **Step Ceiling** value is important. When PSpice simulates a circuit it chooses the time between simulation points to be as large as possible while keeping the error below a specified maximum. The larger the time step, the less time the simulation takes to complete. To keep simulation time to a minimum, leave the **Step Ceiling** value blank. However, with the **Step Ceiling** unspecified, the step size PSpice takes may be too large for you. If this is the case, specify a value for **Step Ceiling**. **Step Ceiling** is the longest time between simulation points PSpice will take. A smaller value of step size will give you more points in your simulation, but will take more time. In our simulation, the voltage source is a 60 Hz sinusoid. Each cycle of the sinusoid takes 16 ms. I would like to see a minimum of 20 points per cycle, so I will choose the value for **Step Ceiling** to be 0.8 ms.

Fill out the dialog box as shown below. Since each cycle takes 16 ms, I will run the simulation for 10 cycles or 160 ms.

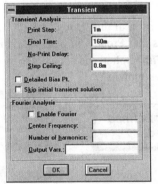

R	D1
Resistor	d1n5401
	D1N5401
	Rectifier diode
C	IN OUT LM7815C U2 GND
Capacitor	**LM7815C**
	+15 volt regulator
BUBBLE	XFMR3 Vin=115 Vo=12
Bubble	IDEAL
	Ideal_XFMR_Vo/Vin
	Ideal transformer
AGND	V1 Amplitude= Frequency=
Ground	**VSIN**
	Sinusoidal voltage source

Click the **OK** button to accept the settings, and then click the **Close** button to return to the schematic. Run PSpice (**Analysis**, **Simulate**). The PSpice simulation window will appear:

```
                    SEC_6E.cir - PSpice
File  Display  Help
Simulating circuit: * C:\PSMAN\SPICE\SEC_6E.sch
Devices: 16  (Types CDEGKLQRSVX)          Memory Used: 62716

                         Transient Analysis

Time step = 800.0E-06    Time = 018         End = 16
```

If you do not see the PSpice simulation window, press the ALT-ESC or ALT-TAB key sequences to toggle the active window. The simulation may take a long time so you should view the PSpice simulation window to observe the progress.

When the simulation is finished, Probe will run. When the Probe window opens, notice that the x-axis is now the time axis. Add trace **V(Vin)** and **V(Vo)** to observe the input and output of the regulator. To add a trace select **Trace** and then **Add** from the Probe menu bar. You will see the following plot:

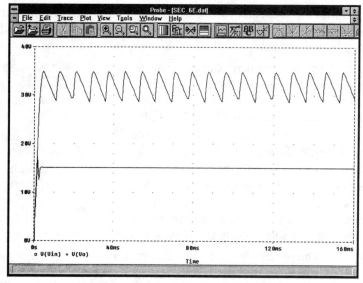

To see what happens at the beginning of the trace we can change the scale of the x-axis. To change the scale, select **Plot** and then **X Axis Settings**:

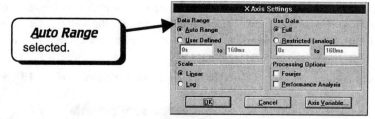

The **Data Range** portion of this dialog box allows you to change the x-axis range. Presently, the **Auto Range** option is selected, which means that Probe will determine the x-axis range. We would like to specify the x-axis range, so we must select the **User Defined** option and specify the range. We would like to set the x-axis to display the first 10 ms of the trace. Fill in the dialog box as shown:

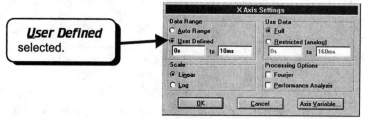

Click the **OK** button to accept the range:

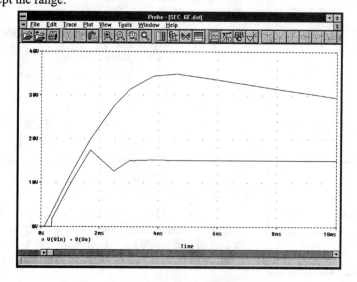

This plot shows us the effect of the **Step Ceiling** in the Transient Analysis. Instead of seeing curved lines, we see straight lines connecting simulation points. If we wish to see a nicer curve for this time scale, we need to reduce the **Step Ceiling**. If we change the Step Ceiling in the Transient Analysis dialog box shown on page 265 to 8 µs rather than 0.8 ms, the simulation will contain 100 times as many points. This will give us much more resolution in the plots. Since we have reduced the step size by a factor of 100, the simulation time will increase. We are interested in more detail for only the first 10 ms, so we will run the simulation for only 10 ms. The modifications to the Transient Analysis dialog box are shown below:

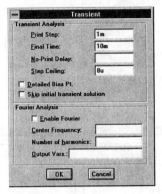

The results of the simulation are shown below:

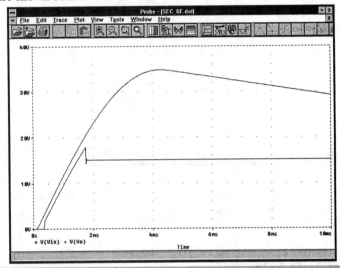

EXERCISE 6-6: Simulate a power supply using a half-wave rectifier and a regulator. Use the same circuit as on page 264, except use a half-wave rectifier instead of a full-wave rectifier. The remainder of the circuit should be the same.

SOLUTION:

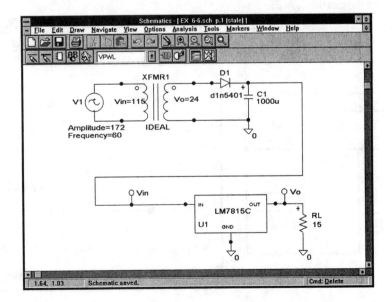

Set up the Transient Analysis to run the simulation for approximately ten cycles of the input source, and use the Step Ceiling to force at least 200 points per cycle:

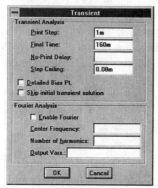

Run the simulation and then display Vin and Vout using Probe :

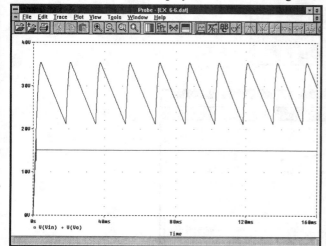 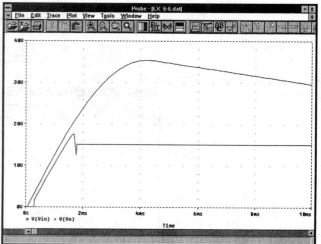

It appears that the circuit still functions properly since the output of the regulator is constant at 15 V. When you look at the input to the regulator, the ripple voltage across the capacitor is about twice the magnitude of the ripple voltage of the full-wave rectifier. This is the expected result.

6.F. Zener Clipping Circuit

In this section we will demonstrate the use of a Zener diode, as well as the piece-wise linear (PWL) waveform voltage source. The PWL source can be used to create an arbitrary voltage waveform that is connected by straight lines between voltage points. Wire the circuit shown:

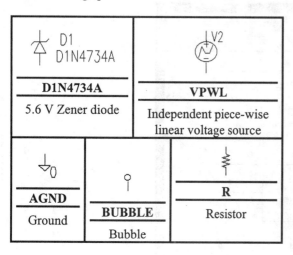 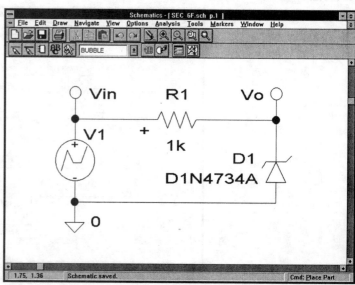

To edit the attributes of the PWL source, double click the *LEFT* mouse button on the PWL graphic, ─◯─. The dialog box below will appear:

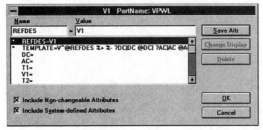

The DC and AC attributes are for the DC and AC analyses. We are running a Transient Analysis, so they will be ignored. The operation of the PWL source is the following: At t1 the voltage is v1; at t2 the voltage is v2; between t1 and t2 the voltage changes linearly from v1 to v2; at t3 the voltage is v3; between t2 and t3 the voltage changes linearly from v2 to v3; at t4 the voltage is v4; and so on. We would like to set up a ±15 V triangle wave. Set the PWL attributes as shown in Table 6-2. When you have made the changes to the attributes of the PWL source, click the **OK** button to accept the changes.

Set up the Transient Analysis to simulate the circuit for 4 ms. To set up the analysis select **Analysis**, **Setup**, and then click the **Transient** button. Fill in the dialog box as shown below:

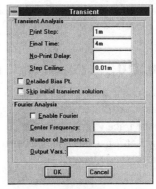

Table 6-2			
t1	t2	t3	t4
0	1m	3m	4m
v1	v2	v3	v4
0	15	-15	0

The analysis is set up to simulate the circuit for 4 ms with a maximum time between simulation points of .01 ms. Thus, this simulation will have a minimum of 400 points (Final Time / Step Ceiling = 4 ms/0.01 ms = 400).

Run the simulation: select **Analysis** and then **Simulate**. When the simulation is finished, run Probe. Add the trace **V(Vin)** (select **Trace** and then **Add**). You will see the triangle wave:

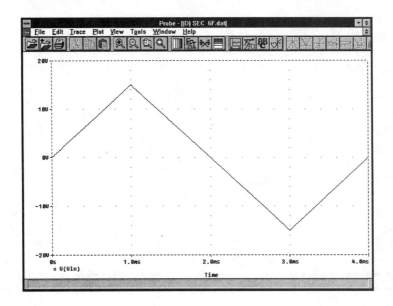

To view the output, add the trace **V(Vo)**:

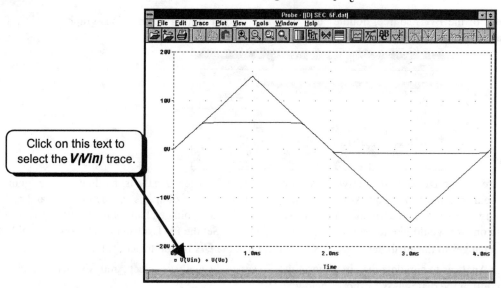

Note that the output of the circuit is limited to a maximum of 5.6 V and a minimum of about –0.7 V.

6.F.1. Plotting Transfer Curves

When you have a clipping circuit, as in the last example, a transfer curve may be desired. This can easily be done by changing the x-axis. A transfer curve usually plots V_o versus V_{in}, so we need to change the x-axis from **Time** to **Vin**. We must first delete the trace **V(Vin)** from the plot above. Click the **LEFT** mouse button on the text **V(Vin)**. The text will appear in red, indicating that it has been selected. From the Probe menu select **Edit** and then **Delete** to delete the trace. You should have a screen with a single trace:

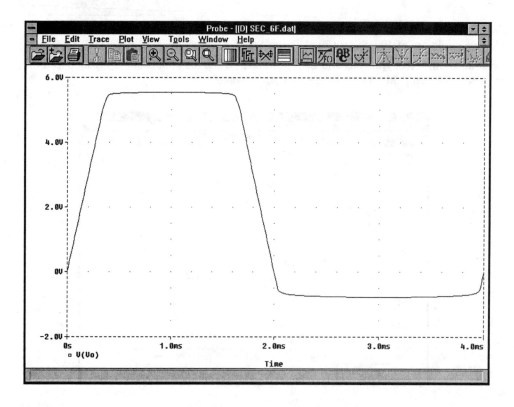

We will now change the x-axis. Select **Plot** and then **X Axis Settings** from the Probe menu bar:

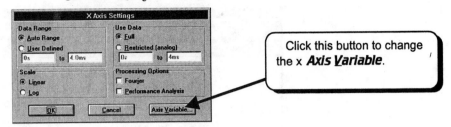

Click the **Axis Variable** button:

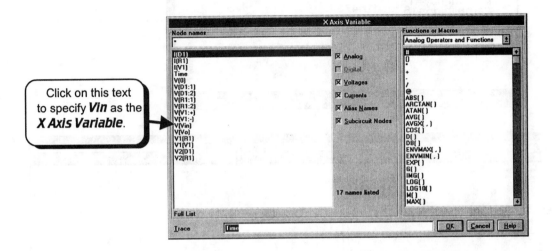

Click the *LEFT* mouse button on the text **V(Vin)** and then click the **OK** button twice. The transfer curve will be displayed:

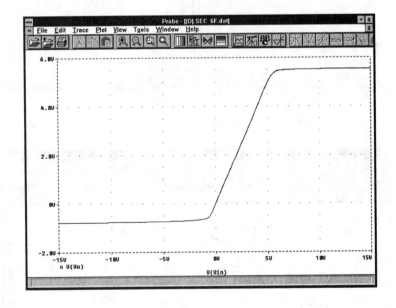

EXERCISE 6-7: Find the output voltage waveform and the transfer curve for the circuit below. Let the input be a ±15 volt triangle wave. Use the source Vtri to create a 1 Hz triangle wave.

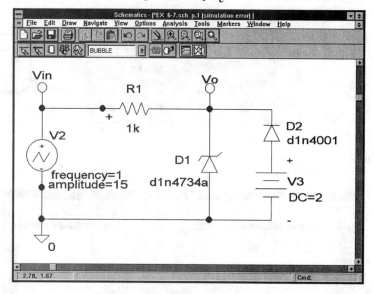

SOLUTION:

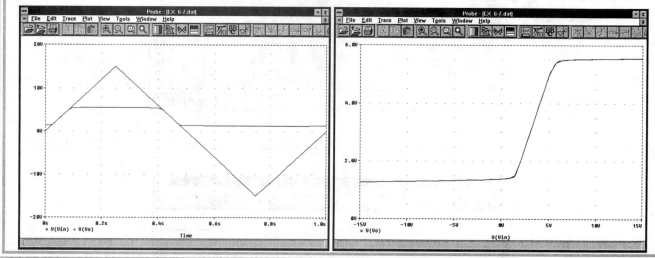

EXERCISE 6-8: For the "Dead-Zone" clipping circuit shown below, find the output voltage waveform and the transfer curve. Let the input be a ±15 volt triangle wave.

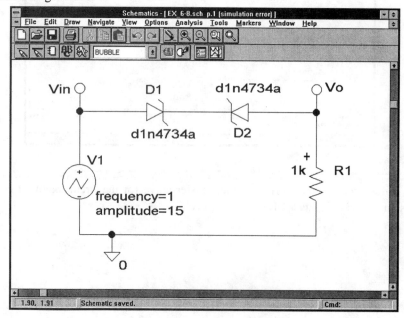

SOLUTION:

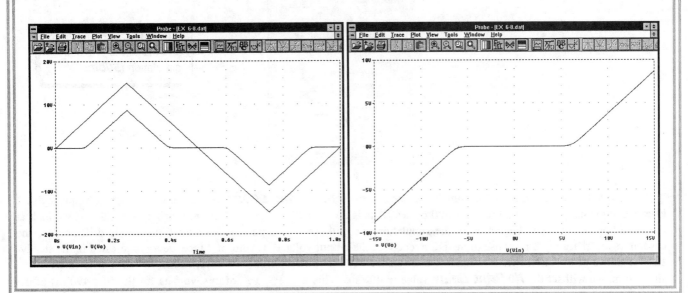

6.G. Amplifier Voltage Swing

In this section we will examine the voltage swing of the transistor amplifier used in Section 5.D. In Section 5.D we found that the gain of the amplifier was about 45 dB, which corresponds to a gain of 193 V/V. We also found that mid-band for this amplifier was from 42 Hz to 6.3 MHz. We would like to see what the maximum voltage swing of this amplifier looks like at mid-band. We will use the same circuit as in Section 5.D, except we will change the signal source from an AC source to a sinusoidal source at a frequency of 1 kHz.

Wire the following circuit:

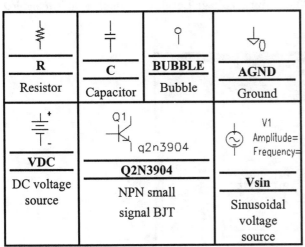

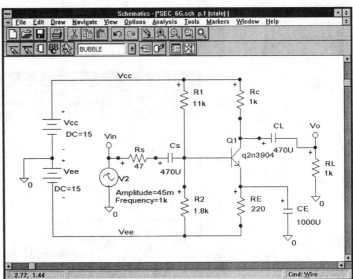

Note that the voltage source part is VSIN, not VAC. The source is now a sinusoidal voltage of amplitude 45 mV and frequency 1 kHz, $V_4 = 0.045\sin(2\pi*1000t)$. The amplitude was chosen to get the maximum swing out of this amplifier. You may have to have to run the simulation several times with different amplitude sine waves for Vin to see what amplitude you need for Vin to observe the maximum swing on the output. Too small an amplitude will cause a small output swing. Too large an output will cause the amplifier to go into saturation or cut-off. The amplitude should be chosen to cause the output of the amplifier to just saturate or cut-off. Set up the Transient simulation dialog box as shown below (**Analysis**, **Setup**, **Transient**):

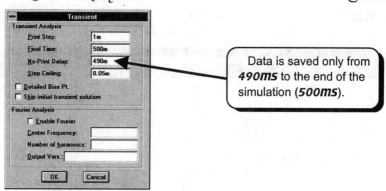

Note that we are running the simulation for 500 ms (**Final Time: 500m**). Since the source is at a frequency of 1 kHz, we are simulating the amplifier for 500 cycles. This is done to allow the amplifier to reach steady-state. If we look at the amplifier output at the beginning of the simulation, we will see the output voltage waveform slowly drift because the capacitors are charging in response to the DC average of the output voltage waveform. Since we are interested in the steady-state response of the amplifier, the simulation is run for 500 ms. Since we are not interested in the data at the start of the simulation, we will set the **No-Print Delay** value to 490 ms. This tells PSpice not to save data for the first 490 ms of the simulation. PSpice will save data only from the last 10 ms of the simulation, 490 ms to 500 ms. If we left the **No-Print Delay** value blank, PSpice would save data from time equals zero to 500 ms, resulting in a huge data file. The **Step Ceiling** is set to 0.05 ms so that there will be 20 points in every 1 kHz cycle. This is unnecessary and increases the run time a fair amount. However, it does make the plotted waveform look a little nicer. To decrease the simulation time you may wish to leave the **Step Ceiling** value blank.

Since this simulation will run for a long time and we have specified a small Step Ceiling, a lot of data will be collected. PSpice normally collects voltage data at every node and current data through every circuit component. This results in a large Probe data file that can take a long time to load and may cause memory problems. Since we are interested only in the input and output voltages, we will tell PSpice to collect data only at the input and output nodes. We will use the markers to specify data collection points. In Schematics select **Markers** and then **Mark Voltage/Level**. Place markers at **Vin** and **Vo** as shown:

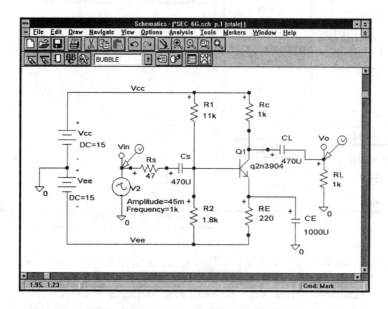

To stop placing markers click the **RIGHT** mouse button. Next, we must specify data to be collected only at the markers. Select **Analysis** and then **Probe Setup** from the Schematics menus:

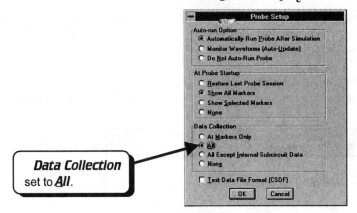

Data Collection set to **All**.

We see that the default setting for **Data Collection** is **All** node voltages and branch currents. We want to specify data collection at the markers only, so click the *LEFT* mouse button on the circle ○ next to the text **At Markers Only**. The circle should fill with a dot ◉:

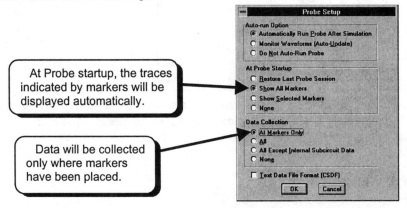

At Probe startup, the traces indicated by markers will be displayed automatically.

Data will be collected only where markers have been placed.

Also note that when Probe starts, the traces indicated by markers will be displayed automatically. Click the **OK** button to accept the settings.

Simulate the circuit: select **Analysis** and then **Simulate** from the Schematics menu bar. When Probe runs, the traces Vin and Vo will be displayed. Note that the time axis only displays times from 490 ms to 500 ms.

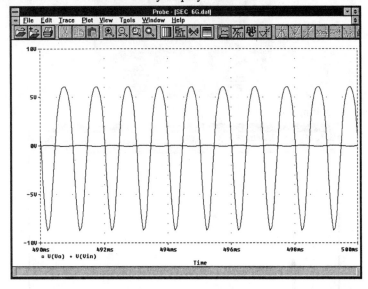

The output is so much larger than the input that you can hardly see any variation on the input. We also notice that the output does look sinusoidal, but also seems to be rounded at the top and peaked at the bottom. The output is not really a pure sine wave and contains some distortion. Distortion means that the output of the amplifier contains frequencies not present in the input. For a distortion-free amplifier, if we input a frequency of 1 kHz, the only frequency contained in the output will be 1 kHz. The output of our amplifier does not look like the input, thus it contains additional frequencies not present in the input. We have two ways to identify these frequencies using PSpice.

6.G.1. Fourier Analysis with Probe

Since we have already spent a good deal of simulation time and we are already in Probe, we can use Probe to view the frequency components of a signal. We will first view the components of the input. Delete the trace **V(Vo)** so that only the trace **V(Vin)** is displayed:

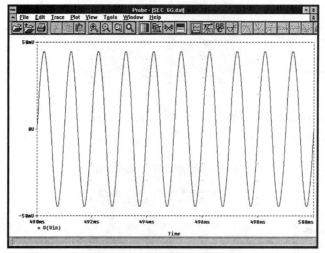

The input should be a sine wave with only one frequency, 1 kHz. To view the frequencies contained in this waveform we need to select the Fourier Processing option. Select **Plot** and then **X Axis Settings** from the Probe menus:

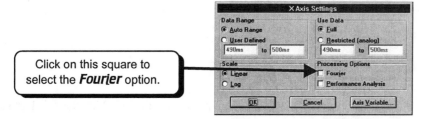

Click the **LEFT** mouse button on the square □ next to the text **Fourier**. The square will fill with an x, ⊠, indicating that the option is selected:

Click the **OK** button to accept the setting:

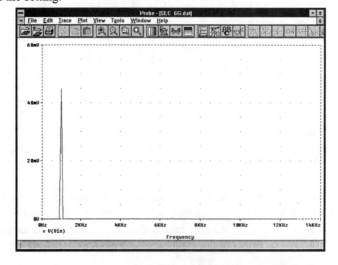

Notice that the signal at **V(Vin)** contains only one frequency at 1 kHz. Since the signal source was a 1 kHz sine wave it should contain only one frequency. Switch back to the time domain by deselecting the Fourier option.

We would now like to look at the frequencies at the output. Delete the trace **V(Vin)** and add the trace **V(Vo)**:

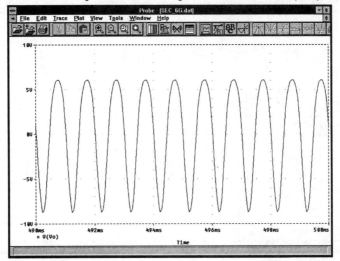

This is what the output looks like versus time. The output is not a perfect sine wave and contains some distortion. To see what frequencies the output waveform contains, we would like to create a second plot that displays the Fourier components of the waveform. Select **Plot** and then **Add Plot** to add a second plot to the same window:

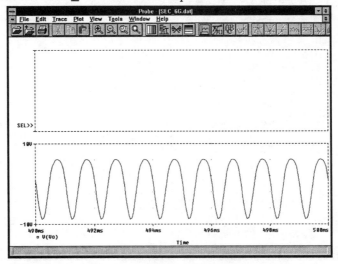

Add the trace V(Vo):

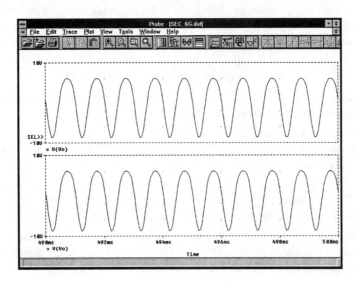

We would like the top plot to display the Fourier components and the bottom plot to display the waveform versus time. For a Fourier plot the x-axis is frequency. For a time plot the x-axis is time. Presently both plots use the same x-axis. To allow the plots to have different x-axes, select **Plot** and then **Unsync Plot**:

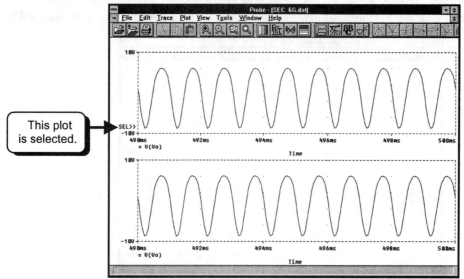

In the screen capture above the top plot is selected. Follow the procedure outlined previously to select the Fourier processing option for the selected plot (select **Plot**, **X** Axis Settings, and then select ***Fourier***):

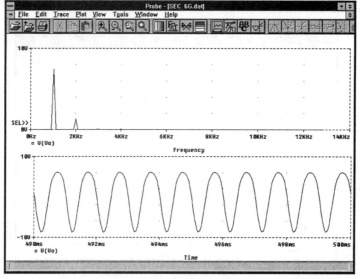

Notice that the largest frequency component contained in the output is at the fundamental (1 kHz), but there is also a large component at the second harmonic (2 kHz). There are other frequencies contained in the output, but they are too small to be seen on the graph. Using the cursors we find that the magnitude of the component at 1 kHz is 7.43 V, and the magnitude of the component at 2 kHz is 1.35 V. An equation for the output voltage would be:

$$Vo=7.43\sin(2\pi \bullet 1000t)+1.35\sin(2\pi \bullet 2000t)$$

This graph tells us that the output contains frequencies of 1 kHz and 2 kHz when the input contains only a single frequency of 1 kHz. This additional frequency is caused by the non-linearity of the amplifier.

   Important Note: To get more accurate results using the Fourier option, you should let the simulation run for many cycles and use a small step size (small value of Step Ceiling in the Transient setup dialog box). This increases the simulation time but will improve the accuracy of your simulation results.

EXERCISE 6-9: Find the magnitude of the harmonics of a ±1 volt, 1 kHz square wave using Probe.

SOLUTION: Use the pulsed voltage source. Set the rise and fall times to 1 μs so that the rise and fall times are much shorter than the pulse width and period of the square wave. Set the print step to a number larger than 20 ns. If you leave the print step small, the simulation will take a large amount of time to complete. Wire the circuit:

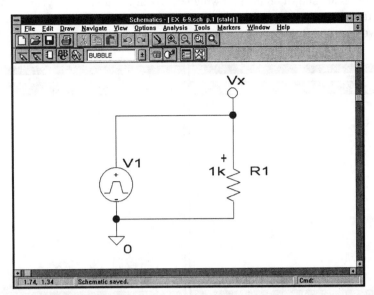

The square wave voltage and Fourier results are:

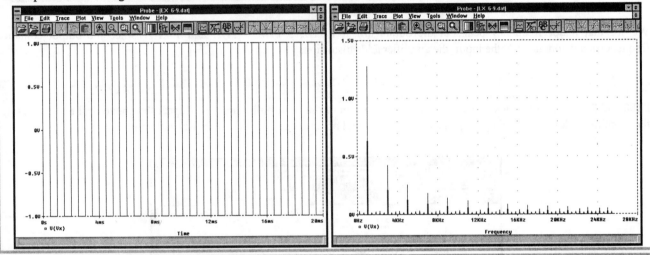

EXERCISE 6-10: Find the harmonics contained in the cross-over distortion of a push-pull amplifier. Let the input to the amplifier be a 2 V amplitude, 1 kHz sine wave. The push-pull amplifier drives an 8 Ω load.

SOLUTION: Wire the push-pull amplifier below. Run a Transient Analysis for several cycles with many points per cycle:

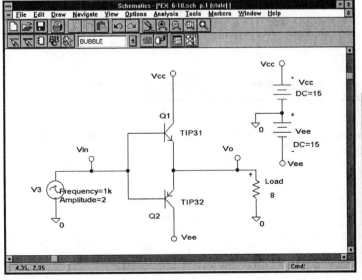

The input and output waveforms are shown on the following left screen capture. Four cycles were simulated, but only one is shown in the following left screen capture to make the distortion easily seen. The right screen capture shows the Fourier components of the output voltage waveform.

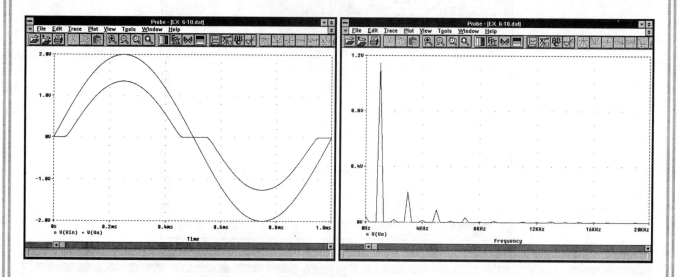

We see that major frequency peaks of the output are at 1 kHz, 3 kHz, 5 kHz, and 7 kHz. Since the output contains frequencies not contained in the input, the amplifier adds distortion.

EXERCISE 6-11: Using feedback, we can reduce the distortion of the push-pull amplifier. Repeat the above drill exercise using the amplifier below. An ideal op-amp is used because a real op-amp cannot drive this push-pull amplifier.

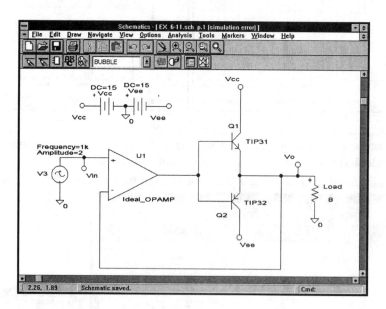

SOLUTION: The input and output waveforms and the Fourier plot of the output of this amplifier are shown below. We see that most of the distortion is removed from the output:

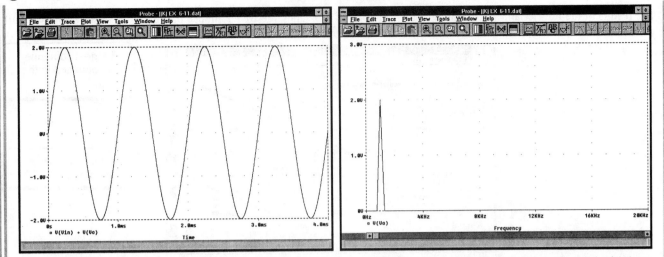

This amplifier has two problems. The first is that most op-amps cannot drive this push-pull amplifier when it is driving an 8 Ω load. The second is that the transistors turn on and off. This adds a phase delay, which can cause the amplifier to oscillate. The Darlington push-pull amplifier on page 283 works much better.

6.G.2. Fourier Analysis with PSpice

The frequency components of a signal can be obtained directly from PSpice by enabling the Fourier option in the Transient Analysis setup dialog box. We will use the common-emitter amplifier circuit shown on page 273. To edit the Transient dialog box select **Analysis**, **Setup**, and then click the **Transient** button. Fill out the dialog box as shown:

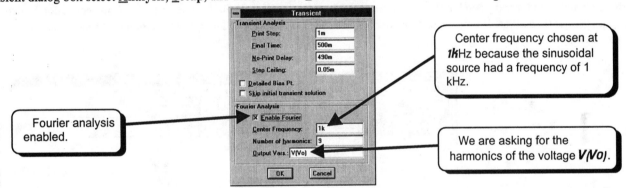

The center frequency is 1 kHz. This frequency was chosen to match the frequency of the sinusoidal input voltage. The harmonics that will be calculated are the first nine: 1 kHz, 2 kHz, 3 kHz, 4 kHz, 5 kHz, 6 kHz, 7 kHz, 8 kHz, and 9 kHz. There may be others, but we want numerical values for only the first nine. The output variable for the Fourier analysis is the voltage at node Vo, **V(Vo)**. This is the output of the amplifier. We could look at the frequency components of any voltage or current, but for this example we are interested only in the output. Click the **OK** button to accept the settings and then run PSpice.

The results of the Fourier analysis will be saved in the output file. When PSpice has finished the simulation, examine the contents of the output file (**Analysis**, **Examine Output**). The Windows Notepad program will run and display the contents of the output file:

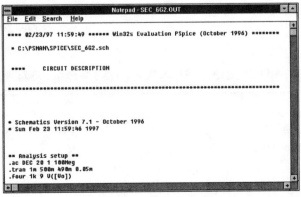

The results of the Fourier analysis are at the end of the output file. Click the **LEFT** mouse button on the down arrow ⬇ until you reach the following text:

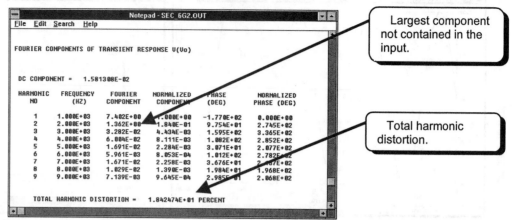

The results of the Fourier analysis show that the magnitude of the sine wave at 1 kHz is 7.402 V, and the magnitude of the sine wave at 2 kHz is 1.362. These results are similar to the results obtained using Probe. This method also gives us the magnitude of the frequency components too small to see on the Probe graph, the phase of each frequency component, and the total harmonic distortion. From the data above, an equation for the output voltage is:

$$V_o = 7.402\sin(2\pi \bullet 1000t - 177.0°) + 1.362\sin(2\pi \bullet 2000t + 97.54°) + 0.0328\sin(2\pi \bullet 3000t + 159.5°) + \ldots$$

EXERCISE 6-12: Find the magnitude of the harmonics of a ±1 volt, 1 kHz square wave using PSpice.

SOLUTION: Use the pulsed voltage source. Set the rise and fall times to 1 μs so that the rise and fall times are much shorter than the pulse width and period of the square wave. Make sure that you set the Print Step in the Transient dialog box to a value greater than 20 ns. If you leave the Print Step small, the simulation will take a large amount of time to complete. Wire the circuit:

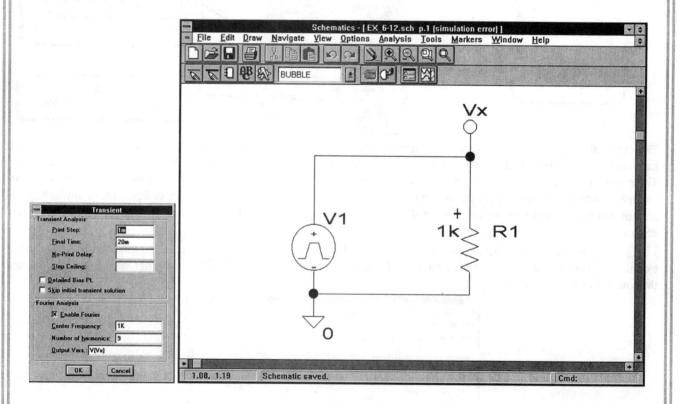

Make sure that you set the appropriate parameters in the "Fourier Analysis" portion of the Transient setup dialog box.

The results are contained in the output file:

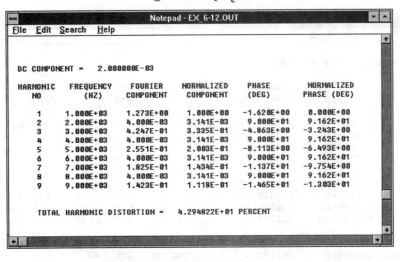

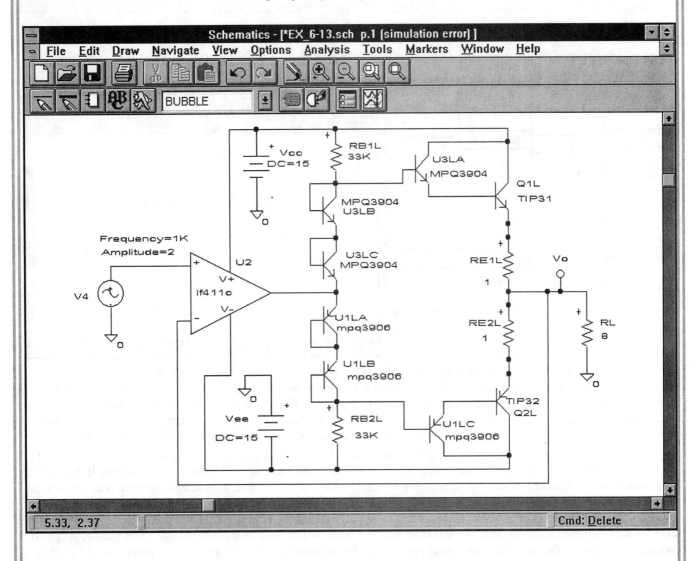

EXERCISE 6-13: Find the distortion of the Darlington push-pull amplifier shown below:

SOLUTION:

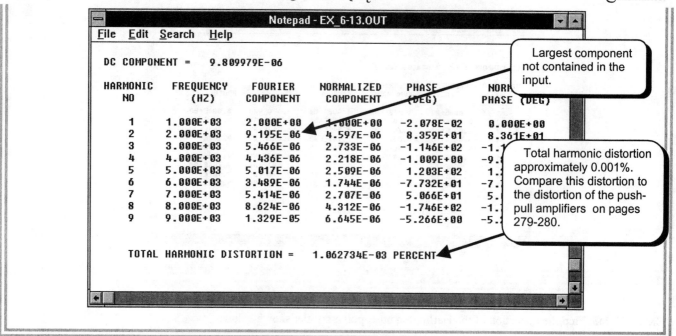

Largest component not contained in the input.

Total harmonic distortion approximately 0.001%. Compare this distortion to the distortion of the push-pull amplifiers on pages 279-280.

6.H. Ideal Operation Amplifier Integrator

In this section we will demonstrate the use of an ideal operational amplifier and the pulsed voltage waveform. Wire the circuit shown below.

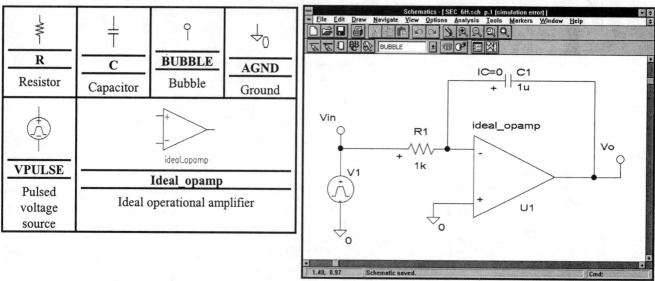

The ideal operational amplifier is very useful in the evaluation version of PSpice. The ideal model has only three components in the subcircuit. This small number of components allows many ideal op-amps to be used before the component limit of the evaluation version is reached. In the evaluation version of PSpice, only two non-ideal op-amp models can be used before reaching the component limit. If you have a circuit with a large number of op-amps, you will be forced to use ideal op-amps in the evaluation version. You should attempt to use the non-ideal op-amp models whenever possible.

The drawback of the ideal op-amp model is that none of the non-ideal properties are modeled. In this example, if a non-ideal op-amp model were used in the simulation, the integrator would not work because of bias currents. If this circuit were tested in the laboratory, it also would not work because of bias currents. Thus, the circuit simulation with a non-ideal op-amp matches the results in the lab, but the circuit simulation with an ideal op-amp does not match the lab results. For this example, the ideal model is not a good choice for simulation because it does not match the results in the lab. We will use it here for demonstration purposes only. See **EXERCISE 6-15** to learn how this integrator performs using non-ideal op-amps. **In general, you should always use the non-ideal op-amp models if possible.** The only reason you should use the ideal op-amp model is if the circuit is too large for the evaluation version of Schematics.

The pulsed voltage source can be used to create an arbitrary pulse shaped waveform. We will use it to create a 1 kHz square wave. The rise and fall times of the square wave will be 1 µs. Double click the **LEFT** mouse button on the pulsed voltage source graphic, to edit its attributes. The following dialog box will appear:

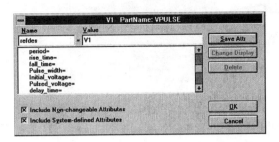

The attributes are described below:

- **period** - The period of the pulse waveform. The pulse shape will repeat itself every **period** seconds. The frequency of the pulse waveform is 1/**period**.
- **rise_time** - The amount of time the voltage source takes to go from the initial voltage to the pulsed voltage, in seconds.
- **fall_time** - The amount of time the voltage source takes to go from the pulsed voltage to the initial voltage, in seconds.
- **Pulse_width** - The amount of time the voltage spends at the pulsed value. It must be true that **Pulse_width** < **period**. For a square wave, **Pulse_width** = **period**/2.
- **Initial_voltage** - The value of the voltage at t=0.
- **Pulsed_voltage** - The value of the voltage source during the **Pulse_width**.
- **delay_time** - At the start of the analysis, the voltage source stays at the initial voltage for an amount of time equal to the **delay_time**. After the **delay_time**, the voltage changes from the initial voltage to the pulsed voltage in the time specified by the **rise_time**.

We would like to create a ±5 V square wave at a frequency of 1 kHz. The rise and fall times will be 1 µs. The dialog box below creates this waveform:

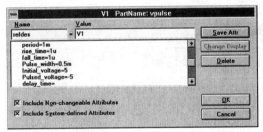

The voltage waveform described by this dialog box will produce the following waveform: From t=0 to 0.0005 s, the voltage will be equal to the initial voltage of 5 V. From 0.0005 s to t=0.001 s, the voltage will be equal to the pulsed voltage of –5 V. At t=0.001 s, the voltage will switch back to the initial voltage and repeat the cycle.

It is important to note that when PSpice calculates the bias to determine the initial condition of the capacitor, it sets the voltage of the pulsed source to its initial voltage. For our pulsed waveform, PSpice will calculate the initial capacitor voltage assuming that Vin = 5 V. This will cause the integrator to saturate initially. To prevent saturation, we must set the initial condition of the capacitor to 0 V. See Section 6.B, pages 251–254, for details on setting the capacitor initial condition.

We would like to simulate the circuit for 10 cycles of the square wave, or 10 ms. Select **Analysis, Setup,** and then click the **Transient** button to obtain the Transient Analysis setup box. Fill in the parameters as shown on the following screen capture. Note that there is an **X** in the box next to the text **Skip initial transient solution**.

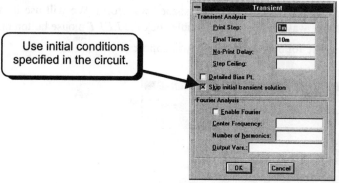

Use initial conditions specified in the circuit.

Run PSpice (**Analysis**, **Simulate**) and then run Probe. In Probe, add the trace **V(Vin)** to observe the square wave voltage source we created:

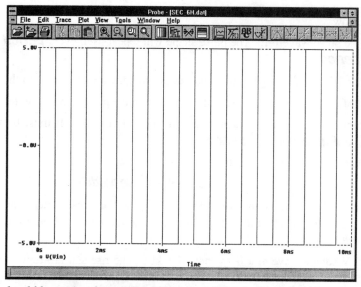

The output of the integrator should be a triangle wave. Add the trace **V(Vo)**. You should see a triangle wave:

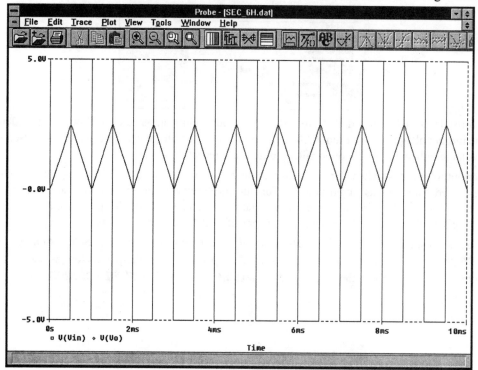

EXERCISE 6-14: Simulate the integrator shown below. Set the initial condition of the capacitor to zero volts. Show the output waveform for a 1 kHz square wave input voltage. Note that this circuit will not work if the resistors are not matched exactly. This is a good circuit for simulation, but never use it in practice. You may also wish to simulate this circuit using non-ideal op-amps to see the effects of bias currents and offset voltages.

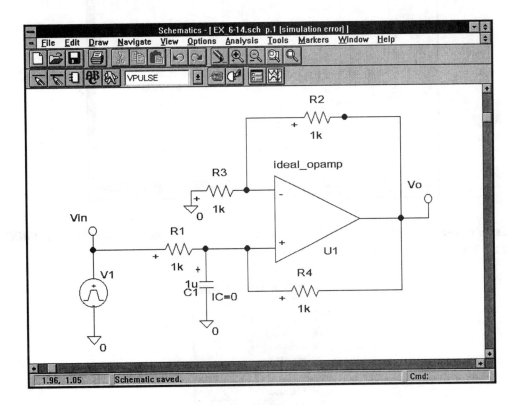

SOLUTION: Use the pulsed voltage source. Set the rise and fall times to 1 µs:

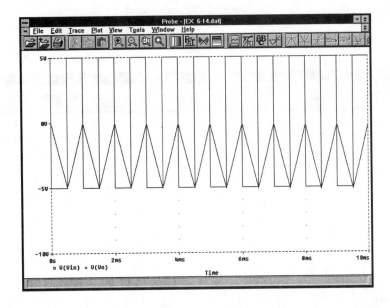

EXERCISE 6-15: Simulate the integrator below using a non-ideal op-amp model like the UA741. For the source, use the square wave voltage source "Vsq."

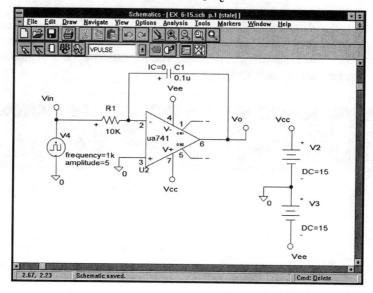

SOLUTION: Make sure you set the initial condition of the capacitor to zero. Remember to check the **Skip initial transient solution** box in the Transient dialog box. First, run the simulation for 10 ms:

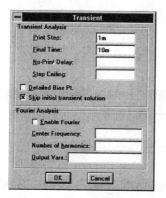

The output voltage is:

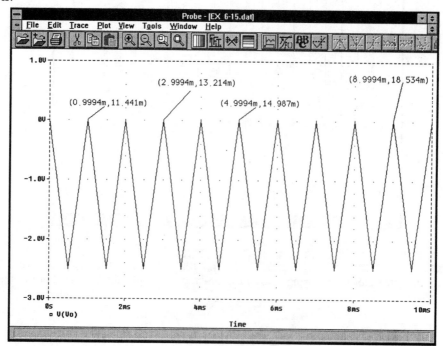

The positive peaks were marked so that we can see that the trace is slowly drifting positive. Set the Step Ceiling to 0.1 ms and then run the simulation again for 100 ms and plot the output:

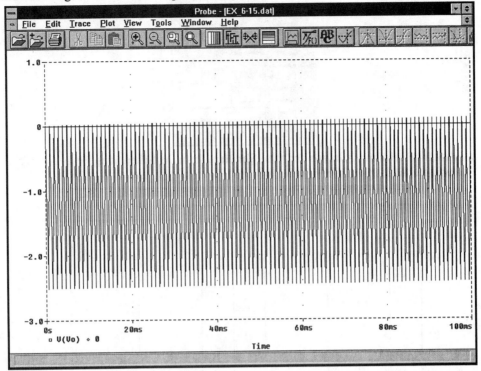

A trace of **0** volts was added so that we can see that the output is slowly drifting positive. If we let the simulation run long enough, the op-amp will eventually saturate and the output will be stuck at the +15 V supply rail. The output drifts up due to the bias currents of the op-amp.

6.I. Multiple Operational Amplifier Circuit

We will now simulate a circuit with three ideal operational amplifiers. With the evaluation version, the component limitation of PSpice limits us to two non-ideal operational amplifiers. Thus, this circuit cannot be simulated using the evaluation version and non-ideal operational amplifiers. The ideal operational amplifier model was created so that a circuit with several operational amplifiers could be simulated using the evaluation version. Wire the circuit shown below.

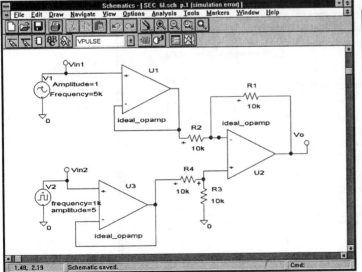

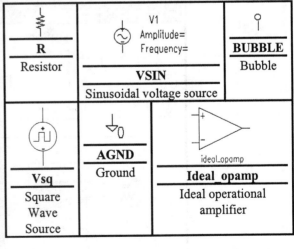

This circuit subtracts a 5 kHz sinusoid from a 1 kHz square wave. We would like to set up a Transient Analysis to simulate this circuit for 5 ms. Select **Analysis**, **Setup**, and then click the **Transient** button to obtain the Transient simulation setup dialog box. The dialog box below is set up to simulate the circuit for 5 ms (**Final Time: 5m**). The maximum step size (**Step Ceiling**) is not specified so that the simulation will finish as soon as possible. The **Print Step** is set at 1 ms so that the integration step will not be affected by the **Print Step**.

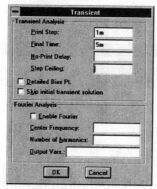

Run the simulation by selecting **Analysis** and then **Simulate** from the Schematics menu bar. When Probe runs, add the trace **V(Vo)**. You will see the following Probe window:

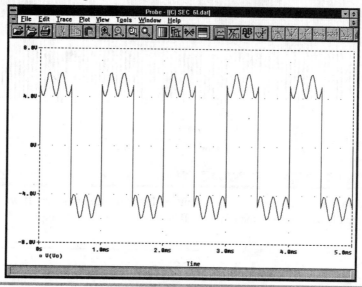

EXERCISE 6-16: Find the output of the circuit below if the input is a 5 V amplitude, 1 kHz square wave. The initial voltage of the capacitor is zero volts.

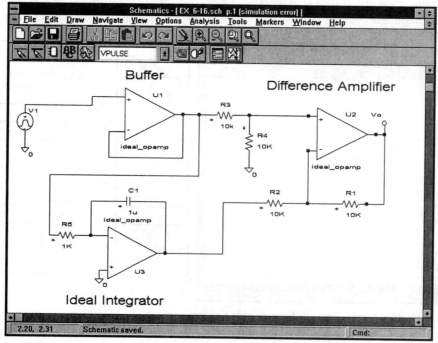

SOLUTION:

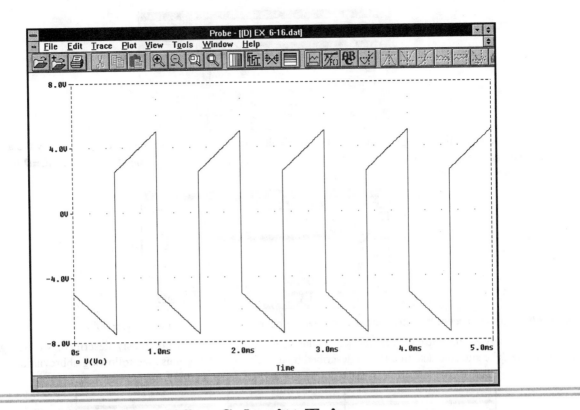

6.J. Operational Amplifier Schmitt Trigger

In this section we will use an operational amplifier to create a Schmitt Trigger. A non-ideal operational amplifier must be used because the ideal op-amp model has trouble converging when it is used as a Schmitt Trigger. Wire the circuit:

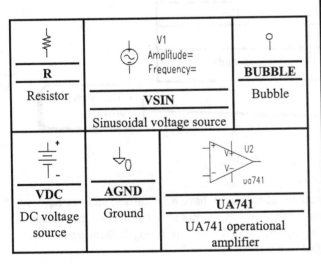

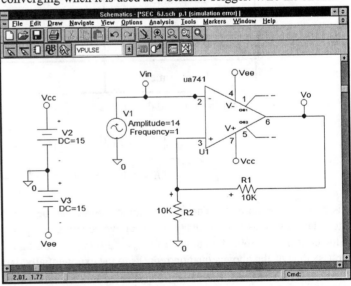

This op-amp circuit is a Schmitt Trigger with trigger points at approximately ±7.5 V. A sinusoidal voltage source will be used to swing the input from +14 V to −14 V and from −14 V to +14 V a few times. The frequency of the source is 1 Hz. This low frequency is chosen to eliminate the effects of the op-amp slew rate on the Schmitt Trigger performance. If you wish to observe slew rate effects, a higher frequency should be chosen. We would also like the sinusoidal source to start at +14 V instead of zero. This can be done by setting the phase of the source to 90°. The phase of the sinusoidal source is not automatically displayed, so it must be changed by editing all the attributes of the sinusoidal source. Double click the *LEFT* mouse button on the sinusoidal voltage source graphic, ⌁. The attribute dialog box for this source will appear:

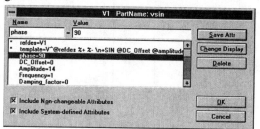

This dialog box shows us all of the possible attributes available for the sinusoidal source. We see that only the amplitude and frequency have been set. We would like to set the phase to 90°, so we must change the attribute **phase=0** to **phase=90**. Set the attributes as shown in the dialog box below:

Click the **OK** button to accept the attributes. The source we have created is: $V_1 = 14\sin(2\pi t + 90°)$.

Since many of the attributes of the sinusoidal source have not been previously described, we will give a brief description here. The sinusoidal source is best described by an equation. We will use the following abbreviations:

Attribute	Symbol	Units
Phase	ϕ	degrees
DC_Offset	B	volts
Amplitude	A	volts
Frequency	F	Hz
Damping_factor	β	sec^{-1}
Delay_time	τ_d	seconds

The equation for the sinusoidal source is:

> This term is a constant.

$$V_{\sin} = \begin{cases} B + A\sin\left(\left(\dfrac{2\pi}{360}\right)\varphi\right) & \text{for} \quad 0 \leq t < \tau_d \\[3mm] B + A\sin\left[2\pi\ F(t - \tau_d) + \left(\dfrac{2\pi}{360}\right)\varphi\right]e^{-\beta(t - \tau_d)} & \text{for} \quad t \geq \tau_d \end{cases}$$

This may not be too clear. The full sinusoidal source can be exponentially damped, can have a DC offset, and can have a time delay as well as a phase delay. In the above equation, the phase (ϕ) is specified in degrees and is converted into radians by the constant $2\pi/360$. We note that for $\tau_d \geq t \geq 0$, $V_{\sin}$ is constant. The sinusoid does not start until $t = \tau_d$. If there are no time or phase delays, the above equation reduces to the exponentially damped sine wave:

$$V_{\sin} = B + A\sin\left[2\pi\ Ft\right]e^{-\beta t}$$

The sources available for use in the Transient Analysis are very flexible to provide the user with many possible waveforms for simulation. Unfortunately, this flexibility can also lead to confusion when using a source for the first time. To make creating waveforms as easy as possible, MicroSim has provided a program for creating and viewing waveforms before running a simulation. The program is called the Stimulus Editor and is discussed in Section 6.M. This program makes understanding the description of the sinusoidal source much easier.

We would now like to set up a Transient Analysis that allows the sinusoidal source to complete two cycles. Since the source has a frequency of 1 Hz, two cycles will take two seconds of simulation time. Obtain the Transient Analysis setup dialog box by selecting **Analysis**, **Setup**, and then clicking the **Transient** button. Fill in the dialog box as shown:

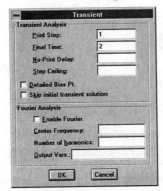

Run PSpice and then Probe. Add the traces **V(Vin)** and **V(Vo)**. You should have the Probe screen shown:

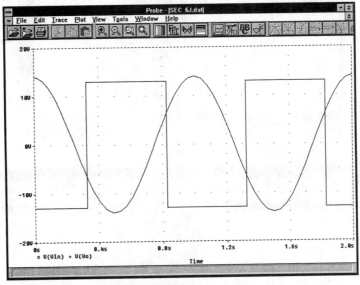

We can see that the Schmitt Trigger changes state at Vin = ±7.5 V.

The next thing we would like to do is plot the hysteresis curve for this Schmitt Trigger. The hysteresis curve is a plot of V_o versus V_{in}. We must delete the trace **V(Vin)**. Click the *LEFT* mouse button on the text **V(Vin)**. It should turn red, indicating that it has been selected. To delete the trace, press the ⌫ DELETE key. You should have the Probe screen shown below:

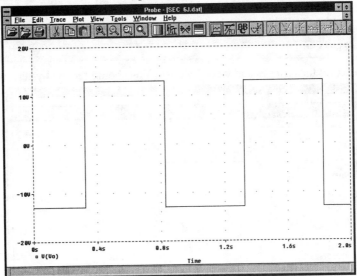

We must now change the x-axis from time to the input voltage, V_{in}. Select **Plot** and then **X Axis Settings** from the Probe menu bar:

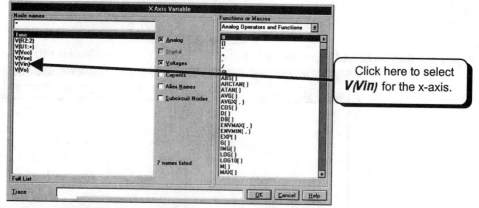

To change the x-axis from time to V_{in}, click the **Axis Variable** button.

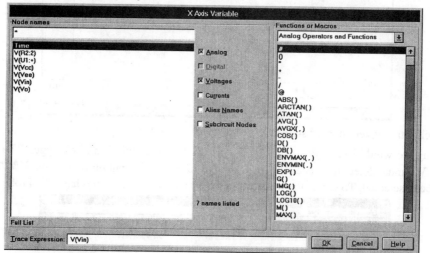

Click here to select **V(Vin)** for the x-axis.

Click the **LEFT** mouse button on the text **V(Vin)** to select the trace:

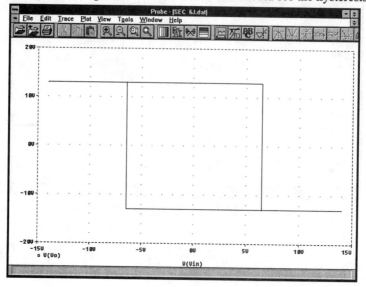

Click the **OK** button twice to accept the changes and view the trace. You should see the hysteresis curve:

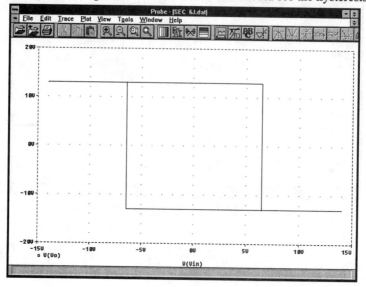

EXERCISE 6-17: Simulate a non-inverting Schmitt Trigger with a ±14 volt sine wave input.

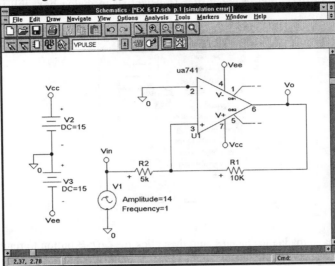

SOLUTION: Use the sinusoidal voltage source. The input and output waveforms are shown on the left screen capture, and the hysteresis curve is shown on the right screen capture.

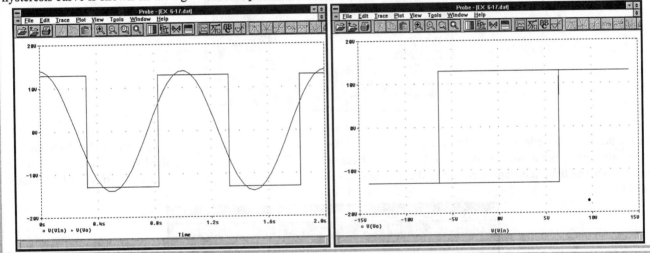

6.K. Parametric Sweep — Inverter Switching Speed

When designing digital circuits we are concerned with how different circuit elements affect the operation of the circuit. In this section we will look at switching speed. First, we will look at a basic BJT inverter and observe its operation. Wire the circuit below:

{}			
R Resistor	**C** Capacitor	**BUBBLE** Bubble	**AGND** Ground
VDC DC voltage source	**Q2N2222** NPN BJT	**VPULSE** Pulsed voltage source	

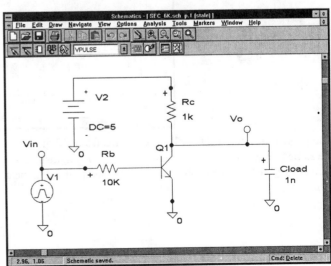

We would like to run a Transient Analysis because we are looking at waveforms versus time. Select **Analysis**, **Setup**, and then click the ***Transient*** button to obtain the Transient setup dialog box. Fill in the Transient dialog box as shown:

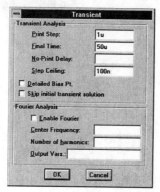

Click the ***OK*** button when you have made the changes, and then click the ***Close*** button to return to the schematic. The input to the inverter will be a short 1 μs pulse. The attributes of the pulsed voltage source are:

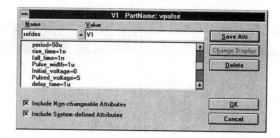

We see that the pulse width is 1 μs. The period is long to let the capacitor come into steady state after the pulse. Note that we are simulating the circuit for only one period. We could have used the PWL source, but the pulsed voltage source is easier to set up. Simulate the circuit and then run Probe. We will first look at the input to see if the pulse is correct. Add the trace **V(Vin)**:

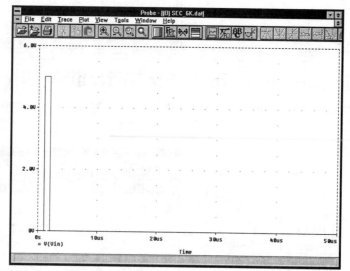

The pulse looks correct. Usually waveforms from a digital circuit have edges close together. If we plot both Vin and Vo on the same graph, the two traces may be hard to distinguish. To see both waveforms clearly, we will add a plot. Select **Plot** and then **Add Plot** from the Probe menus. A second plot will appear on the same window. Add the trace **V(Vo)**:

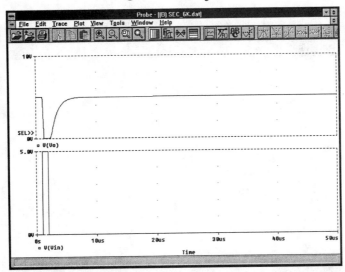

Now that we know what the input and output waveforms look like, we will see how changing the collector resistor **Rc** affects the waveform. Return to the schematic. The Parametric Sweep can be used to change the value of any circuit parameter. First we must define the parameter we want to change. Get a part called "PARAM" and place it in your circuit:

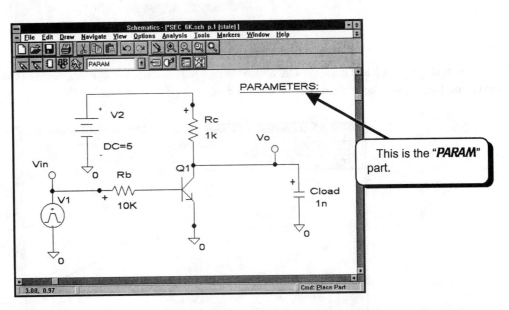

To edit the attributes of the **PARAM** part, double click the **LEFT** mouse button on the text **PARAMETERS**. The attributes allow us to set up three parameters. Each parameter has a name and a value. We must define the name of the parameter and its value. We will define a parameter called **R_val** and give it an initial value of **1k**:

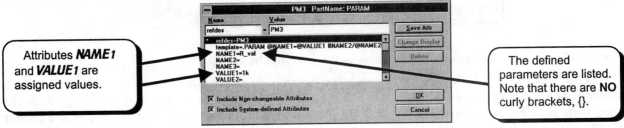

Click the **OK** button to accept the attributes:

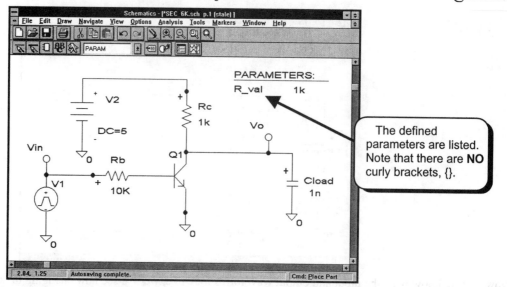

In a schematic a parameter is treated as a value. We must now change the value of **RC** from **1k** to the name of the parameter, **R_val**. Double click on the text **1k** next to resistor **RC**:

Type in the text **{R_val}** and click the **OK** button. **When parameters are used as component values they are enclosed in curly brackets.** The value of the resistor in the schematic should change to **{R_val}**:

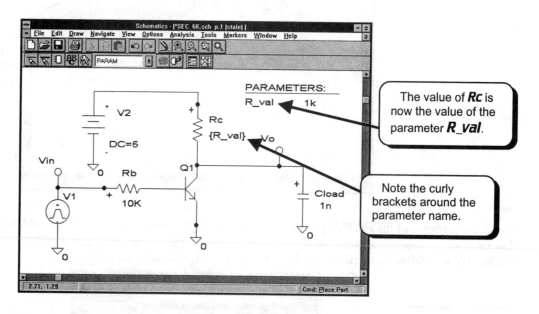

When the circuit is simulated, the value of **RC** is the value of the parameter **{R_val}**.

We will now set up a Parametric Sweep to change the value of the parameter. Suppose we want to see the performance of the circuit for values of **RC** from 1 kΩ to 10 kΩ. We need to set up a Parametric Sweep. Select **Analysis** and then **Setup** from the Schematics menus, and then click the ***Parametric*** button. A parameter is referred to as a ***Global Parameter*** in this dialog box. Fill in the dialog box as shown:

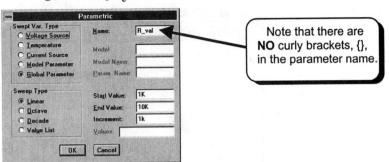

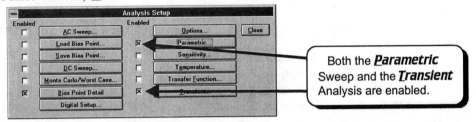

Note that there are no curly brackets around the parameter. Click the **OK** button to accept the values. You will return to the *Analysis Setup* dialog box. Both the *Parametric* Sweep and the *Transient* Analysis must be enabled. Note that the squares next to these two buttons are filled with **X**'s, ⊠.

Click the **Close** button to accept the setup. Logically, the Transient Analysis executes inside the Parametric Sweep. That is, for each value of the parameter, the Transient Analysis is run. Thus, for this setup, ten Transient Analyses will be run.

Run the simulation (**Analysis, Simulate**). When the simulation is complete Probe will run:

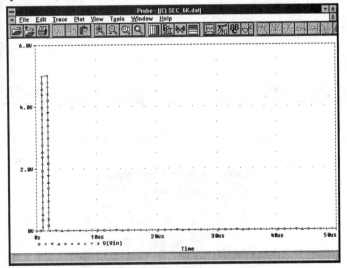

Since the Transient Analysis was run several times, Probe is asking which of the runs we want to view. We would like to see all results on the same graph. By default, all of the runs are selected, so click the **OK** button. Next, add the trace **V(Vin)**:

Ten traces are shown because the Transient Analysis was run ten times and *V(Vin)* was in each simulation. Next, we will add a plot to make seeing the traces easier. Select **Plot** followed by **Add Plot**, and then add the trace **V(Vo)**:

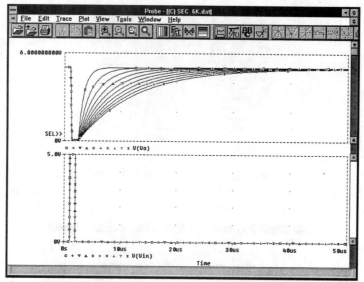

Ten traces are plotted, one for each value of R_val.

EXERCISE 6-18: For the BJT inverter, simulate the circuit to observe how **Rb** effects switching speed. Let **RC** remain constant at 1 kΩ.

SOLUTION: Add a second parameter called **RB_val**. Change the value of **Rb** from 10k to **{RB_val}**:

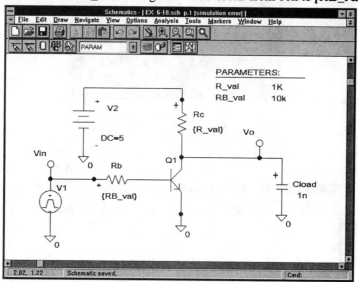

Modify the Parametric Sweep to sweep parameter **RB_val** from 5 kΩ to 30 kΩ in 5 kΩ steps:

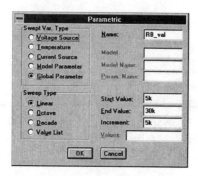

Modify the Transient setup to run the simulation for 10 μs and use a Step Ceiling of 10 ns. The results of the simulation are:

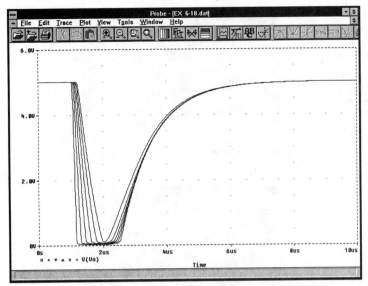

The results show that the base resistance affects the fall time and the rise time. See **EXERCISE 6-19** to generate plots of fall time versus *Rb*.

6.L. Performance Analysis — Inverter Rise Time

The Performance Analysis capabilities of Probe are used to view properties of waveforms that are not easily described. Amplifier bandwidth, rise time, and overshoot are examples. To calculate the bandwidth of a circuit, you must find the maximum gain, and then find the frequency where the gain is down by 3 dB. To calculate rise time, you must find the 10% and 90% points, and then find the time difference between the points. The Performance Analysis gives us the capability to plot these properties versus a parameter or device tolerances. The Performance Analysis is used in conjunction with the Parametric Sweep to see how the properties vary versus a parameter. The Performance Analysis is used in conjunction with the Monte Carlo Analysis to see how the properties vary with device tolerances. In this section we will plot the rise time of a BJT inverter versus the value of the collector resistor. See Section 9.G to learn how to use the Performance Analysis in conjunction with the Monte Carlo Analysis.

Suppose that for the example of Section 6.K we would like to see a plot of how the rise time is affected by the value of the collector resistor. This plot can be accomplished using the Performance Analysis capabilities of Probe. Repeat the procedure of Section 6.K. When Probe runs, select all of the runs and add the trace **V(Vo)**:

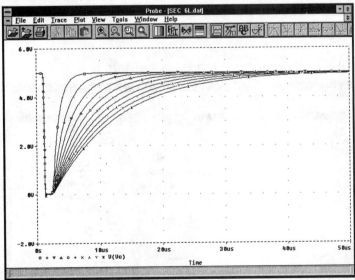

We see that all of the traces reach the final value of 5 V. To plot the rise time versus the parameter {R_val} we must select the Performance Analysis. From the Probe menu select **Plot** and then **X Axis Settings**:

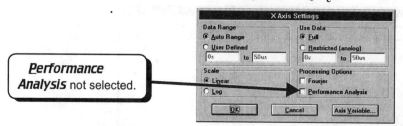

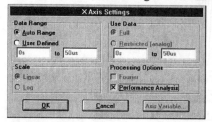

Under **Processing Options** we see that the square □ next to **Performance Analysis** does not have an **X** in it, indicating that it is not enabled. To enable the **Performance Analysis** click the *LEFT* mouse button on the text **Performance Analysis**. The square should fill with an **X**, indicating that the **Performance Analysis** is enabled:

Click the **OK** button to return to Probe. You should have two plots on the screen:

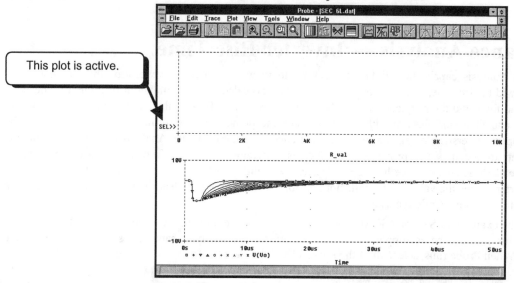

Notice that the top plot has the text **SEL>>** next to it. This means that the top plot is the active plot. Any traces that are added will appear on the active plot.

We now wish to add the rise time plot. Select **Trace** and then **Add**:

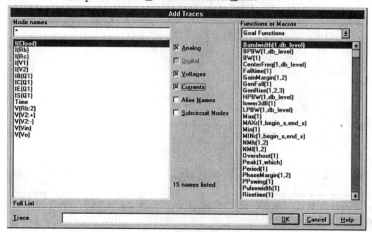

The left pane contains the normal voltage and current traces. The right pane contains goal functions, such as **upper3dB**, **lower3dB**, and **Risetime**. These functions are defined using the Performance Analysis capabilities of Probe. To see how the functions are defined, use the Windows Notepad program to edit the file c:\msimev71\msim.prb. Near the center of the file you will see the text:

```
Risetime(1) = x2-x1
*

*#Desc#* Find the difference between the X values where the trace first
*#Desc#* crosses 10% and then 90% of its maximum value with a positive
*#Desc#* slope.
*#Desc#* (i.e. Find the risetime of a step response curve with no
*#Desc#* overshoot. If the signal has overshoot, use GenRise().)
*

*#Arg1#* Name of trace to search
*

* Usage:
*           Risetime(<trace name>)
*

  {
     1|Search forward level(10%, p) !1
      Search forward level(90%, p) !2;
  }
```

The name of the function is **Risetime**. It has **1** input argument. **1|Search forward level** means search the first input forward and find a level. The level we are looking for is the 10% voltage level. 0% is defined as the minimum level of the trace; 100% is defined as the maximum. The **p** means find the specified level when the trace has a positive-going slope. When the point is found, the text **!1** designates its coordinates as x1 and y1. **Search forward level(90%, p) !2** means search the first input forward and find a point on the positive-going slope that is at the 90% level. When the point is found, the text **!2** designates its coordinates as x2 and y2. The function returns **x2-x1**, which is the time difference between the two points. Since the x-axis is time in a Transient Analysis, **x2-x1** is the time difference between the two points, or the rise time. A second function is :

```
Falltime(1) = x2-x1
  {
  1|Search forward level(90%, n) !1
  Search forward level(10%, n) !2;
  }
```

The n indicates a negative-going slope.

This function is similar to the Risetime function, except that it finds the time between the 10% and 90% points when the points lie on the negative-going slope.

Type in the text **Risetime(V(Vo)):**

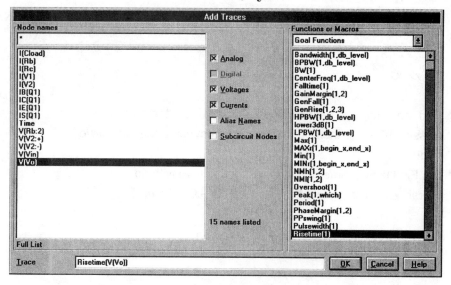

Click the **OK** button to view the plot:

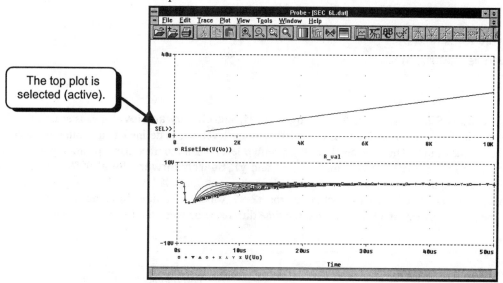

Depending on your setup the plots may or may not be easy to read. We will delete the lower plot so that the top plot will fill the screen. At the moment the top plot is active. Click the **LEFT** mouse button on the bottom plot to select it. The text **SEL>>** should toggle to the bottom plot:

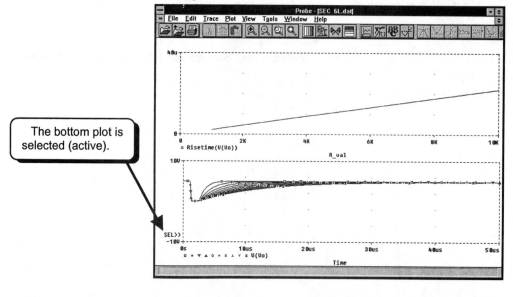

To delete the active plot, select **Plot** and then **Delete Plot** from the Probe menus:

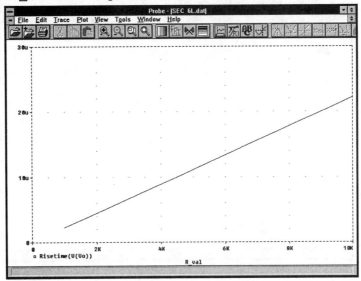

This plot is a little easier to read. For this type of plot it may be necessary to show the individual data points on the line. Thus, we would like to display the markers. Select **Tools** and then **Options** from the Probe menus:

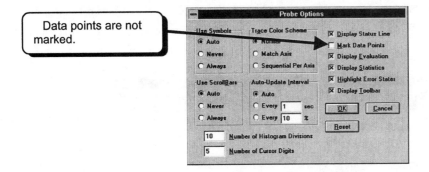

We notice that the individual data points are not displayed. To mark the points, click the *LEFT* mouse button on the text **Mark Data Points**. The square next to the text **Mark Data Points** should fill with an *X*, ⊠, indicating that the option is enabled:

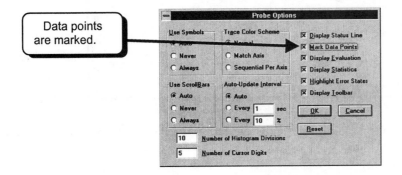

Click the **OK** button to change the setting. **Important note: By clicking the OK button, you will change the settings for all future plots. All plots in this session as well as plots in the future will mark the data points. If you do not want to make the setting permanent, you must change the setting back before you exit Probe.**

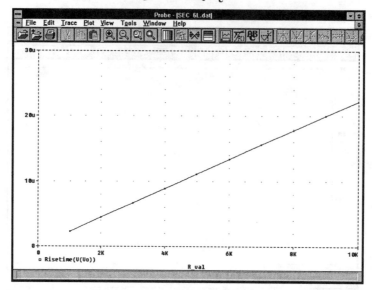

We can now see the individual points. This plot shows us that the rise time of the circuit is a linear function of the collector resistor value.

EXERCISE 6-19: Continuing with **EXERCISE 6-18**, we would like to see how the base resistor **Rb** affects the rise and fall time.

SOLUTION: Rerun **EXERCISE 6-18**. When Probe runs, enable the Performance Analysis. First add the trace `Falltime(V(Vo))`:

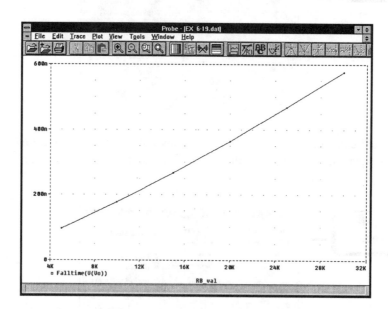

We see that the fall time is approximately a linear function of the base resistor value. Delete the trace and add the trace `Risetime(V(Vo))`:

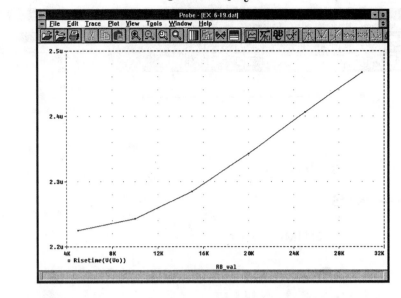

6.M. Stimulus Editor

The Stimulus Editor is a tool that allows us to create and view a signal source before we run a simulation. It can be used with analog voltage sources, analog current sources, and digital sources. Here, we will demonstrate its use with an analog voltage source. The Stimulus Editor is a very useful tool for creating waveforms if you are not familiar with the various sources available with PSpice. It can be used to create sources such as VPWL, VSIN, VPULSE, VSFFM, and VEXP. Unfortunately, the evaluation version of the Stimulus Editor is limited to sinusoidal voltages. However, it is still a very useful tool.

The parts used for analog voltages and currents are called **ISTIM** and **VSTIM** and are located in the **source2.slb** library:

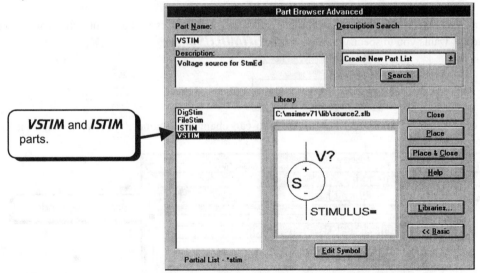

Place the **VSTIM** part in your schematic with a resistor load:

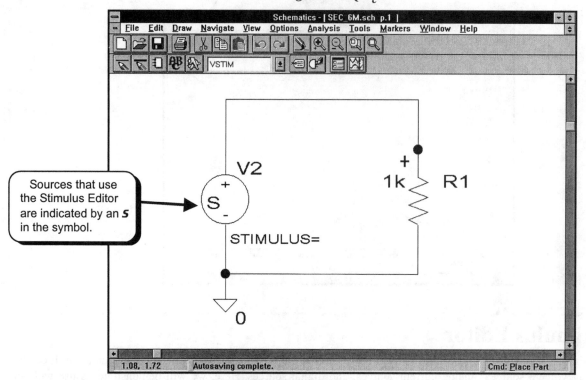

Notice that the graphic has the attribute **STIMULUS=**. When you create a waveform with the Stimulus Editor, you will give the waveform a name. The name specified with the Stimulus Editor will be specified in the schematic using the **STIMULUS=** line.

Before you can use the Stimulus Editor, you must save the schematic. See Section 1.H if you are unfamiliar with saving a schematic. We will save the schematic with the name **sec_6m.sch**.

Notice that the name of the stimulus is still unspecified. To create the waveform for the part, double click the **LEFT**

mouse button on the VSTIM graphic, 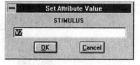. Before the Stimulus Editor runs, Schematics will ask you what you wish to call the waveform. In other words, it wants you to specify a value for the **STIMULUS=** line:

The default name is the name of the source, **V2** for my schematic. You can change the name if you wish. Click the **OK** button to run the Stimulus Editor:

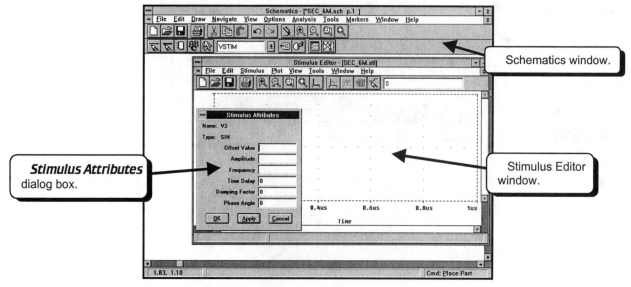

The Stimulus Editor window allows us to view the waveform. The **Stimulus Attributes** dialog box allows us to change the properties of the waveform. Since the evaluation version of the Stimulus Editor allows only the creation of sinusoidal sources, the **Stimulus Attributes** dialog box automatically comes up with a waveform type of **SIN**:

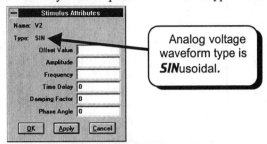

We have a lot of flexibility since we can specify the **Offset Value**, **Amplitude**, **Frequency**, **Time Delay**, **Damping Factor**, and **Phase Angle**. The dialog box below specifies a 1 kHz sine wave with a 1 V amplitude and 1 V offset:

To view the waveform with the Stimulus Editor click the **Apply** button:

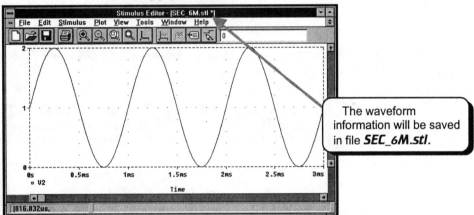

The waveform information will be saved in file **SEC_6M.stl**. If you copy this circuit to a floppy, make sure you copy the files **SEC_6M**.sch and **SEC_6M.stl**. If you do not copy the file **SEC_6M.stl**, you will lose the waveform information when you copy the files to a floppy disk.

Suppose we wish to see the effect of the other parameters. You may change any of the parameters in the **Stimulus Attributes** dialog box and then click the **Apply** button to view the waveform. Modify the parameters as shown below:

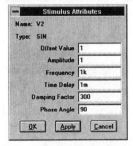

The additional parameters add a time delay, a phase angle, and exponential damping to the waveform. Click the **Apply** button to view the waveform:

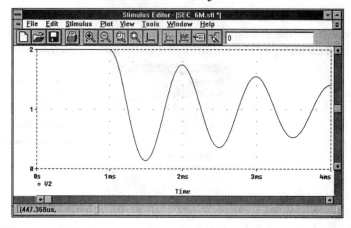

The Stimulus Editor allows us to see the effect of each of the parameters. As a last example, change the parameters as shown:

We will accept these parameters as the final waveform settings, so click the **OK** button. The waveform will be displayed, and the **Stimulus Attributes** dialog box will disappear:

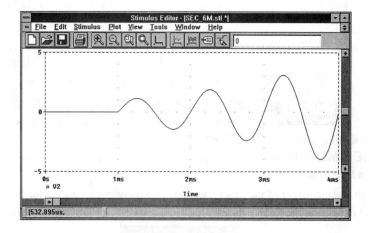

If you need to make further changes to the waveform, select **Edit** and then **Attributes** from the Stimulus Editor menus. The **Stimulus Attributes** dialog box will reappear:

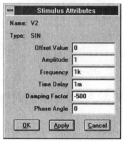

When you are finished making changes, click the **OK** button. The updated waveform will be displayed:

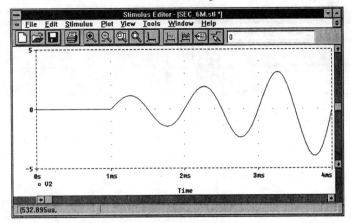

To apply the waveform displayed in the Stimulus Editor window to the V2 part in the schematic and return to the schematic, select **File** and then **Exit** from the Stimulus Editor menu bar:

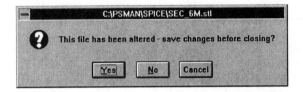

Click the **Yes** button to save the changes. You will return to the schematic:

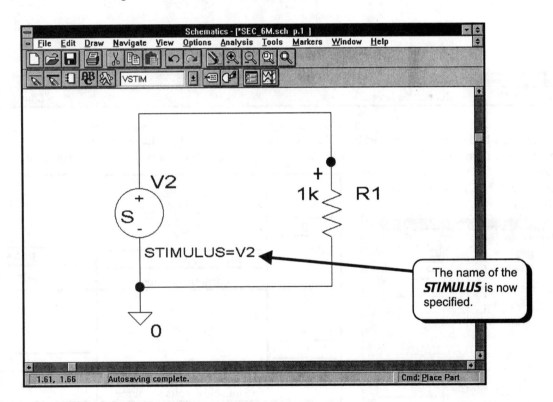

Notice that the name of the stimulus, **STIMULUS=V2**, is now specified in the schematic. If you run a Transient Analysis and plot the voltage across R1, the voltage will be the same as we saw in the Stimulus Editor.

6.N. Temperature Sweep — Linear Regulator

A temperature sweep can be used in conjunction with most of the analyses. Here we will show how temperature affects the performance of a linear voltage regulator. Create the circuit shown below:

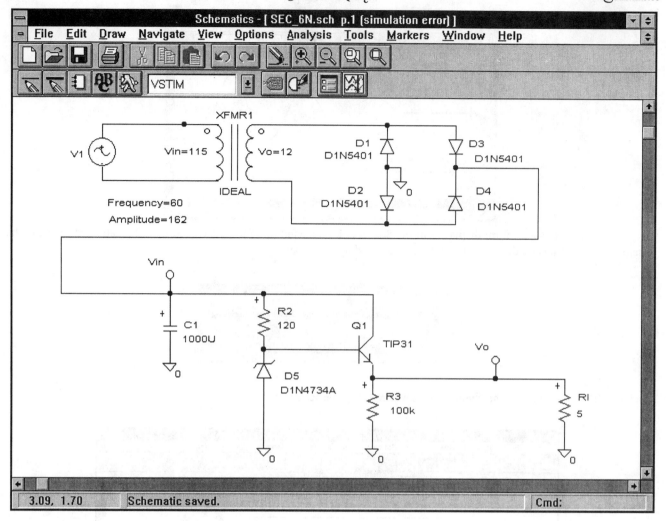

This circuit is a bridge rectifier followed by a filter capacitor to produce a DC voltage with ripple at Vin. After Vin is a linear regulator made from a Zener voltage reference and an NPN pass transistor. We will first run a Transient Analysis to see the operation of the circuit at room temperature (27 °C). Set up a Transient Analysis as shown:

Transient

Transient Analysis
- Print Step: `1m`
- Final Time: `160m`
- No-Print Delay:
- Step Ceiling: `10u`
- ☐ Detailed Bias Pt.
- ☐ Skip initial transient solution

Fourier Analysis
- ☐ Enable Fourier
- Center Frequency:
- Number of harmonics:
- Output Vars.:

[OK] [Cancel]

R Resistor	**Ideal_XFMR_Vo/Vin** Ideal transformer	**D1N5401** Rectifier Diode
C Capacitor	**VSIN** Sinusoidal voltage source	**TIP31** Power BJT
BUBBLE Bubble	**AGND** Ground	**D1N4734A** Zener diode

Run the simulation and plot Vin and Vo:

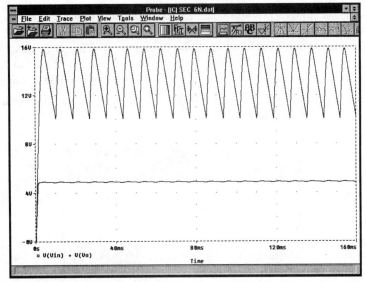

We will zoom in on the Vo trace to see it more closely:

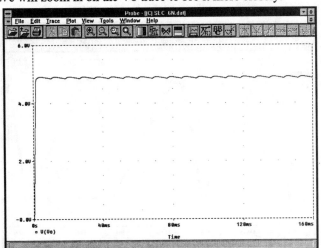

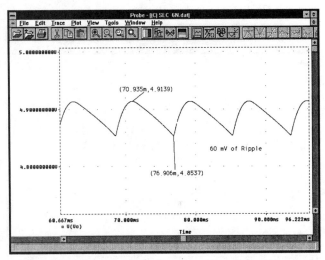

We see that the output has about 60 mV of ripple.

Next we will see how the output changes with temperature. We will run a Transient Analysis at –25 °C, 25 °C, and 125 °C. Return to the schematic and select **Analysis**, **Setup**, and then click the ***Parametric*** button. Fill in the dialog box as shown:

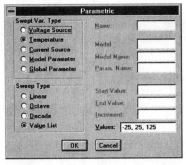

Click the ***OK*** button:

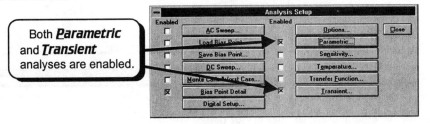

Both ***Parametric*** and ***Transient*** analyses are enabled.

Note in the dialog box above that both the **Transient** and **Parametric** analyses are enabled. Click the **Close** button and then run the simulation. It will take three times as long as the previous simulation because the Transient Analysis will run three times. Logically, the Transient Analysis runs inside of the Parametric sweep. For this example, the temperature will be set to –25 °C, and then the Transient Analysis will be run. Next, the temperature will be set to 25 °C, and then the Transient Analysis will be run. Finally, the temperature will be set to 125 °C, and then the Transient Analysis will be run.

When the three simulations are complete, Probe will run automatically. You will have a choice of which of the runs you would like to view. By default, all of the runs are selected:

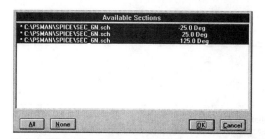

Click the **OK** button to select all of the runs. First we will plot the input voltage waveform. Add the trace **V(Vin)**:

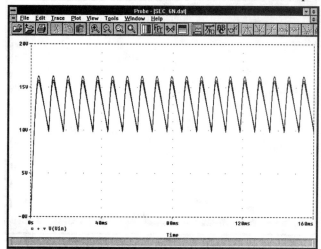 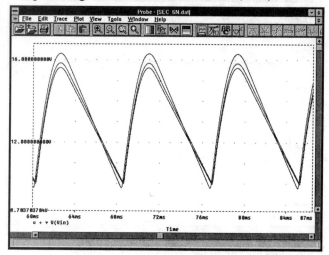

Three traces are shown, one for each temperature. The result shows that temperature does not have too much of an effect on the input voltage. Remove the trace and plot the output voltage:

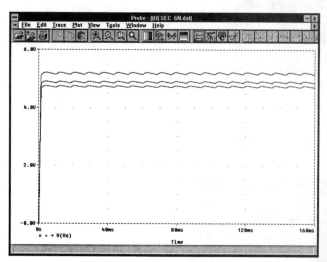 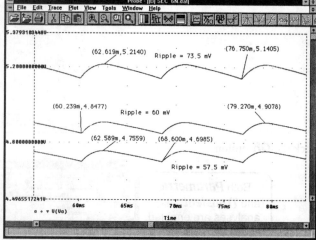

The top trace has 73.5 mV of ripple and the bottom trace has 57.5 mV of ripple.

The question may arise as to how to tell which trace is at what temperature. We can find out more information about each trace. Double click the *LEFT* mouse button on the square, as shown below:

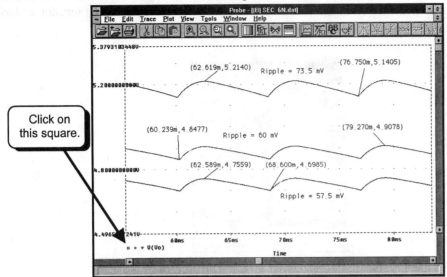

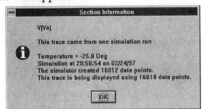

If you double click fast enough, a dialog box will appear:

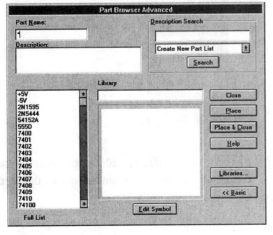

We see that the trace marked by squares is at a temperature of –25 °C. We can use this procedure to determine the identity of each trace.

6.O. Analog Behavioral Modeling

The Analog Behavioral Modeling (ABM) parts give you an easy way to include function blocks in your circuit without having to create a circuit that implements the function. You can implement simple math functions like addition, subtraction, multiplication, and division of waveforms. More complicated functions include power, log, sin, and absolute value. Circuit functions are also provided such as gain with limits and Butterworth and Chebyshev filters. These blocks are usually used instead of creating a circuit to perform the function. In the evaluation version, they can be used to replace circuit blocks with a large number of components by a single part. This enables the simulation of more complicated circuits by reducing the component count.

6.O.1. Examples of ABM Parts

The ABM parts are located in library abm_port.slb in the libraries that accompany this text, and in library abm.slb in the libraries that are standard with the Evaluation version. We will look at some of the parts. Type **CTRL-G** to get a new part and then click the **Libraries** button. Select the **abm_port.slb** library:

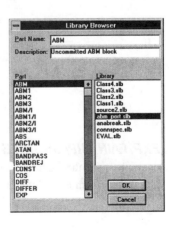

The first few parts are uncommitted ABM parts for general use. These blocks do not perform a specific function. To see what functions can be performed, refer to the ABM section of the reference manuals provided on the CD-ROM. Further down the list we see some specific functions such as **ABS**, **ARCTAN**, **BANDPASS**, and so on. To see what function a block performs, click the **LEFT** mouse button on the name of the block to select it. For example, click on **DIFF**:

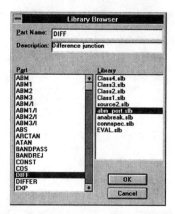

The description of this block says that it is a **Difference junction**; that is, this block will subtract two waveforms. Next, select the part named **DIFFER**:

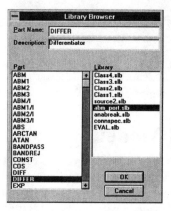

The block says that this is a **Differentiator**. The output of this block is the time derivative of the input. Scroll down the part window to see more parts:

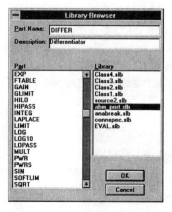

We see such parts as **EXP**, **GAIN**, **HIPASS**, **INTEG**, **MULT**, **SIN**, and **SQRT**. The blocks perform the stated function on the input waveform. For example, with the **SQRT** function, the output voltage is the square root of the input voltage. With the **INTEG** function, the output waveform is the integral over time of the input waveform.

These blocks can be used together with any circuit components in a Transient Analysis. We will demonstrate the function of a few blocks. Create the circuit below. The parts are listed in Table 6-3 on page 318:

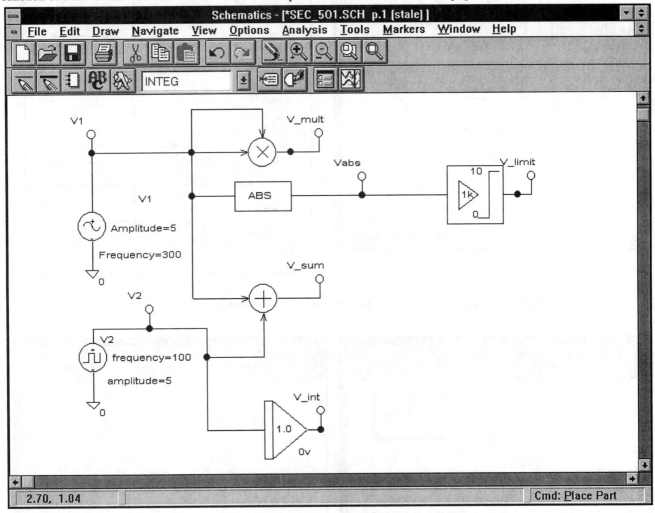

Part *glimit* is a gain block with output voltage limits. The part placed has a gain of *1k* with the output voltage constrained between *0* V and *10* V. This block can be viewed as a single sided op-amp with one supply pin tied to ground and the other to 10 V. We can change the limits and the gain. To change the gain, double click on the text *1k*. A dialog box will open:

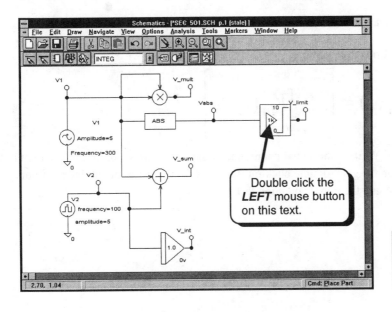

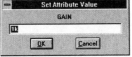

Type **5** and click the **OK** button to accept the change and return to the schematic. The gain will now be displayed as **5**:

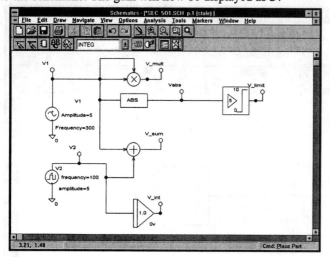

Table 6-3		
Mult Multiplier	**Sum** Summing junction	**AGND** Ground
ABS Absolute value	**Glimit** Gain with voltage limits	**Integ** Integrator
Vsin Sinusoidal voltage source	**Vpulse** Pulsed voltage source	**BUBBLE** Bubble

To change the limits, double click on the **0** or the **10**. Change the limits to –15 and 15:

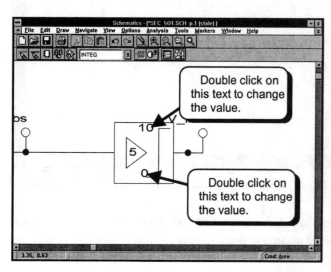
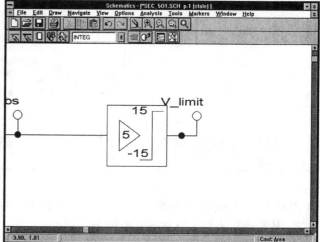

This block can now be viewed as an ideal amplifier with a gain of 5 and ±15 V supplies.

The integrator block performs the function

$$V_{out}(t) = \left[A \int_{t=0} V_{in}(t)dt \right] + V_{in}(0)$$

where A is the gain of the integrator and $V_{in}(0)$ is the initial condition. In the block shown on the schematic the gain is **1.0** and the initial condition is **0V**. To change the gain, double click on the text **1.0**; to change the initial condition, double click on the text **0V**. I will change the gain to 5 and the initial condition to 1 V.

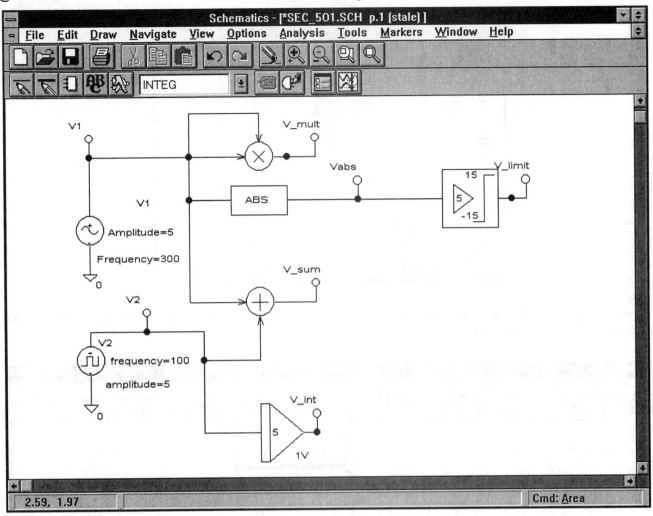

We will now simulate the circuit with a Transient Analysis for 50 ms:

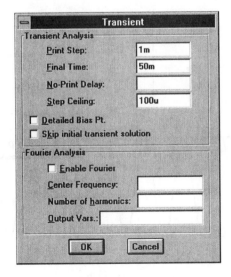

Run the simulation and plot the sine wave and square wave on separate plots. To add plots to the Probe window, select **Plot** and then **Add Plot** from the Probe menus.

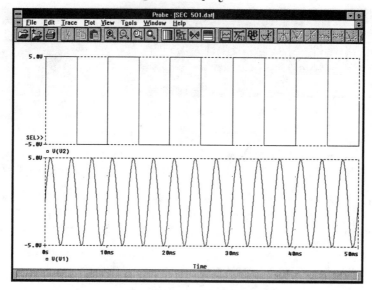

We will now look at each output individually and compare them to the two input waveforms. The output of the multiplier should be the square of a sine wave:

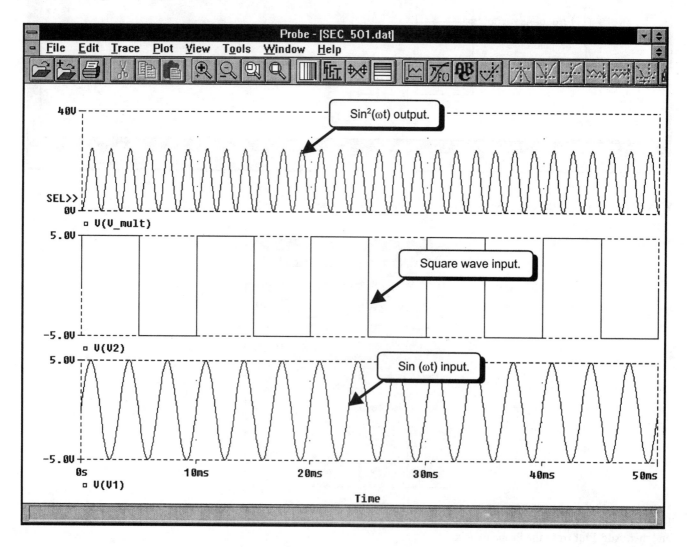

Note that the output of the multiplier is always positive. The output of the ABS function is:

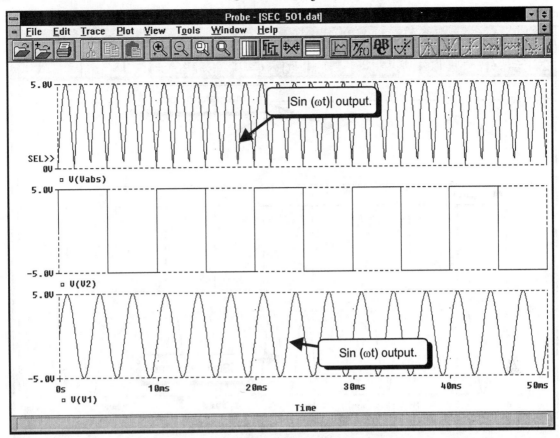

The output of the Glimit part is:

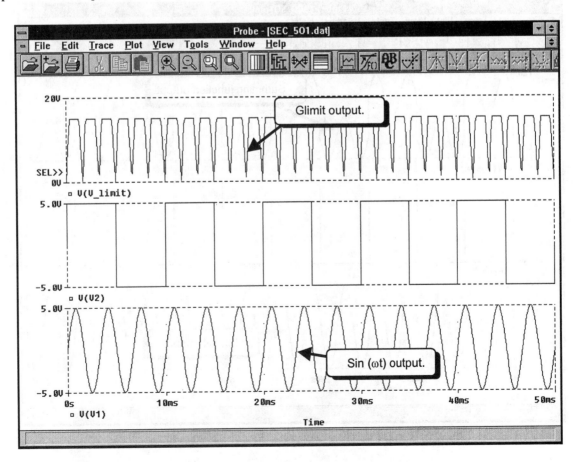

The output of the integrator is:

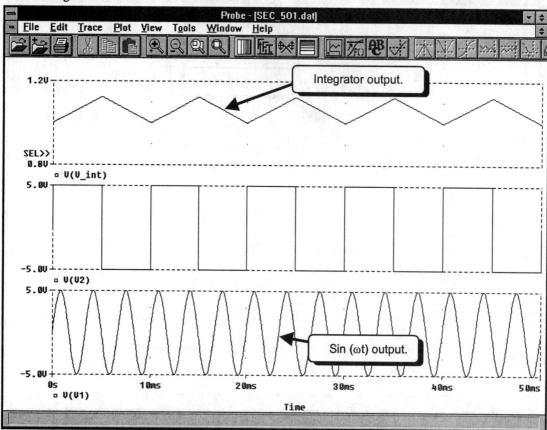

And the output of the summing junction is:

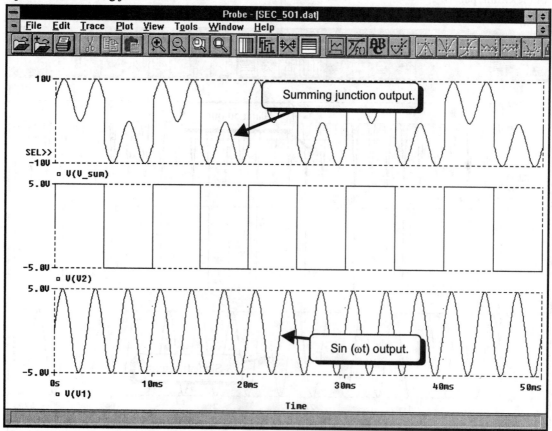

6.O.2. Modeling the Step Response of a Feedback System

In this example we will show the performance of a feedback system using only ABM parts. Wire the circuit below:

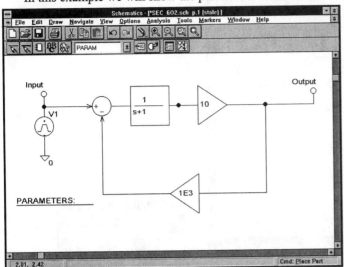

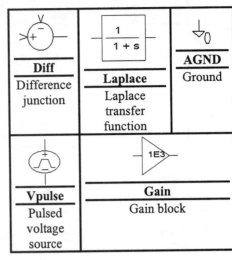

We would like to see how the output step response changes for different feedback gains. Presently the feedback gain is **1E3** or 1000. We would like to vary the feedback gain to see how the feedback affects the operation of the system. We will use a parameter to change the value of the feedback gain. Double click on the text **PARAMETERS:** to edit the attributes of the part:

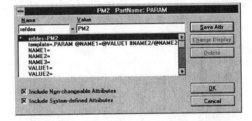

Assign the attributes as shown:

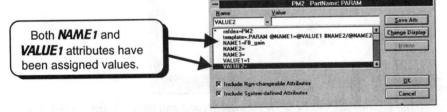

Both **NAME1** and **VALUE1** attributes have been assigned values.

We have defined a parameter named **FB_gain** and given it an initial value of **1**. Click the **OK** button to accept the changes:

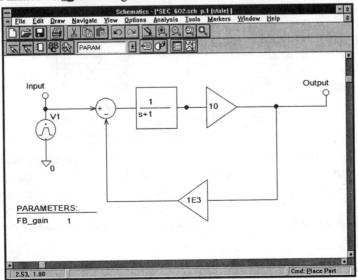

Double click on the text **1E3** of the bottom gain block to change the gain of the feedback block:

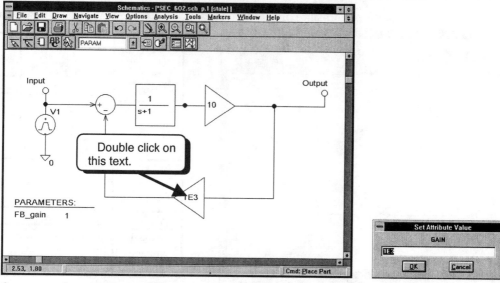

Type the text {**FB_gain**} and click the **OK** button to accept the value. In the screen capture below, the text {**FB_gain**} has been moved to a location that makes it easier to read.

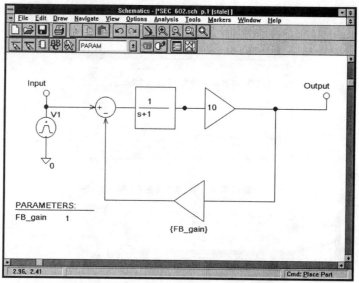

We would like the input to be a 0 to 5 V step input. The step will be 0 for the first second and then 5 V for the remainder of the simulation. This waveform can be constructed with a pulsed voltage source. Set the attributes of the VPULSE source as shown:

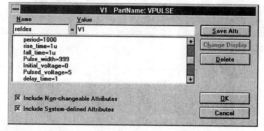

A **_period_** of **_1000_** seconds and a **_Pulse_width_** of **_999_** seconds make this a waveform that is at the pulsed voltage for most of the time. The **_delay_time_** of **_1_** second keeps the voltage at the initial voltage for the first second of the simulation. Thus, when the simulation starts, the voltage will stay at the initial voltage of 0 for the first second. Then the source will change to the pulsed voltage of 5 volts for 999 seconds. The simulation will end well before 999 seconds, so for purposes of this example, the pulsed source is a step from 0 to 5 volts. The waveform of this source is shown below:

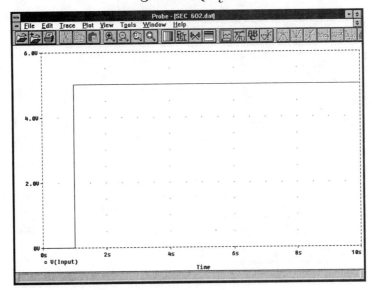

We would like to simulate the response of the system for 10 seconds. Set up a Transient Analysis with the following parameters:

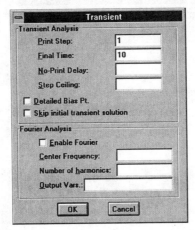

We would like to observe the step response for several different values of the feedback gain. We will use a Parametric Sweep to vary the value of the parameter FB_gain. Fill in the Parametric dialog box as shown:

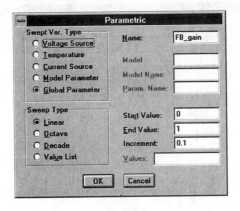

We will be running the Transient Analysis 11 times. The first time the feedback gain will be zero, the second time the feedback gain will be 0.1, the third time the feedback gain will be 0.2, and so on. Run the simulation and plot the output:

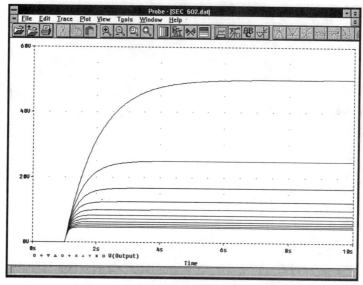

Just for fun, we can simulate more complicated transfer functions. We will change the transfer function 1/(s+1) to (s−1)/(s²+2s+1). To change the numerator, double click the **LEFT** mouse button on the text **1**:

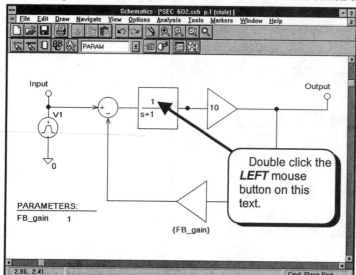

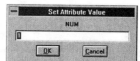

Type the text **s-1** and press the **ENTER** key:

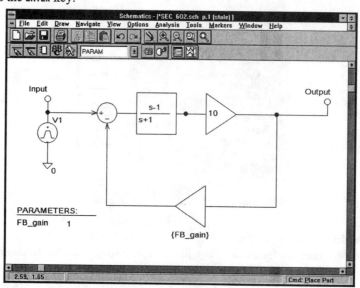

To change the denominator, double click on the text **S+1**:

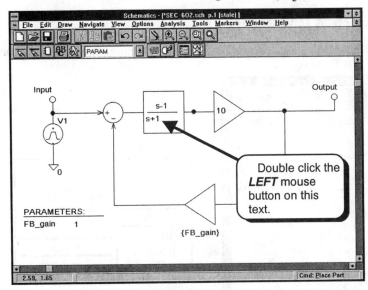

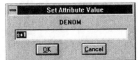

Type the text **s*s+2*s+1** and press the ENTER key:

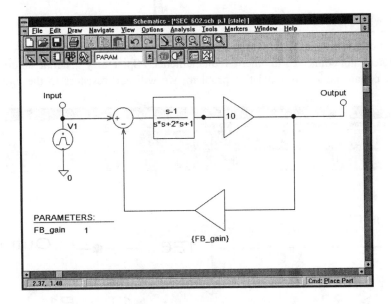

Simulate the circuit and plot the output for the first 5 seconds:

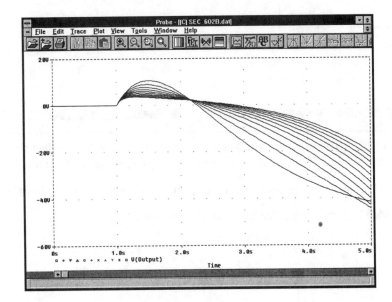

6.O.3. Op-Amp Models with ABM Parts

It is easy to create ideal parts using ABM parts. The circuit below has infinite input impedance and a gain of 1,000,000. The output has no supply limits:

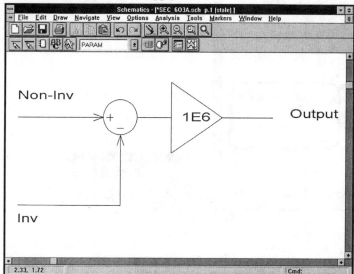

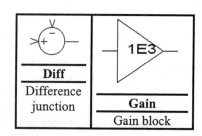

If we make a subcircuit out of this circuit, we could run into problems when the circuit is used. A problem that may result is that, if any of the terminals are not connected to other circuit elements, PSpice may generate an error message that only one element is connected to a certain node. To avoid this problem, we will add resistors to the circuit that do not affect its operation:

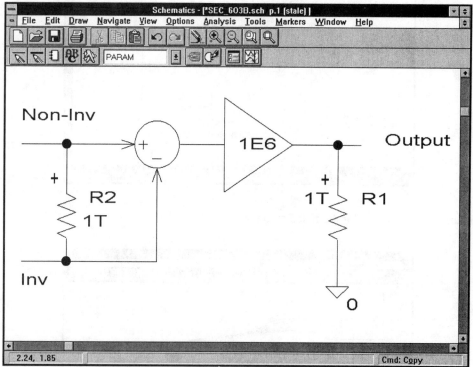

The resistors have a value of 1T or 10^{12} Ω. These resistors are so large that they will not affect the operation of the circuit in most cases, and they avoid the problem of only one circuit element being connected to each node.

The circuit below is an improved model for an op-amp. It has a low frequency gain of $10^6/30$ and infinite input impedance, but we have added frequency dependence and supply limits of ±15 V:

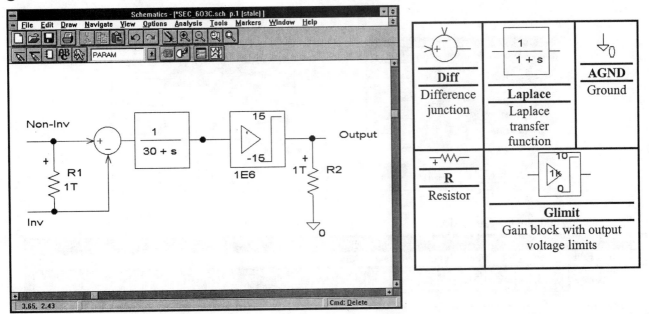

The output voltage swing is limited to ±15 V and we have added a pole at 30 rad/sec. This is approximately the frequency response of a 741 op-amp. Note that this circuit is still ideal because it does not include many of the other non-ideal characteristics of a 741 op-amp such as bias currents, offset voltages, and slew rate. In Section 7.G, we will show how to make this circuit into a subcircuit, add it to the libraries, and create a graphic symbol for it. We will then be able to use the subcircuit in other circuits.

6.O.4. AC Sweep with ABM Parts

Next, we will use the op-amp circuit created in the previous section to demonstrate an AC Sweep. We created an op-amp with frequency dependence in the previous section. We will now show how the frequency response varies with feedback. Wire the circuit below:

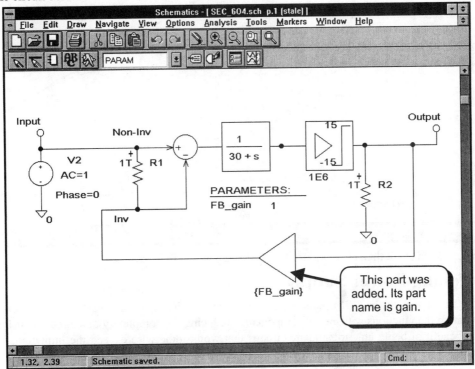

Set up the AC and Parametric sweeps as shown:

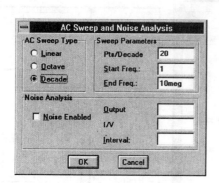

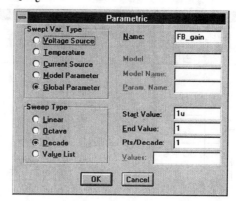

Run the simulation and plot the gain in decibels:

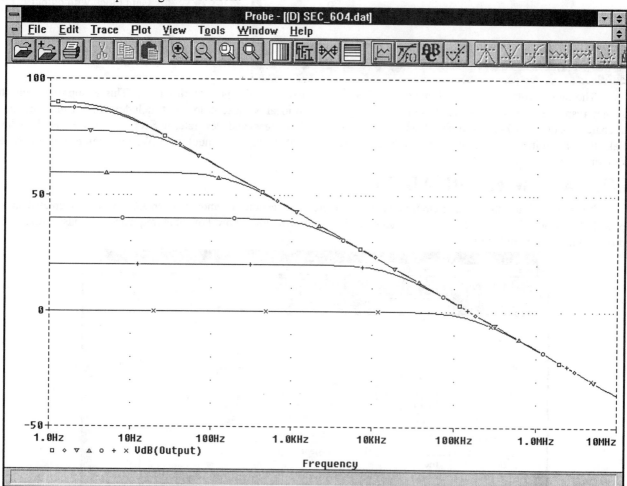

The plot shows the expected result that, as we reduce the gain, the bandwidth increases.

6.O.5. Switching Power Supply

One of the advantages of the ABM parts is that we can create circuits with complicated functions easily and with fewer parts than creating the functions with circuit components. Reducing the complexity of a circuit by using the ABM parts gives us the added benefit of being able to simulate circuits that would otherwise exceed the limitations of the evaluation version. An example would be the switching power supply below:

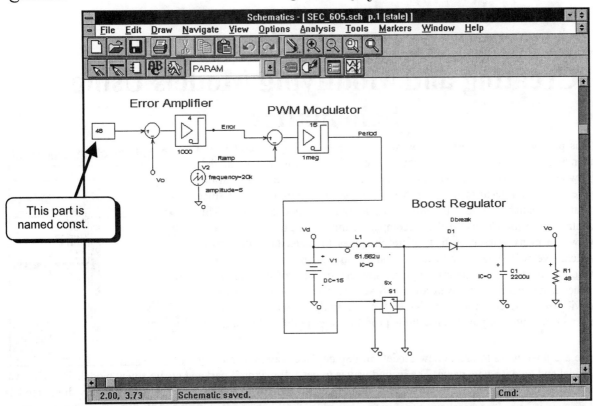

To simulate this circuit with real components we would need to create an error amp with output limits and then a ramp generator with output limits. Notice that the limits are not traditional supply limits and would require limiting circuits. This circuit would take at least two op-amps plus some limiting circuits, which could be Zeners or active clamps that also use op-amps. A real circuit of this complexity could easily exceed the limits of the evaluation version. The purpose of this simulation was not to accurately simulate the control circuit, but to create an "ideal controller" so that we could test the theory behind the switching circuit. The circuit also uses the PSpice part Vswitch rather than a MOSFET or IGBT. Vswitch is an ideal switch. Using this switch rather than a MOSFET or IGBT eliminates the circuitry needed for driving the MOSFET.

We will not simulate the circuit here because it would take more than an hour to complete. The circuit is given as an example of how the ABM parts can be used to simplify a complicated control circuit.

6.P. Summary

- The Transient Analysis is used to look at plots of voltages and currents versus time. This simulation displays waveforms as you would see them on an oscilloscope screen.
- Use Probe to view the results graphically.
- Use the Transient Analysis with sources such as Vsin, Isin, Vpulse, Ipulse, VPWL, IPWL, etc.
- The AC sources VAC and IAC are set to zero.
- DC sources keep their specified value.
- Use the Transient Analysis to view inductor and capacitor transient responses.
- Use Transient Analysis to observe an amplifier's voltage swing.
- The Fourier components of a time signal can be viewed graphically with Probe by selecting **Axis** and then **Fourier** from the Probe menu.
- The Fourier components of a time signal can be displayed in the output file by enabling the Fourier option in the Transient Analysis dialog box.
- The Parametric Sweep can be used to see how values of devices affect the performance of a circuit.
- Goal Functions can be used to obtain numerical data from Probe graphs. The result of evaluating a Goal Function is a single numerical value.
- The Performance Analysis is the use of a Goal Function in conjunction with a Parametric Sweep. The result of the Goal Function is plotted versus the swept parameter.
- The user can create complex waveforms with the Stimulus Editor without knowing all of the parameters of a particular source.

PART 7
Creating and Modifying Models Using Schematics

In this part we will demonstrate how to modify existing PSpice models and how to create new models. We will assume that the user is familiar with PSpice models and knows how he or she would like to modify the models. A discussion of the various models requires too much detail to be given here. The user is referred to the PSpice Reference Manual available from MicroSim Corporation for model details. This manual is contained on the CD-ROM that accompanies this text. You will probably need to review the many references that MicroSim gives to understand the model parameters. Here, we will show how to make changes to existing models or create simple new models. Section 7.F contains simplified models for some of the commonly used parts. The model parameters given are for first-time users. For more accurate models, you will need to refer to more detailed texts covering Spice models. If you are more familiar with the models, you can use these procedures to modify all parameters in a model.

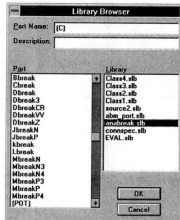

There are three ways to create new models in PSpice. One way is to modify an existing model and give it a new name. The second way is to get a "breakout" part and create a new model. The third way is to create a new model using MicroSim's Parts program. Since the evaluation version of Parts creates only diode models, we will not discuss it here. The breakout parts are contained in the library called *anabreak.slb*. This library contains graphic symbols for all parts available in Schematics. As you scroll down through the part window you will notice many different parts:

Model	Description
Cbreak	Capacitor model.
Dbreak	Generic diode.
DbreakZ	Zener diode model. The PSpice models used to create a diode and a Zener diode are the same. The only difference is in the graphic symbols of the parts.
JbreakN	N-type junction FET.
JbreakP	P-type junction FET.
Lbreak	Inductor model.
MbreakN MbreakN4	N-channel MOSFET. Both graphic symbols have 4 terminals. The fourth terminal is the substrate.
MbreakN3	N-channel MOSFET. The graphic symbol has only three terminals available. The substrate terminal is tied to the source in the graphic.
MbreakP MbreakP4	P-channel MOSFET. Both graphic symbols have 4 terminals. The fourth terminal is the substrate.
OP-AMP_breakout	Operational Amplifier. The default model is equivalent to a 741 op-amp. This part is in library Class2.slb.
MbreakP3	P-channel MOSFET. The graphic symbol has only three terminals available. The substrate terminal is tied to the source in the graphic.
QbreakN	NPN bipolar junction transistor.
QbreakP	PNP bipolar transistor.
Rbreak	Resistor model.

To demonstrate changing models, we will use the circuit components below. We will not bother wiring the parts since we are interested only in creating models. The only purpose of this example is to demonstrate how to change models.

Q2 QbreakN **QbreakN** NPN BJT	D3 Dbreak **DbreakZ** Zener diode	M1 L= W= MbreakN **MbreakN** N-channel MOSFET	R1 Rbreak **Rbreak** Resistor
D5 D1N4001 **D1n4001** Rectifier diode	Q1 q2n3904 **Q2N3904** NPN small-signal BJT	D1 d1n4734A **D1n4734A** Zener diode	

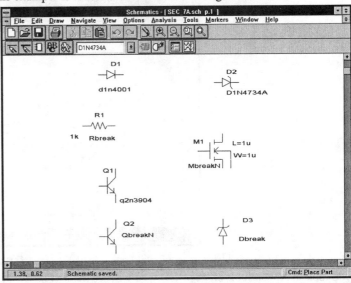

Before we change any of the models we must save the schematic. I will save this schematic as "sec_7A.sch." Refer to Section 1.H if you are unfamiliar with saving a schematic. You will not be able to modify or create new models unless you save your schematic first.

7.A. Changing the Model Reference

Click the **LEFT** mouse button on the graphic symbol for the breakout Zener diode **D3**, ⎯▷⎯. It should turn red, indicating that it has been selected. Select **Edit** and then **Model** from the Schematics menus[*]:

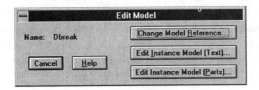

This dialog box allows us to change the model reference, create a model with the Parts program, or edit the model. The model reference for **D3** was **Dbreak**. If I want to change the reference to another model already defined in our libraries, we can click the **LEFT** mouse button on the **Change Model Reference** button:

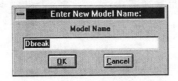

The dialog box tells us that the model for this diode is **Dbreak**. Suppose we want both Zener diodes in the circuit to use the model D1n4734A. Type the text **D1n4734A** and click the **OK** button to accept the change. In the schematic, the text Dbreak will change to **D1n4734A**:

---

[*]If the text for **Model** appears grayed out (Model), you have either not selected a part or the part you have selected does not have a model. Attempt to select the part again, or select a different part.

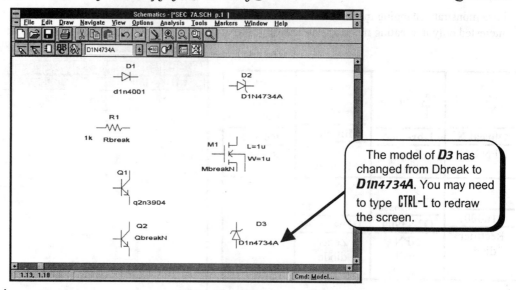

You may need to type **CTRL-L** to redraw the screen. The model **D1n4734A** must be defined in one of the PSpice libraries. If you look in Appendix E on page 466, you will see the text below in the file class.lib:

```
.model D1N4734A D(Is=1.085f Rs=.7945 Ikf=0 N=1 Xti=3 Eg=1.11 Cjo=157p M=.2966
+      Vj=.75 Fc=.5 Isr=2.811n Nr=2 Bv=5.6 Ibv=.37157 Nbv=.64726
+      Ibvl=1m Nbvl=6.5761 Tbv1=267.86u)
*      Motorola     pid=1N4734     case=DO-41
*      89-9-19 gjg
*      Vz = 5.6 @ 45mA, Zz = 40 @ 1mA, Zz = 4.5 @ 5mA, Zz = 1.9 @ 20mA
```

Lines that begin with a + sign are continuation lines and are part of the model definition. Lines that begin with an * are comment lines and are ignored by PSpice. This text defines model D1N4734A. The graphic symbol does not define the model. The graphic symbol refers to a model named D1N4734A and this model must be defined in a library. In this case, the library is file class.lib.

If you add new model libraries (files named ????????.lib) you can change model references, rather than define new graphic symbols for each part in the library. The professional versions of Schematics and PSpice have several thousand parts. Instead of creating a new graphic symbol for each part, you would get a breakout part and then change the model reference to a model in the ".lib" file. You are encouraged to look through all files named with the suffix ".lib." These files define the models available to you. Not all of these models have predefined graphic symbols that refer to them. You may use the models in the ".lib" files by changing the model reference of a breakout part.

As a second example of changing a model reference, we will change diode **D1**. Note that **D1** is a part with a pre-defined model, **d1n4001**. Click the **LEFT** mouse button on the graphic for diode **D1**, ▷┝. It should be highlighted in red, indicating that it has been selected. First, we will look at the attributes of the diode. Select **Edit** and then **Attributes**:

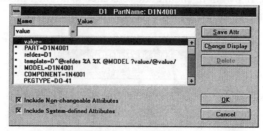

We see that the **MODEL** is **D1N4001**, the **COMPONENT** is **1N4001**, and the **PART** is **D1N4001**. If you use the **Change Display** button to find out which attributes are displayed on the screen, you will find that the **PART** attribute is displayed and not the **MODEL** attribute. Thus, when we change the model, the change will not be displayed on the screen. Click the **Cancel** button to return to the schematic. We will now change the model. **D1** should still be highlighted in red. Select **Edit** and then **Model**:

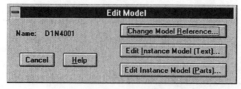

Click the *LEFT* mouse button on the ***Change Model Reference*** button:

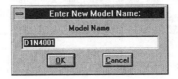

The current model is the ***D1N4001*** model. Enter the text **Dx** to specify a model named "Dx" and click the ***OK*** button. You will be returned to the schematic:

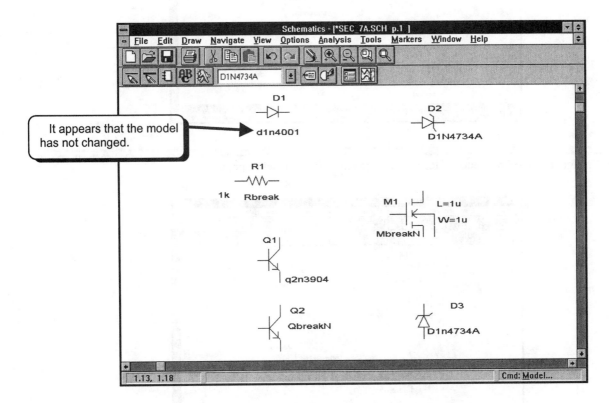

Note that the model Dx does not appear in the schematic. This is because, with predefined models like the ***D1N4001***, the text ***D1N4001*** is the part name and not the model name. If you double click the *LEFT* mouse button on the ***D1*** graphic, ─▷⊢, you will see the attributes for ***D1***:

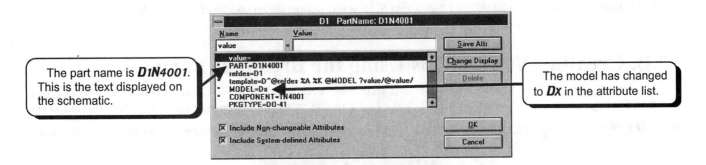

We see that the ***MODEL*** attribute has been changed to ***Dx*** in the attribute list so PSpice will use model Dx in the simulation. The part name is still ***D1N4001*** and this is the text displayed on the screen.

The model is displayed on the screen with breakout parts. When the model is changed, the changed model name appears on the schematic. This is because Schematics assumes that you wish to change the breakout model to a different model, and that you wish to identify that model on the screen. Schematics assumes that when you place a breakout part, you will change it. If you know you will be changing the model reference of a part, you should use the breakout parts.

Repeat the procedure to change the model reference of **D3** back to **Dbreak**:

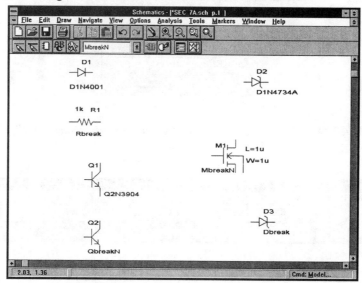

EXERCISE 7-1: Change the model reference of R1 to the 5 percent resistor model R5pcnt.

SOLUTION:

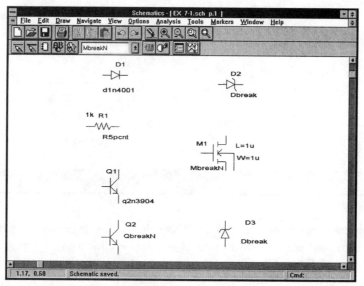

We will continue this example in **EXERCISE 7-3**. See Part 9, page 401 for details on creating models with tolerance.

7.B. Creating New Models Using the Breakout Parts

For this section, the schematic of the previous section has been saved as sec_7A. This was done by selecting **File** and then **Save As** from the Schematics menus. Click the **LEFT** mouse button on the graphic symbol for the breakout Zener diode **D3**, ⊶ . It should turn red, indicating that it has been selected. Select **Edit** and then **Model** from the Schematics menus[*]:

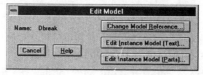

_____

[*]If the text for **Model** appears grayed out (Model), you have either not selected a part or the part you have selected does not have a model. Attempt to select the part again or select a different part.

To edit the existing model for Dbreak in order to create a new model, click the *LEFT* mouse button on the **Edit Instance Model (Text)** button. This button will allow us to modify the model associated with **Dbreak**:

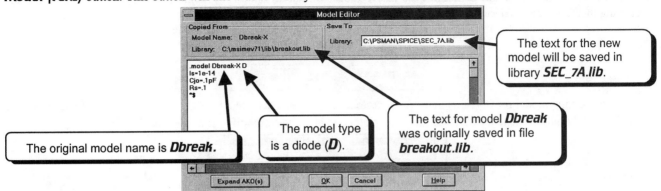

This window is a text editor and allows us to edit this model. Notice that in the upper right corner we see the text **Save To**. The new model will be saved in file **SEC_7A.lib**. When we create a new model, we are also creating a new library file for the schematic. The schematic was named sec_7A.sch so the library file will be named **SEC_7A.lib**. When you create new models for a schematic, a new library file is created with the name of the schematic and the extension ".lib." You can change the library name if you wish.

Let us create a new Zener diode with a breakdown voltage of 3 volts. We will name this new Zener model DZ3V. The model parameter in PSpice that controls the breakdown voltage is called BV. Change the model as shown:

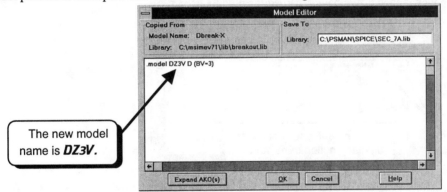

The only model parameter we are defining is BV. All other parameters will be left at the PSpice defaults. Click the **OK** button to accept the model:

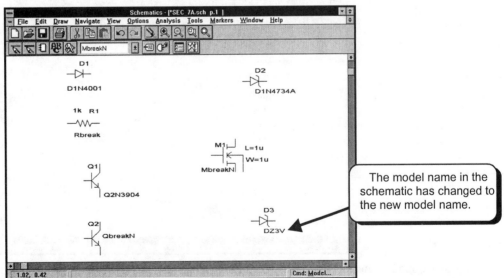

Notice in the schematic that the model reference for **D3** has changed from Dbreak to **DZ3V**. Thus, **D3** will now use the newly defined model **DZ3V**.

Next, suppose we wish part **Q2** to have an H_{FE} of 33. Chances are, no model in the library has this parameter. We will define a new model for this part. Click the **LEFT** mouse button on the NPN graphic for **Q2**, ⊬. It should turn red, indicating that it has been selected. Select **Edit** and then **Model** from the Schematics menus:

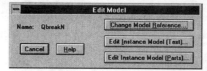

To edit the existing model for QbreakN in order to create a new model, click the **LEFT** mouse button on the **Edit Instance Model (Text)** button. The text editor window will appear and allow us to modify the model **QbreakN**:

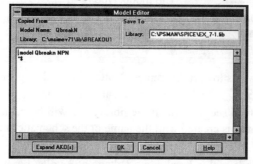

This is the model for **QbreakN**. This model uses all default model parameters. The PSpice parameter that represents H_{FE} is the parameter BF. Change the model as shown. We have renamed the model QBF33:

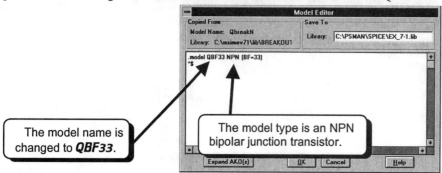

Click the **OK** button to accept the model:

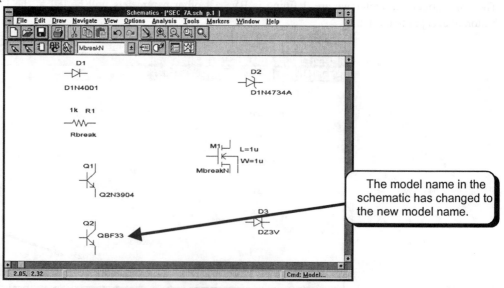

Notice that the model reference has changed from QbreakN to **QBF33**. If you use the Windows Notepad to edit the file sec_7A.lib, you will see the models you have just created:

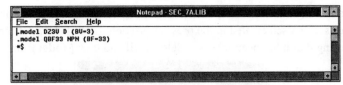

This procedure can be used to create a model using any of the breakout parts. Before you create any new parts, you must know the definition of the model parameters. The user is strongly urged to refer to the MicroSim Circuit Analysis Reference Manual [1] for the definitions of the model parameters. This manual is contained on the CD-ROM that accompanies this text.

EXERCISE 7-2: Change the model parameters of M1 in the circuit above to define an enhancement MOSFET with the parameters K = 20 μA/V^2 and V$_T$ = 3 V. MOSFET operation in the saturation region is governed by the equation:

$$I_D = K(V_{GS} - V_T)^2$$

In PSpice, the equation that defines the MOSFET operation in the saturation region is:

$$I_D = \frac{K_P}{2}(V_{GS} - V_{TO})^2$$

SOLUTION: From the two equations, we see that K$_P$ = 2K and V$_{TO}$ = V$_T$. Thus, in our model we will define K$_P$ = 40 μA/V^2 and V$_{TO}$ = 3 V. I will rename the model to **Mx**:

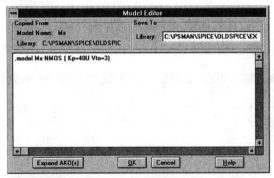

When we return to the schematic, **MbreakN** should be renamed **Mx**:

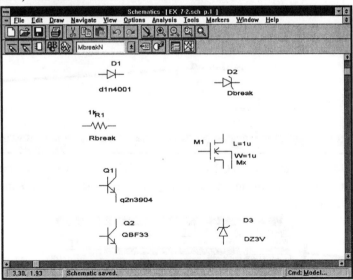

7.C. Modifying Existing Models

When you edit an instance of a model, Schematics makes a copy of the original model and saves it in a new library, leaving the original model unchanged. This feature allows you to edit entire existing models and modify the parameters you wish.

We would like to use the Q2N3904 model, but we would like to set H$_{FE}$ to its minimum value of 50. We will continue with the circuit named sec_7A.sch created in the previous section. Click the **LEFT** mouse button on the **Q1** graphic, ⌐⌐. It should turn red, indicating that it has been selected. Select **Edit** and then **Model** from the Schematics menus.*

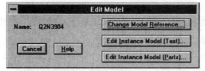

Select **Edit Instance Model (Text)** to edit the **Q2N3904** model:

This screen gives all of the model parameters for the 2n3904 model. To distinguish the new model we are about to create from the original model, the model has been renamed **Q2N3904-X**. This new model will be saved in file **SEC_7A.lib**. The original model will remain unchanged. Change the model parameters that you want and then click the **OK** button. You will return to the schematic:

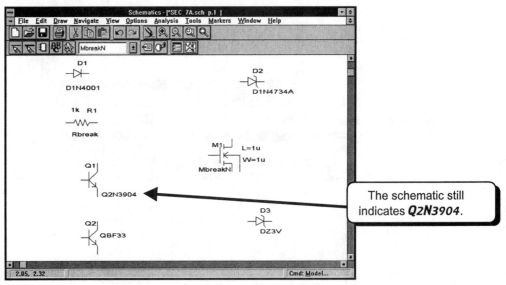

We see that the text **Q2N3904** has not changed to Q2N3904-X. This is because, with predefined models like the **Q2N3904**, the text **Q2N3904** is the part name and not the model name. If you double click the **LEFT** mouse button on the **Q1** graphic, ⌐⌐, you will see the attributes for **Q1**:

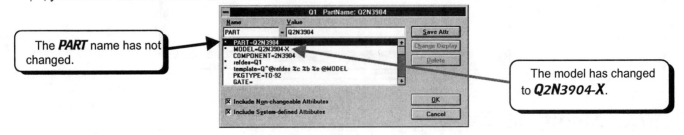

---

*If the text **Model** appears grayed out (Model), you have either not selected a part or the part you have selected does not have a model. Attempt to select the part again or select a different part.

We see that the model has been changed to **Q2N3904-X** in the attribute list. When PSpice runs it will use model **Q2N3904-X**. Note that the **PART** name has not changed, and it is the **PART** name that is displayed in the schematic. This is different from the breakout parts. With breakout parts, the new model name is displayed because the breakout part assumes that you wish to create a new part and to identify that part on the schematic. When you modify an existing model, the original model name is displayed because Schematics assumes that you are only making small modifications to the original model. The modified model is still basically the original part.

If you use the Windows Notepad program to display the contents of the file sec_7A.lib, you will see the new model **Q2N3904-X**:

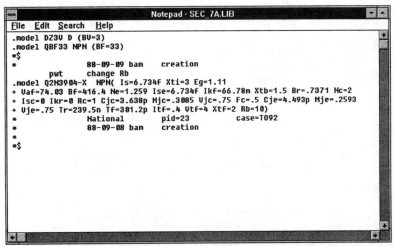

We see that the original model for the 2n3904 has been copied, with the exception of the changes we made to the model.

EXERCISE 7-3: Continuing with **EXERCISE 7-1**, edit the 5 percent model of **R1** and change the model to a 10 percent model. The model reference of R1 was changed from Rbreak to **R5pcnt** in **EXERCISE 7-1**.

SOLUTION: Use the Edit Instance Model button to edit the resistor model **R5pcnt**:

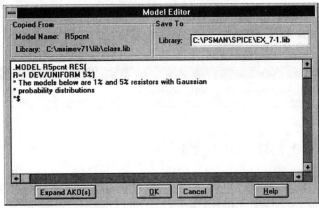

Change the name of the model to **R10pcnt** and change the **5%** tolerance to **10%**:

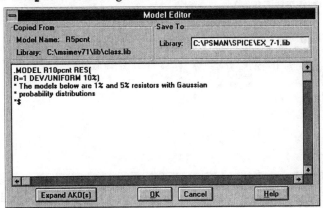

The model name of R1 should change in the schematic:

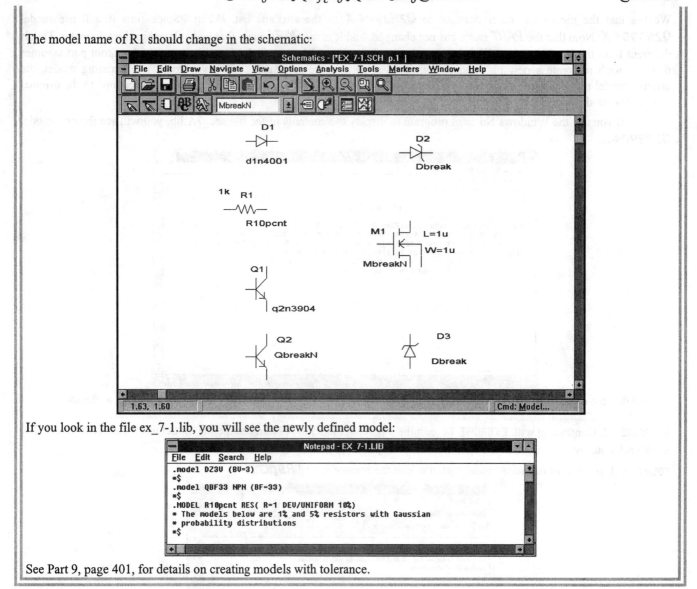

If you look in the file ex_7-1.lib, you will see the newly defined model:

See Part 9, page 401, for details on creating models with tolerance.

7.D. Changing the Library Path

Suppose you have created a number of new models and you would like to use them in another circuit. This can be done by adding the library name to the library path. The current file we are using is **SEC_7A.sch**. The new models created in this circuit are stored in the file called sec_7A.lib. Use the Windows Notepad program to edit the file sec_7A.lib.

This file contains all of the models we created for the current schematic. Close the Notepad and return to the schematic. We will now look at the current library path. Select **Analysis** and then **Library and Include Files** from the Schematics menus:

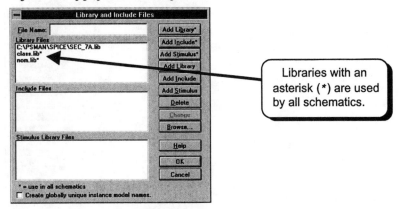

This screen lists all libraries used with this schematic. The libraries **class.lib** and **nom.lib** are the default libraries used by all schematics. The library **SEC_7A.lib** was created for use by the present schematic.

What if we would like to create a new circuit but use the models QBF33 and DZ3V? This can be done by modifying the **Library and Include Files** dialog box. Return to the schematic by selecting the **Cancel** button. Create a new schematic by selecting **File** and then **New.** Add the breakout parts shown below and then save the schematic as sec_7D:

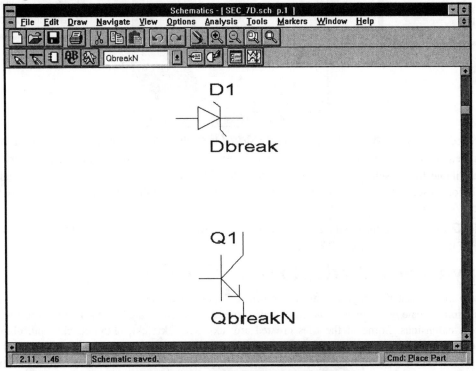

To view the model libraries for this circuit select **Analysis** and then **Library and Include Files** from the Schematics menus:

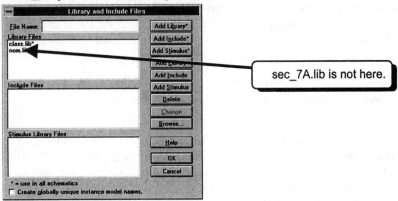

Only the default libraries are present. If we would like to add sec_7A.lib to this list, type **sec_7A.lib** in the text window next to **File Name**:

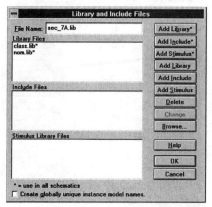

Click the *LEFT* mouse button on the *Add Library* button. The new library will be added to the list:

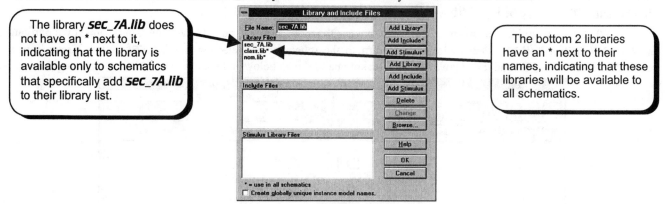

The library *sec_7A.lib* does not have an * next to it, indicating that the library is available only to schematics that specifically add *sec_7A.lib* to their library list.

The bottom 2 libraries have an * next to their names, indicating that these libraries will be available to all schematics.

There is a difference between the *Add Library** and the *Add Library* buttons. The *Add Library* button adds the library for the current schematic only. Note that there is no asterisk (*) next to the text *sec_7A.lib*. An asterisk indicates that the library will be available for all schematics. If you use the *Add Library** button, the library will be added to the list of libraries available for all schematics. All libraries shown in the window above that have an asterisk (*) are libraries available for all schematics.

Click the *OK* button to return to the schematic. You can now change the model references as shown in Section 7.A. You can use any of the models in file *sec_7A.lib*.

7.E. Copying Schematics to New Machines

When you make a schematic, you create several files. Suppose we create a new schematic and call it "ckt5." When we run the simulations, create new models, and run Probe, several files are created. All start with the text "ckt5" but there are many different extensions. Some of the files created are: ckt5.sch, ckt5.ckt, ckt5.net, ckt5.dat, ckt5.lib, ckt5.stl, and ckt5.als. The question is, when you copy files to a floppy disk, which files do you need to copy? Some of the files are very large and if you have several circuits, they could easily fill a floppy disk. We would like to copy the minimum number of files. Only three files are needed; **you must copy ckt5.sch, ckt5.stl, and ckt5.lib**. All files that have the suffix ".sch" are the actual circuit drawings. All files that have the suffix ".lib" are the new model libraries you created. All files that have the suffix ".stl" contain waveforms created with the Stimulus Editor. If you forget to copy the ".lib" file and you created new models, you will get an error message when you try to simulate your circuit because Schematics cannot find the models.

In summary, the files you need to copy are described below:

1.   If you created new models, as shown in this part of the manual, you need to copy all files with the ".lib" suffix.

2.   If you used the Stimulus Editor to create waveforms, you will need to copy all files with the ".stl" suffix.

3.   You must always copy the files with the ".sch" suffix.

Note about the Windows File Manager: If you need to rename a file, use the Schematics **Save As** command. Do not rename the files using the File Manager. For example, suppose we have files named ckt5.sch, ckt5.lib, and ckt5.stl. Inside file ckt5.sch are pointers to its support files, ckt5.lib and ckt5.stl. Suppose we use the Windows File Manager to rename the files to xx.sch, xx.lib, and xx.stl. File xx.sch contains the same information as file ckt5.sch. Thus, inside file xx.sch are pointers to the old files ckt5.lib and ckt5.stl. Unfortunately, the files ckt5.lib and ckt5.stl no longer exist because we renamed them. Thus, when we run the file xx.sch, Schematics will generate an error. It is not a good idea to rename files using the File Manager unless you change all of the pointers inside the ".sch" file.

7.F. Model Parameters for Commonly Used Parts

This section gives brief descriptions of model parameters for most of the parts mentioned in this manual. Only a few of the parameters for each model are given to allow the user to create simple models. To create more detailed models, the reader is referred to *The Design Center Circuit Analysis - Reference Manual* which is contained on the CD-ROM that accompanies this manual. The semiconductor models presented here include only the simple DC parameters because the models get complicated very quickly when dynamics are included. Temperature dependence parameters for some models are discussed.

CAPACITOR

The capacitor model is used to specify capacitor voltage dependence and tolerance. The main use of the capacitor model in this manual is to specify capacitor tolerance. See Part 9, page 402, for an example of using this model to specify tolerance.

Model Parameter	Description	Units	Default Value If Not Specified
C	Capacitance multiplier	none	1
VC1	Linear voltage coefficient	volt^{-1}	0
VC2	Quadratic voltage coefficient	volt^{-2}	0
TC1	Linear temperature coefficient	°C^{-1}	0
TC2	Quadratic temperature coefficient	°C^{-2}	0

The value of the capacitance used in the simulation is:

Capacitance Value =

$$(\text{SCH\_VAL}) \cdot \textbf{C} \cdot \left[1 + \left(V_c \cdot \textbf{VC1}\right) + \left(V_c^2 \cdot \textbf{VC2}\right)\right]\left[1 + \textbf{TC1}(T - Tnom) + \textbf{TC2}(T - Tnom)^2\right]$$

The variables in this equation are:

- Capacitance Value – The value of capacitance used in the simulation.
- SCH_VAL – The value of capacitance specified in the schematic.
- V_C – The voltage across the capacitor, determined during the simulation.
- T – temperature at which the simulation is run.
- Tnom – The nominal temperature, 27 °C.

Note that if a capacitor model is not specified, all model parameters are set at their default value and the value in the simulation is the value specified in the schematic.

EXAMPLE CAPACITOR MODELS:

.Model Cideal CAP (C=1)	– Equivalent to not using a model.
.Model Clin CAP (VC1=5)	– Capacitor with voltage dependence.
.Model Cdouble CAP (C=2)	– The capacitance used in the simulation is twice the value specified in the schematic.

DIODE

The diode model is used to create rectifier, signal, and Zener diodes.

Model Parameter	Description	Units	Default Value If Not Specified
IS	Saturation current.	amp	1×10^{-15}
N	Emission coefficient.	none	1
BV	Diode reverse breakdown voltage. Used to set the diode breakdown voltage in rectifier diodes, and the Zener breakdown voltage in Zener diodes.	volt	∞
RS	Diode parasitic resistance. Can be viewed as the series resistance of the diode.	ohm	0

The diode equation using these parameters is:

$$\text{Diode Current} = \textbf{IS}\left[\exp\left(\frac{V_D}{\textbf{N} \cdot V_T}\right) - 1\right]$$

EXAMPLE DIODE MODELS:

.Model Dx D (IS=1.5e–15 RS=3 BV=100)

.Model Dideal D (IS=1e–15)

.Model Dzener D (BV=4.7)

JUNCTION FETs

Two different models are given to specify n- and p-type jFETs. Both models use the same parameters.

Model Parameter	Description	Units	Default Value If Not Specified
VTO	Threshold voltage.	volt	0
BETA	Transconductance coefficient.	amp/volt^2	1×10^{-4}
LAMBDA	Channel-length modulation.	volt^{-1}	0

The equation for the drain current using these parameters is:

$$\text{Drain Current} = \begin{cases} \textbf{BETA} \cdot (1 + \textbf{LAMBDA} \cdot V_{DS}) \cdot \left[2 \cdot (V_{GS} - \textbf{VTO})V_{DS} - V_{DS}^2\right] & - \textit{Linear Region} \\ \textbf{BETA} \cdot (1 + \textbf{LAMBDA} \cdot V_{DS}) \cdot (V_{GS} - \textbf{VTO})^2 & - \textit{Saturation Region} \end{cases}$$

Some readers may be familiar with a jFET described by the equation:

$$\text{Drain Current} = I_{DSS} \cdot \left(1 - \frac{V_{GS}}{V_P}\right)^2 \qquad\qquad - \textit{Saturation Region}$$

This is an equivalent description with LAMBDA equal to zero, VTO $= V_P$, and BETA $= \dfrac{I_{DSS}}{V_P^2}$.

EXAMPLE jFET MODELS:

.Model jxx NJF (BETA=250u VTO=–3)   – n-type jFET model

.Model Jy PFJ (BETA=100u VTO=2)   – p-type jFET model

INDUCTOR

The inductor model is used to specify inductor voltage dependence. The value of the inductance used in the simulation is:

Model Parameter	Description	Units	Default Value If Not Specified
L	Inductance multiplier.	none	1
IL1	Linear current coefficient.	amp^{-1}	0
IL2	Quadratic current coefficient.	amp^{-2}	0
TC1	Linear temperature coefficient.	$°C^{-1}$	0
TC2	Quadratic temperature coefficient.	$°C^{-2}$	0

Inductance Value =

$$(\text{SCH\_VAL}) \cdot \mathbf{L} \cdot \left[1 + \left(I_L \cdot \mathbf{IL1}\right) + \left(I_L^2 \cdot \mathbf{IL2}\right)\right]\left[1 + \mathbf{TC1}(T - Tnom) + \mathbf{TC2}(T - Tnom)^2\right]$$

The variables in this equation are:

- Inductance Value – The value of inductance used in the simulation.
- SCH_VAL – The value of inductance specified in the schematic.
- I_L – The current through the inductor, determined during the simulation.
- T – temperature at which the simulation is run.
- Tnom – The nominal temperature, 27 °C.

Note that if an inductor model is not specified, all model parameters are set at their default value and the value in the simulation is the value specified in the schematic.

EXAMPLE INDUCTOR MODELS:

    .Model Lideal IND (L=1)   – Equivalent to not using a model.

    .Model L_lin IND (IL1=5)   – Inductor with current dependence.

    .Model Ldouble IND (L=2)   – The inductance used in the simulation is twice the value specified in the schematic.

MOSFET

Two different models are given to specify n- and p-type MOSFETs. Both models use the same parameters. These models are used to specify both enhancement and depletion type MOSFETs.

Model Parameter	Description	Units	Default Value If Not Specified
VTO	Threshold voltage.	volt	0
KP	Transconductance coefficient.	$amp/volt^2$	2×10^{-5}
L	Channel length.	meter	100×10^{-6}
W	Channel width.	meter	100×10^{-6}
LAMBDA	Channel-length modulation.	$volt^{-1}$	0

The equation for the drain current using these parameters is:

$$\text{Drain Current} = \begin{cases} \left(\dfrac{\mathbf{W}}{\mathbf{L}}\right)\dfrac{\mathbf{KP}}{2} \cdot \left(1 + \mathbf{LAMBDA} \cdot V_{DS}\right) \cdot \left[2 \cdot \left(V_{GS} - \mathbf{VTO}\right)V_{DS} - V_{DS}^2\right] & - \textit{Linear Region} \\[2ex] \left(\dfrac{\mathbf{W}}{\mathbf{L}}\right)\dfrac{\mathbf{KP}}{2} \cdot \left(1 + \mathbf{LAMBDA} \cdot V_{DS}\right) \cdot \left(V_{GS} - \mathbf{VTO}\right)^2 & - \textit{Saturation Region} \end{cases}$$

EXAMPLE MOSFET MODELS:

    .Model M1 NMOS (KP=25u LAMBDA=0.01 VTO=3)   –Enhancement type NMOS

.Model M2 NMOS (KP=20u LAMBDA=0.1 VTO=–2 W=10u L=5u)   –Depletion type NMOS

.Model M3 PMOS (KP=8u LAMBDA=0.01 VTO=–2)   –Enhancement type PMOS

.Model M4 PMOS (KP=6u LAMBDA=0.05 VTO=1.5 W=10u L=2u)   –Depletion type PMOS

BIPOLAR JUNCTION TRANSISTORS

Two different models are given to specify n- and p-type BJTs. Both models use the same parameters.

Model Parameter	Description	Units	Default Value If Not Specified
BF	Ideal maximum forward current gain.	none	100
BR	Ideal maximum reverse current gain.	none	1
RE	Emitter ohmic resistance.	ohm	0
RC	Collector ohmic resistance.	ohm	0
RB	Base ohmic resistance.	ohm	0
VAF	Forward Early voltage.	volt	∞

EXAMPLE BJT MODELS:

.Model Qnpn1 NPN (BF=50)   –NPN model

.Model Qpnp PNP (BF=25)   –PNP model

RESISTOR

The resistor model is used to specify resistor tolerance and temperature dependence. We will not discuss temperature dependence here. The main use of the resistor model in this manual is to specify resistor tolerance. See Part 9, page 401, for examples of using this model to specify tolerance.

Model Parameter	Description	Units	Default Value If Not Specified
R	Resistance multiplier	none	1
TC1	Linear temperature coefficient	$°C^{-1}$	0
TC2	Quadratic temperature coefficient	$°C^{-2}$	0
TCE	Exponential temperature coefficient	%/°C	0

The value of the resistor used in the simulation is:

$$\text{Resistance Value} = \begin{cases} (\text{SCH\_VAL}) \cdot \mathbf{R}\left[1 + \mathbf{TC1}(T - Tnom) + \mathbf{TC2}(T - Tnom)^2\right] & \text{if } \mathbf{TCE} \text{ is not defined} \\ (\text{SCH\_VAL}) \cdot \mathbf{R}\left[1.01^{\mathbf{TCE}(T-Tnom)}\right] & \text{if } \mathbf{TCE} \text{ is defined} \end{cases}$$

The variables in this equation are:

- Resistance Value – The value of resistance used in the simulation.
- SCH_VAL – The value of resistance specified in the schematic.
- T – temperature at which the simulation is run.
- Tnom – The nominal temperature, 27 °C.

Note that if a resistor model is not specified, all model parameters are set at their default value and the value in the simulation is the value specified in the schematic.

EXAMPLE RESISTOR MODELS:

.Model Rideal RES (R=1)   – Equivalent to not using a model.

.Model Rdouble RES (R=2)   – The resistance used in the simulation is twice the value specified in the schematic.

7.G. Creating a Subcircuit

In this section we will show how to take a schematic and change it into a subcircuit so that you can use it with other circuits. In order to follow this section you must have access to the libraries and configuration files for Schematics. If you are running Schematics on a network, you may not have access to these files. If you are running this example on your own computer, you should be able to follow the example with no problems. We will illustrate two examples of creating subcircuits. Both circuits will be added to the same library. We will start with the op-amp circuit created from ABM parts in Section 6.O.3.

7.G.1. Creating a Subcircuit and Adding It to the Symbol Library

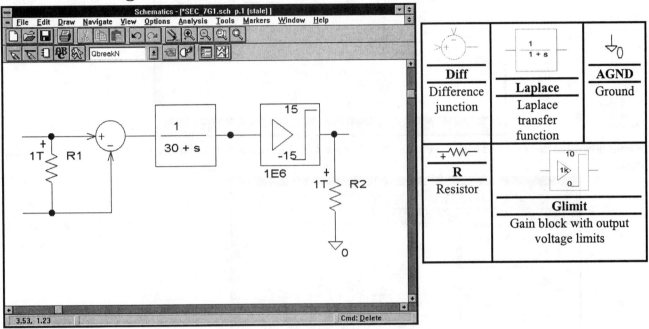

Note that none of the wires are labeled in this example. The text labels in the circuit of Section 6.O.3 were removed.

We must now add input and output interface ports. We will first add the input ports. Get a part named IF_IN and place it on the two inputs of the op-amp circuit:

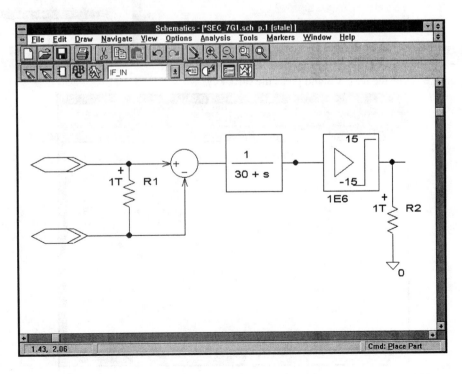

Next, get a part named IF_OUT and place it on the output of the op-amp circuit:

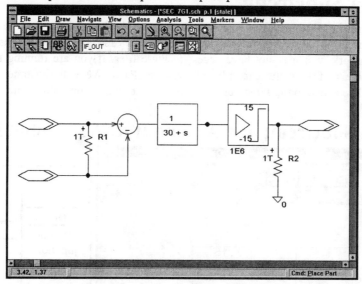

We must now label the input and output ports. Double click the **LEFT** mouse button on the graphic for the input port on the + input:

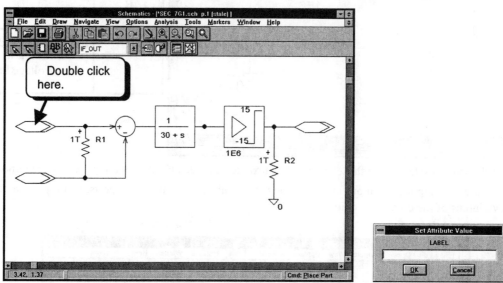

We will label this port Vp. Type the text **Vp** and press the **ENTER** key:

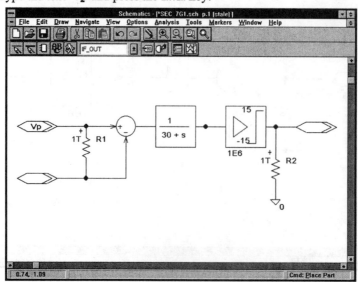

Label the other input port Vm and the output port Out using the same method:

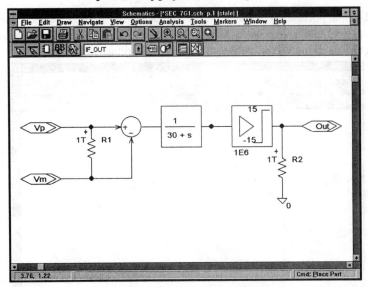

We can now create a subcircuit. Select **Tools** and then **Create Subcircuit**. A netlist of the subcircuit will be created. The file will be saved with the extension ".sub". Since my file was named sec_7G1.sch, the subcircuit netlist file will be saved with the name sec_7G1.sub. We will now look at this file. Select **Analysis** and then **Examine Netlist**. We did not create a netlist for this circuit; we created a subcircuit. Thus, a netlist file does not exist and we get an error message:

Click the **NO** button. This will open the Windows Notepad with an empty window. We can now open the subcircuit file. Select **File** and then **Open**:

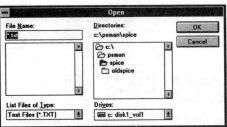

Type ***.sub** and press the **ENTER** key to display all files with the ".sub" extension:

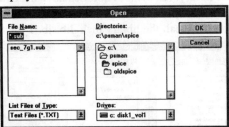

Select file **sec_7g1.sub** and click the **OK** button. The netlist for the subcircuit will be displayed:

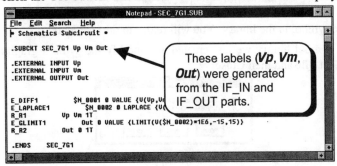

We notice that the input and output ports are now listed on the **.SUBCKT** line. These are the three nodes that will be used to connect to this circuit. We will not modify this file. We just opened it so that you would know where to find the file and how to observe its contents. Select **File** and then **Exit** to close the notepad and return to Schematics:

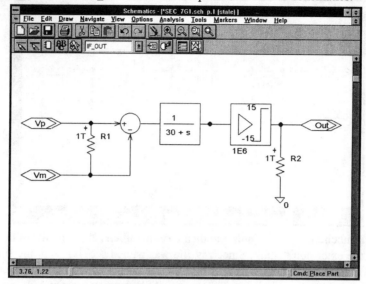

     Next we need to create a graphic symbol for the subcircuit. Select **File** and then **Symbolize** from the Schematics menus:

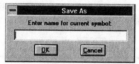

We need to give a name to the symbol. This will be the name we use to place the part in a schematic. I will call it OP-Ideal. Type **OP-Ideal** and press the **ENTER** key:

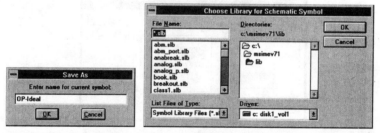

We must now place the symbol in a symbol library. All symbol libraries have the extension ".slb". If you are doing this for a class, your instructor may have a name of a library that everyone should use. If you are doing this on your own computer, you can name the library whatever you like. I will call the library meh_lib.slb. Type the name of the library:

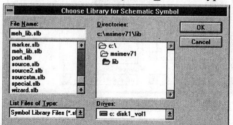

Click the **OK** button to save the symbol in the library. You will return to the schematic.

     We will now edit the symbol library we just created and modify the symbol graphics to our tastes. First we will close the current schematic. Select **File** and then **Close**:

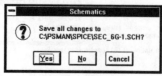

Click **Yes** to save the changes. A blank Schematics screen will appear:

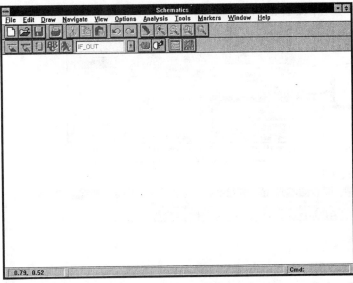

Closing the previous file was not a required step, but it may remove some confusion that might have occurred later in this example. Presently there are no open schematics. We will now open the library file. Select **File** and then **Edit Library**:

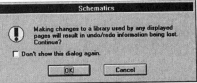

Since we closed all files, any changes we make will not cause any problems. You may not have received the dialog box above if someone using Schematics previously had selected the ***Don't show this dialog again*** option in the dialog box. Click the ***OK*** button:

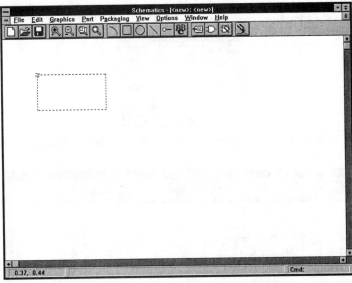

Notice in the above screen capture that the menus are a little different. This is the configuration of the Schematics window for editing graphics symbols. We must open the symbol library we created previously. Select **File** and then **Open**:

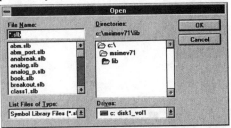

Select the file we just created. In my case, it is meh_lib.slb. It is not shown in the window pane so I will need to scroll the pane down a bit:

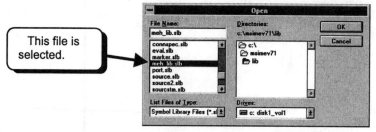

Click the **OK** button to open the file:

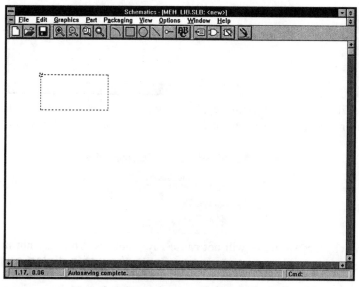

A symbol library can contain many parts. Note that the evaluation version is limited to 20 parts in each library. We must now get the part we wish to modify. Select **Part** and then **Get**:

Since I just created this library it contains only one part. Select the part by clicking the *LEFT* mouse button on the text **OP-Ideal**, and then click the **Edit** button:

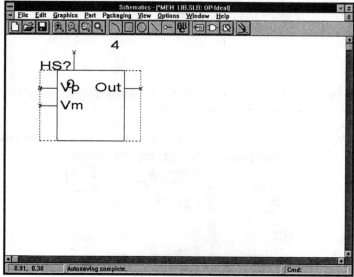

By default, when we create a symbol using the method shown, the symbol is always a square with the pins connected to it. Note that this symbol has 4 pins labeled **Vp**, **Vm**, **Out**, and **O**. Pin **O** is ground and it is a hidden pin. The ground pin is shown here, but it will be hidden when we place this part in a schematic. The other pins correspond to the IF_IN and IF_OUT ports we placed in the subcircuit.

The labels tell us which pin is which. Instead of labels **Vp** and **Vm**, I would like to use + and −. However, the labels **Vp** and **Vm** are necessary to identify the pins and relate the pins in the symbol to the calling nodes in the netlist of the subcircuit. From the previous screen capture, we know that the top left pin is the + pin and the bottom left pin is the − pin. We will first hide the labels. Double click the *LEFT* mouse button on the text **Vp**:

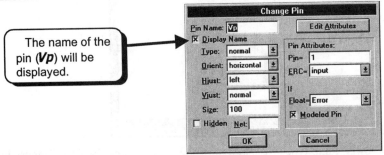

The name of the pin (**Vp**) will be displayed.

Notice that the square next to the text **Display Name** has an x in it ⊠. This means that the name of the pin is displayed in the symbol. We wish to hide the name. Click on the square to remove the x from the square ☐:

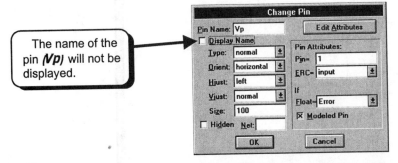

The name of the pin (**Vp**) will not be displayed.

Click the **OK** button:

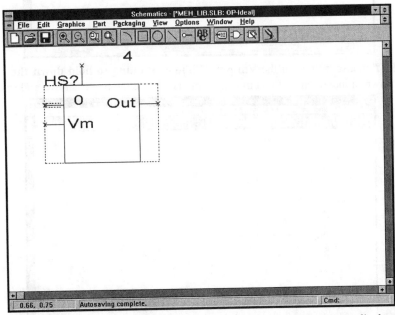

The text Vp is no longer displayed. The pin is still named Vp; however, the name is not displayed. **Note that we do not want to change the name of the pin.** If you change the name of the pin, it will no longer correspond to a calling node in the netlist of the subcircuit. When you use the symbol in a schematic and then run PSpice, an error will be generated by PSpice. We will add text later to place a + next to the pin.

Use the same procedure to hide the text **Out** and **Vm**:

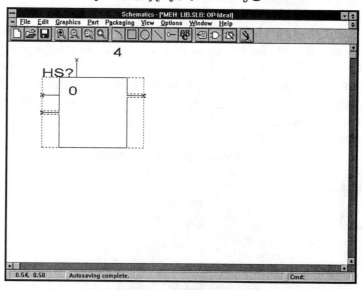

We will now move the Vm pin a bit lower. First, type **CTRL-L** to redraw the screen. Click the *LEFT* mouse button on the Vm pin. It should be enclosed in a dashed box:

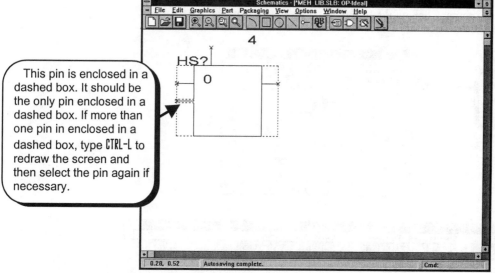

This pin is enclosed in a dashed box. It should be the only pin enclosed in a dashed box. If more than one pin in enclosed in a dashed box, type **CTRL-L** to redraw the screen and then select the pin again if necessary.

Click and **HOLD** the *LEFT* mouse button on the Vm pin. **While continuing to hold down the button**, move the mouse down. The pin will jump down in increments as you move the mouse. Place the pin as shown and release the mouse button:

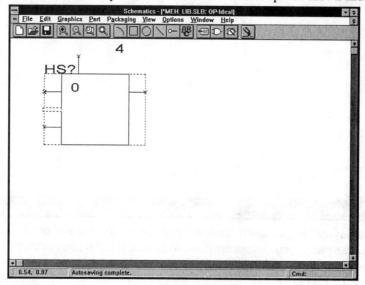

Using the same method, move the Out pin down and place it as shown:

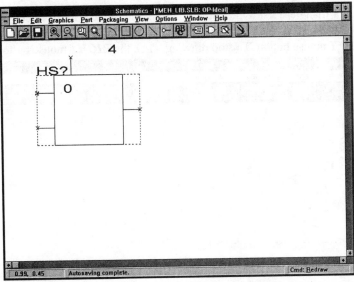

We now wish to delete the box and draw a triangle. Click the *LEFT* mouse button on the square to select it. It should turn red, indicating that it has been selected.

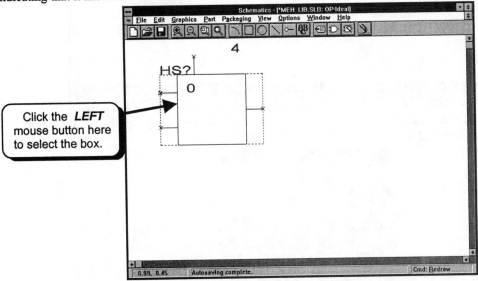

Once the box is selected, press the DELETE key to delete the box:

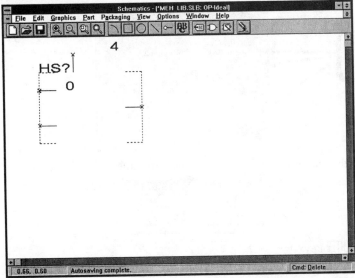

The box disappears. If it does not work the first time, repeat the procedure until the box disappears.

Next we need to draw a triangle. First, type **CTRL-L** to redraw the screen. Next, select **Graphics** and then **Line**. The cursor will be replaced by a pencil. Draw a triangle symbol. You can draw lines the same way you draw wires in Schematics. To place a corner, click the **LEFT** mouse button. To stop drawing, click the **RIGHT** mouse button:

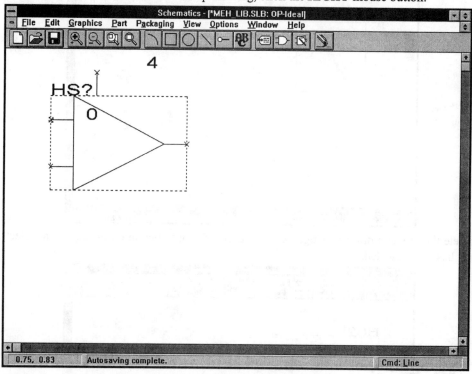

The last thing we need to do is add text for the plus and minus signs. We will not place this text on grid increments so we will first disable the grid. Select **Options** and then **Display Options**:

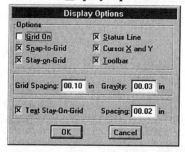

Change the options as shown:

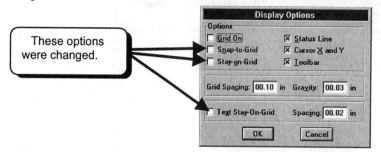

Click the **OK** button to accept the settings. You will return to the schematic. Select **Graphics** and then **Text** from the menus:

Type the **+** key and press the **ENTER** key:

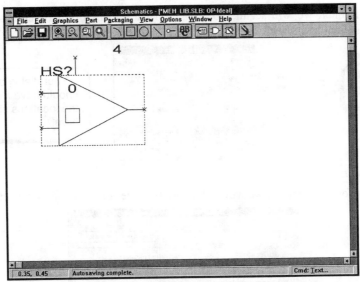

The cursor is replaced by a box that outlines the text string we just entered. Place the box next to the top left pin and click the *LEFT* mouse button to place the text:

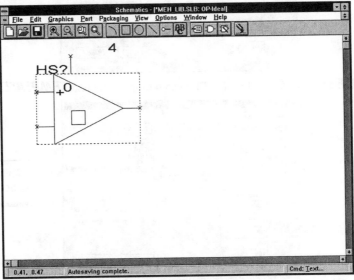

We note that the outline is still attached to the mouse. This indicates that we can place multiple copies of the text. To stop placing text, click the *RIGHT* mouse button. The box should disappear. Use the same procedure to place a minus sign next to the lower pin:

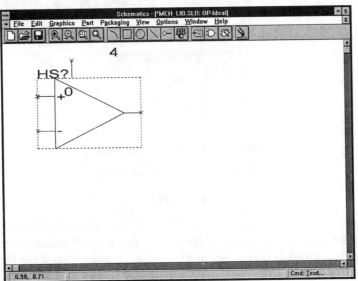

Before we leave the symbol editor, we will set the configuration back to its original settings. Select **Options** and then **Display Options**. Select the options as shown:

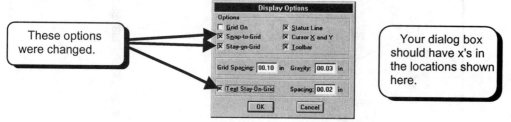

Click the **OK** button to accept the changes.

We are now done. Before we continue we will look at the attributes of this symbol. Select **Part** and then **Attributes**:

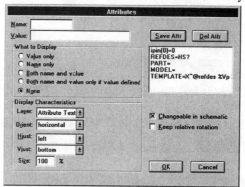

We will look at the **TEMPLATE** attribute. Click the **LEFT** mouse button on the text **TEMPLATE** to select it:

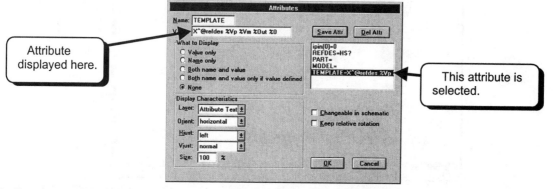

This is the line that specifies the connections for the subcircuit. Note that the calling nodes are **%Vp**, **%Vm**, **%Out**, and **%0**. The nodes have the same names and are in the same order as we saw in the netlist of the subcircuit. The names and order specify how Schematics connects the subcircuit to the netlist when you use this part. We will not change anything here. Click the **Cancel** button to return to the Symbol Editor.

We will now save the changes and exit the Symbol Editor. Select **File** and then **Close**:

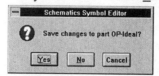

Click the **Yes** button to save the changes:

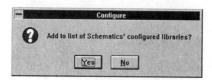

Schematics asks if we want to add the library meh_lib.slb to the list of standard symbol libraries used by Schematics. Click the **Yes** button. If you select **No**, you will not be able to retrieve the part using Schematics unless you add the library at a later time. The library will not be harmed. However, Schematics will not browse through the library when looking for parts. We can add the library later, but selecting **Yes** here is the easiest way to add it. After clicking **Yes**, you will return to Schematics with no schematic open.

We will now look at how the configuration of libraries has changed. Select **File** and then **New** to start a new schematic:

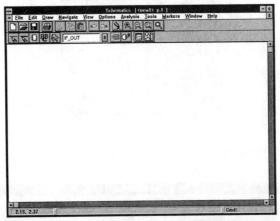

Select **Options** and then **Editor Configuration**:

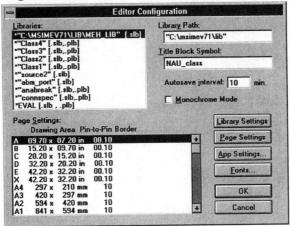

Notice that library **MEH_LIB** has been added to the list of libraries. **Important Note:** We see that the new library has been added at the top of the list of libraries. The limitation on libraries for the evaluation version is 10 libraries. If you add another new library, it will appear at the top of the list. Only the first 10 libraries will be used. The result will be that the last library in the list will not be browsed when searching for parts. **If you add more than one new library, you will no longer be able to access the parts in the standard libraries that come with the evaluation version.**

7.G.2. Using the New Symbol in a Circuit

We will now create a circuit using the symbol we just created to test that it works. Create the circuit below. It uses the voltage source Vac and the part we created, OP-Ideal.

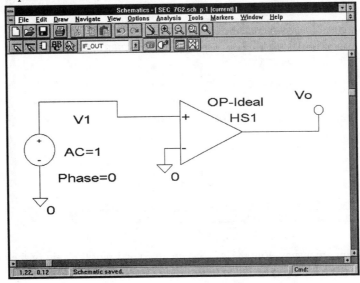

We will use this circuit to plot the open loop gain versus frequency. Set up the AC analysis below:

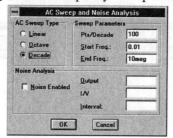

Run the analysis and plot the output in decibels:

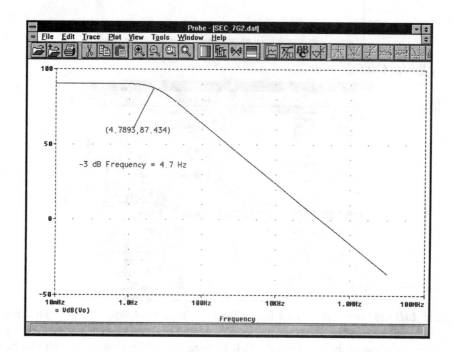

We will now run a circuit with feedback:

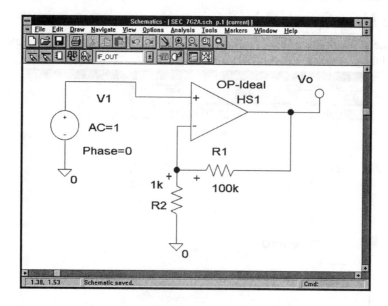

We will now plot the gain versus frequency. Run the AC Sweep again and plot the gain of the circuit. It should have a low frequency gain of 100 and an upper 3 dB frequency of 1.59 kHz:

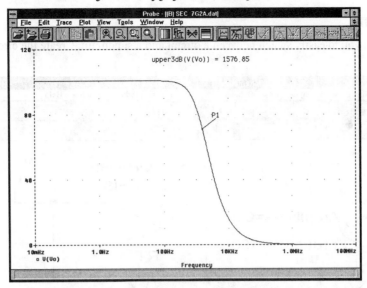

The circuit has the correct frequency response (1576 Hz is close to the theoretical value of 1.59 kHz). Last, we will test the circuit in a Transient Analysis to test the output voltage limits of ±15 V. The circuit uses the voltage source VSIN:

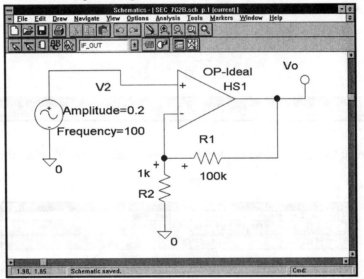

Set up the Transient Analysis below and plot the output waveform:

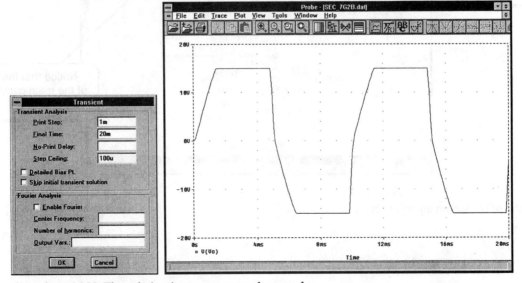

The output is clipped at ±15 V. The subcircuit appears to work properly.

7.G.3. Navigation

Schematics allows you to easily switch between a circuit and a subcircuit. Create the circuit below:

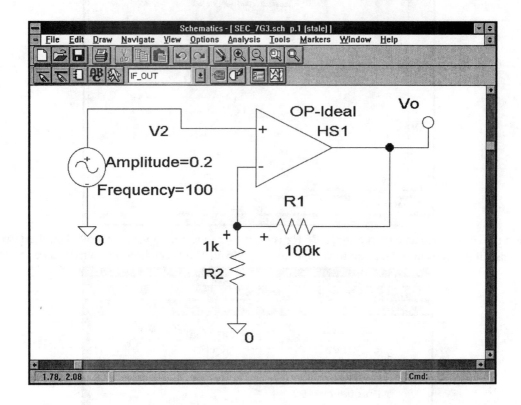

This circuit uses the subcircuit OP-Ideal that we created. To access the circuit, double click the **LEFT** mouse button on the subcircuit icon:

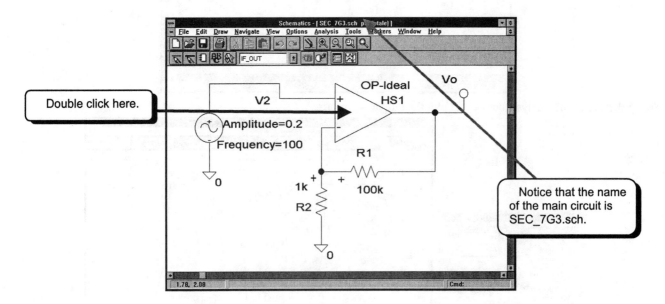

The schematic of the subcircuit will open:

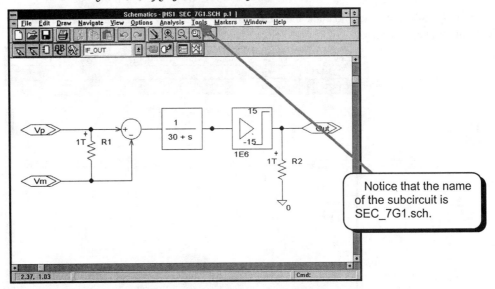

Notice that the title of the circuit is **SEC_7G1**. This is the same title under which we saved the subcircuit. To return to the main circuit, select **Navigate** and then **Pop**. The subcircuit will be closed and the original circuit will be displayed:

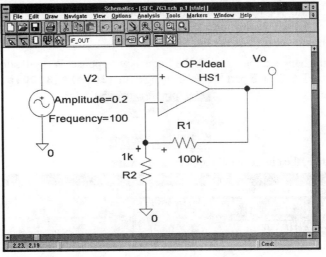

7.G.4. Modifying the Subcircuit

To modify a subcircuit we need to open the schematic for the subcircuit. We can do this by using the menus and selecting **File** and then **Open**, or we can use the navigation tools provided by Schematics. We will use the Navigation tools to open the circuit. We will start with the circuit we used in Section 7.G.2.

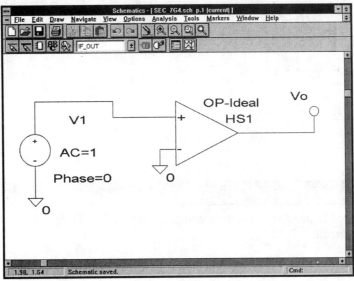

To access the subcircuit, double click the *LEFT* mouse button on the graphic for the part OP-Ideal. The schematic of the subcircuit will be opened:

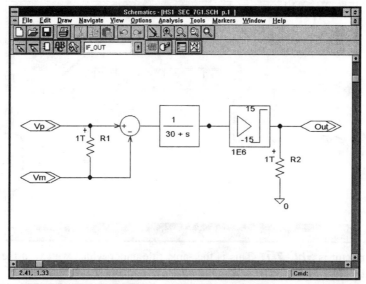

We can now modify the subcircuit. We will not add any more input or output pins. If we add pins, we must change the graphic for the subcircuit. Since we will change only the internal operation of the subcircuit, the graphic will have the same number of input and output pins and we can use the same graphic.

The only thing we will change in the circuit is the frequency response. We will add a second high frequency pole at 3000 r/s. The denominator of the LAPLACE part has been changed to **(30+s) * (3000+s)**:

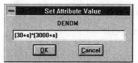

When you have made the change, the circuit should appear as shown:

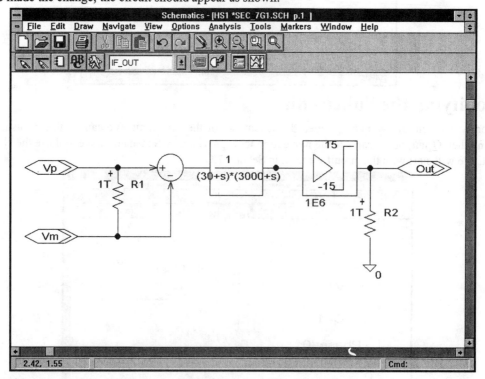

The last thing we need to do is re-create the subcircuit netlist. Select **Tools** and then **Create Subcircuit**. This will rewrite the netlist for the subcircuit:

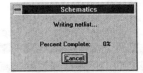

When the netlist is complete, you will return to the schematic of the subcircuit:

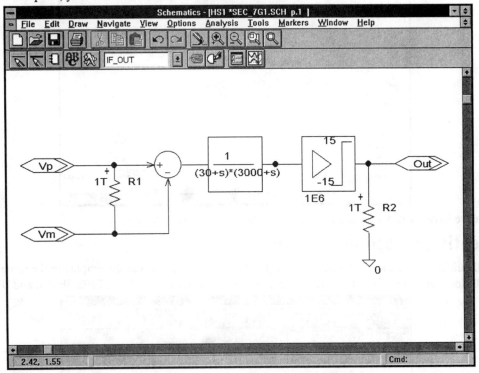

We will now return to the schematic of the main circuit. Select **Navigate** and then **Pop**:

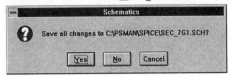

Click the **Yes** button to save the changes to the subcircuit:

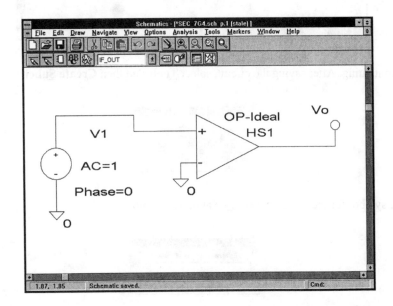

We will now run the open loop frequency response to see if the second pole has been added:

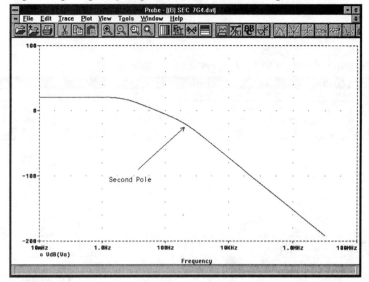

We see that once we have created a subcircuit, modifying the subcircuit is very easy.

7.G.5. Creating a Second Symbol

We will quickly create a second subcircuit and graphic to show how to add the graphic to the same library where we placed the the OP-Ideal graphic. Create the S-R latch below. The parts in this circuit are 7402, IF_IN, and IF_OUT.

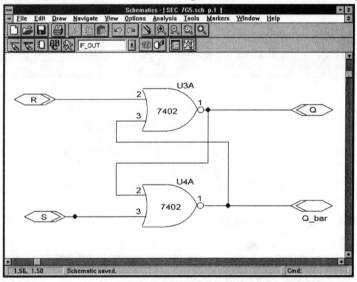

Save the circuit before continuing. After saving the circuit, select **Tools** and then **Create Subcircuit**:

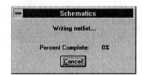

Next, we must create the symbol for the circuit. Select **File** and then **Symbolize**:

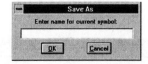

Name the symbol **SR-Latch** and click the *OK* button to continue:

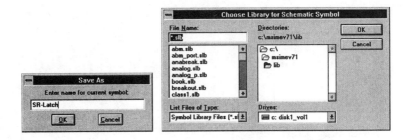

We must now specify in which library we would like to place the graphic. I placed the last symbol in library file meh_lib.slb. I will select this library:

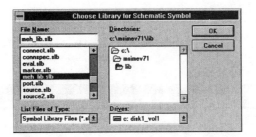

Select the library and click the *OK* button. The graphic is now saved in the library. We must edit the library and modify the graphic. The procedure for modifying the graphic was shown on pages 352-360. I will edit library meh_lib.slb and get the part named SR-Latch:

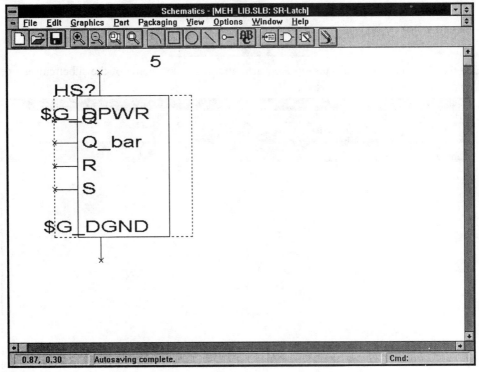

Modify the symbol as shown:

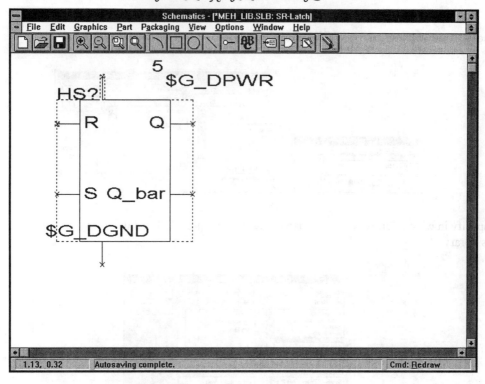

Pins **$G_DGND** and **$G_DPWR** are hidden pins and will not be displayed when the part is placed in a schematic.

   After making the changes, select **File** and then **Close**:

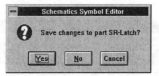

Select **Yes** to save the changes. You will return to the schematic. If the schematic of the subcircuit is open, select **File** and then **Close** to close the schematic of the subcircuit:

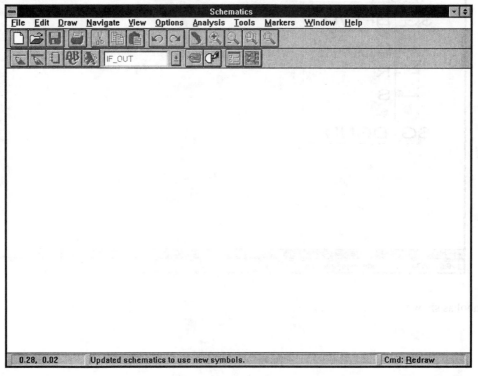

We will now create a new circuit and place the part SR-Latch in it. Select **File** and then **New**. Place the part SR-Latch in the circuit:

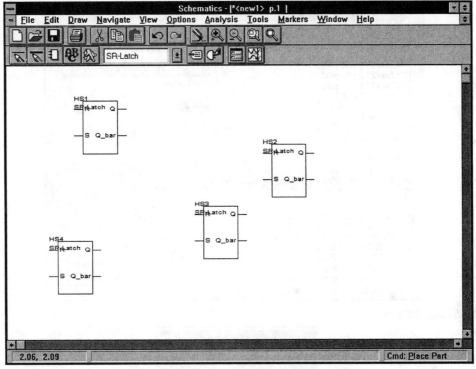

You should now test the subcircuit to see that it works properly.

7.G.6. Restoring Schematics' Library Configuration

The number of symbol libraries you can use in the evaluation version of Schematics is limited to 10. If many users create a subcircuit, create a graphic for the subcircuit, add the graphic to a symbol library with a unique name, and then tell Schematics to add the new library to the list of configured libraries, the limit of 10 libraries will be easily reached. When we follow the procedure outlined in Section 7.G.1 and add a new symbol library to the list of configured libraries, the new library is added at the beginning of the list. Each time we add a new library, it is added at the beginning, and the number of configured libraries grows. The problem with the evaluation version is that Schematics will use only the first 10 libraries in the list. This means that the libraries that will be used are the ones we created ourselves, and some of the standard libraries shipped with the evaluation version will be unavailable to us because they are at the end of the list of configured libraries. The result is that the parts in those libraries will be unavailable to us.

The problem is easily fixed by manually configuring the libraries. We will start with the Schematics window:

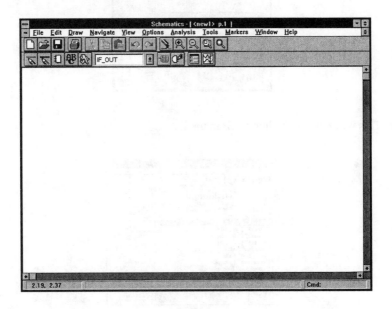

Select **Options** and then **Editor Configuration**:

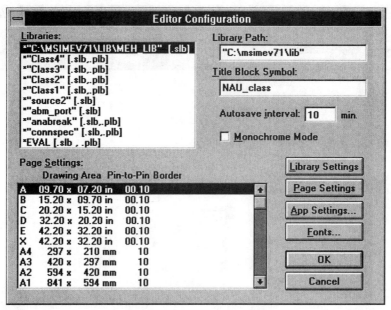

Click the **Library Settings** button:

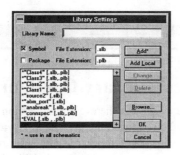

This window lists the configured libraries. Even though more than 10 libraries may be listed in this window, only the first 10 libraries will be accessible in the evaluation version. We can delete any of the libraries in the list. To demonstrate, I will delete the library I created, **meh_lib.slb**. Select library **meh_lib.slb**:

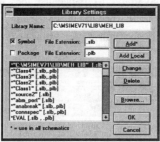

Click the **Delete** button. This will remove the library from the list.

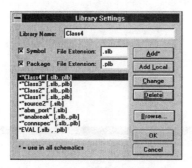

You may have more libraries than are listed in the screen capture above, but the libraries listed above are the standard libraries that come with this book. To restore Schematics to its original configuration, you should have only the libraries listed in the above screen capture.

Note that the procedure given above removes the library from the list of configured libraries. It does not actually remove the file from your hard disk. Click the **OK** button twice to return to the schematic, and then select **File** and **Exit** to terminate Schematics and update the changes we made to the msim_evl.ini file.

7.H. Summary

- Changing the model reference does not create a new part. It tells PSpice to use a model that is already located in one of the library files (all files ending with the .lib suffix). You must know the name of the model and the name of the library file.

- Editing an instance of a model creates a new model by changing the contents of an older model. The original model is left unchanged. This method is used to create new models with slightly different characteristics from the original model.

- The breakout models are used to create completely new models.

- If Schematics or PSpice has trouble finding a model, check the library path to see if the files listed in the library exist.

PART 8
Digital Simulations

PSpice can simulate pure analog, mixed analog/digital, and pure digital circuits. This part describes how to run pure digital and mixed analog/digital simulations. The circuits given are fairly simple, but the examples can be applied to larger systems. Although the digital components can be used with any of the previously described simulations, digital circuits are usually simulated with the Transient Analysis because we are interested in a gate's output at a particular time. Only Transient simulations will be demonstrated here. If you are not familiar with the Transient Analysis, review Part 6 before proceeding. This part also assumes that you are familiar with displaying traces using Probe. If this is your first time using Schematics and PSpice, you should review Part 1 for instructions on drawing a circuit, Part 2 for instructions on using Probe, and run a few of the examples in Part 6 to become familiar with using the Transient Analysis.

The library "eval.slb" shipped with the evaluation version contains over 150 logic circuits ranging from a 7400 to a 74490. This library is rich enough for students to simulate most circuits found in first- and second-semester logic classes.

NOTE ON CIRCUIT LIMITATIONS

The limitations of the evaluation version of Schematics and PSpice on digital simulations are the number of parts you can draw on a page (50 parts), the number of logic transitions (10,000), and the number of parts contained in a netlist that can be read by PSpice (70). Some of the digital circuits are fairly complicated and contain a large number of internal parts. Even though a circuit may appear small on the screen, when it is compiled into its subcircuits, it may have a large number of internal parts and may reach the limit of 70 parts. The limitations on the mixed analog/digital simulations are the same as on pure analog simulations. However, when simulating mixed analog/digital circuits, PSpice must add interface circuits between the analog and digital components. The interface circuits increase the part count and can cause a circuit to reach the part limitation sooner than if the circuit contained only digital parts. These limitations are fairly generous and allow for the simulation of fairly large circuits.

8.A. Digital Signal Sources

Although a digital circuit could have only analog inputs, digital signal sources are available to provide a digital signal to the digital components. If an analog voltage source is used to provide a signal to a digital gate, PSpice will insert analog-to-digital conversion circuits between the source and the gate. These conversion circuits could result in longer simulation times. For example, you could use the analog source Vpulse to generate a 0 to 5 volt square wave for use as a clock. This would require longer simulation times than if you used the digital sources described below.

8.A.1. Digital Signal

The "Digital_signal" source allows us to set a single wire to any bit sequence in time. This part is located in the library **Class4.slb**. To get this part, select **D**raw, **G**et New Part, and then **L**ibraries.

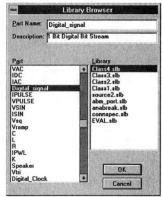

The part **Digital_signal** is located near the top of the **Class4.slb** library. Select the **Digital_signal** part and click the **OK** button. The mouse pointer will be replaced by the graphic for the **Digital_signal** part, ⌐⌐⌐⌐>. Place the part in your schematic by clicking the *LEFT* mouse button. To stop placing parts, click the *RIGHT* mouse button. Edit the part's attributes by double clicking the *LEFT* mouse button on the **Digital_signal** graphic. The attributes of this source are:

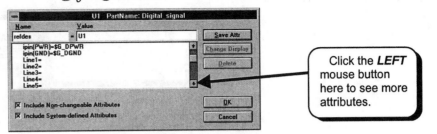

This source has thirteen attributes that you can define, *Line 1* to *Line 13*. To see the remaining attributes, click the *LEFT* mouse button on the down arrow, 🔽. Not all of the lines must contain data, but they must be filled in sequentially. At a minimum, you must define the attribute *Line 1*. These lines may be defined in several different ways, but only three will be presented here. For a complete description, see *The Design Center Circuit Analysis-User's Guide*. This manual is included on the CD-ROM that accompanies this text.

ABSOLUTE TIME

The screen capture below shows the attributes defined using absolute time:

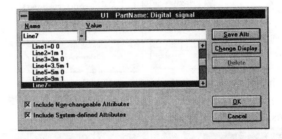

These attributes define the source as the following bit stream: At time zero the source has a value of logic zero. The source remains zero until time equals 1 ms. At 1 ms, the source changes to a logical one. The source remains a one until time equals 3 ms. At 3 ms, the source changes to a zero. It remains at zero until time equals 3.5 ms. At 3.5 ms, it switches back to a one. It remains a one until time equals 5 ms, when it changes back to a zero. The source remains at zero until time equals 9 ms, when it changes back to a logical one. For time greater than 9 ms, the source will remain at logical one.

From this example we see that *Line 1* through *Line 13* consist of time and transition pairs. The output of the source remains constant until time reaches one of the times specified in the attributes. At that time, the source makes the transition from its present value to the value specified by the attribute. You need to define only as many attributes as necessary. In this example only six of the thirteen "Line" attributes were needed. The waveform this source describes is:

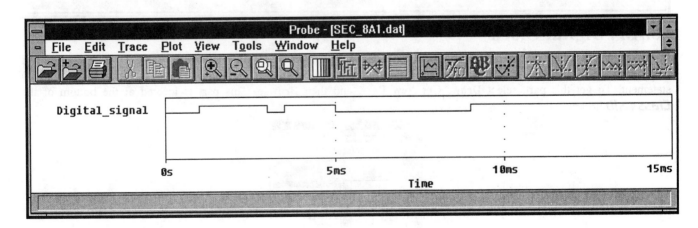

RELATIVE TIME

We will now define a source that produces the same waveform as above, but we will use relative timing. Relative timing is similar to absolute time, except that the time-transition pairs specify the amount of time from the last transition. The following attributes specify the same waveform specified above using relative time. Note the plus (+) signs before the time values.

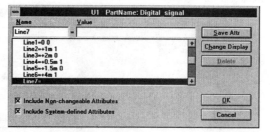

At time equals zero the source is a logic zero. It remains at zero for 1 ms and then changes to a one. It remains a one for 2 ms and then changes to a zero. It remains a zero for 0.5 ms and then changes to a one. It remains a one for 1.5 ms and then changes to a zero. It remains a zero for 4 ms and then changes to a one. It remains at one for the remainder of the simulation.

REPEATED LOOPS

Since many digital waveforms are periodic, a GOTO statement is provided for looping. The attributes below produce a clock frequency of 1 kHz for four cycles and then 500 Hz for an infinite number cycles:

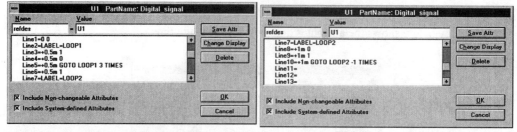

Two copies of the attribute box are shown since not all the attributes can be shown in a single screen capture. The line **GOTO LOOP1 3 TIMES** means execute the **GOTO** statement 3 times. The line **GOTO LOOP2 -1 TIMES** means execute the **GOTO** statement an infinite number of times. These attributes describe the waveform:

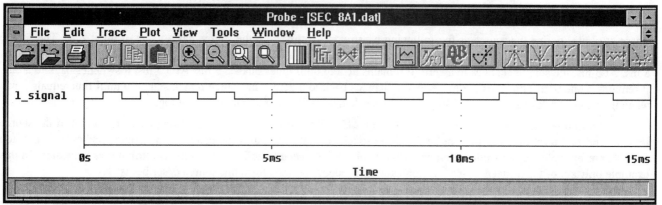

8.A.2. Digital Clock

Since most digital systems use a clock with a constant frequency, a special part called "Digital_Clock" is provided to generate clock waveforms. This part is nothing more than a special case of the "Digital_signal" part using the GOTO statement. To get this part, select **Draw**, **Get New Part**, and then **Browse**. This part is located at the bottom of the **Class4.slb** library.

The graphic for the **Digital_Clock** is ⌐CLK⊃. Notice that the frequency is indicated on the schematic:

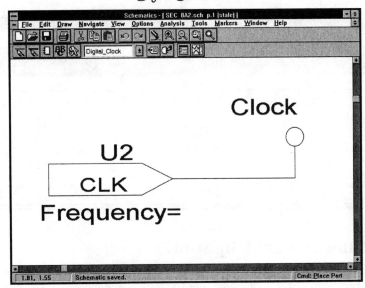

The BUBBLE was added to view the waveform using Probe. To set the frequency, double click on the text **Frequency=**:

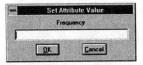

Type in the value for the frequency:

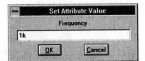

Click the **OK** button. The updated value of the frequency will be shown on the schematic:

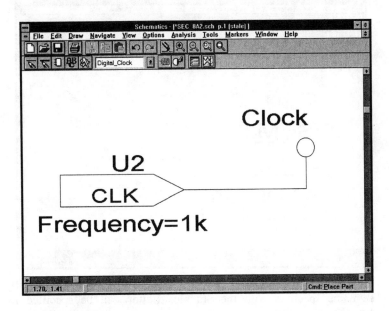

The Digital_Clock part will produce the specified frequency with a 50% duty cycle (the time high equals the time low). The 1 kHz clock waveform is shown below:

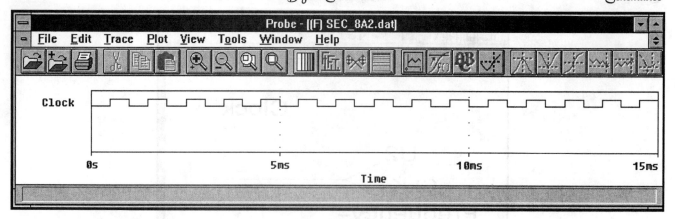

8.A.3. Digital Stimulus Part (DigStim)

The Stimulus Editor is a tool that allows us to create and view a signal source before we run a simulation. It can be used with analog voltage and current sources, and digital sources. Here, we will demonstrate its use with a digital signal. The Stimulus Editor is a very useful tool for creating waveforms if you are not familiar with the various sources available with PSpice. Unfortunately, for digital signals the evaluation version is limited to clock signals.

The part **DigStim** is located in the **source2.slb** library:

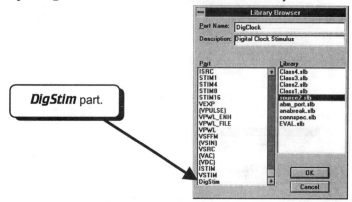

Place the DigStim part in your schematic:

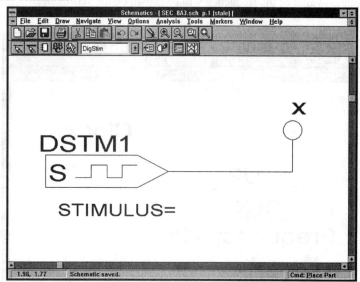

Notice that the graphic has the attribute **STIMULUS=**. When you create a waveform with the Stimulus Editor, you will give the waveform a name. The name specified with the Stimulus Editor will be specified in the schematic using the **STIMULUS=** line.

Before you can use the Stimulus Editor, you must save the schematic. See Section 1.H if you are unfamiliar with saving a schematic. We will save the schematic with the name **SEC_8A3.sch**.

A bubble was added to the circuit with the label *X* so that we can view the waveform after we run the simulation. Notice that the name of the stimulus is still unspecified.

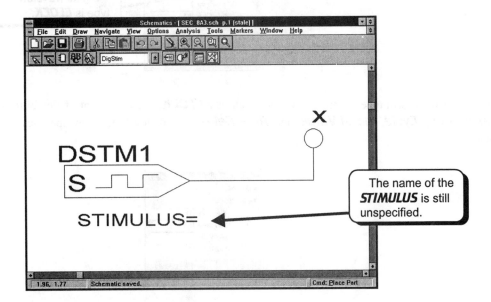

To create the waveform for the part, double click the *LEFT* mouse button on the DigStim graphic, ⌐s⎍▷—. Before the Stimulus Editor runs, Schematics will ask you what you wish to call the waveform. In other words, it wants you to specify a value for the ***STIMULUS=*** line:

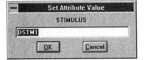

The default name is the name of the source, ***DSTM1***. You can change the name if you wish. Click the ***OK*** button to run the Stimulus Editor:

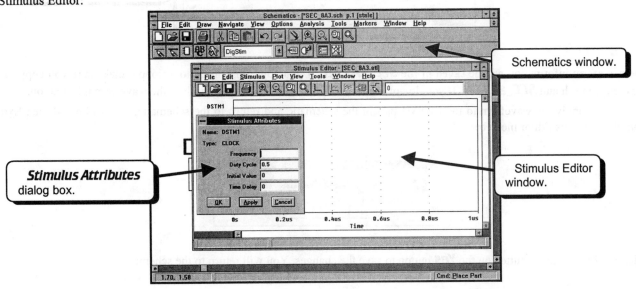

The Stimulus Editor window will allow us to view the waveform. The ***Stimulus Attributes*** dialog box allows us to change the properties of the waveform:

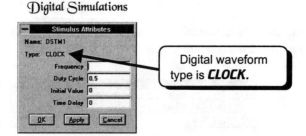

The only digital waveform type we are allowed to create is a **CLOCK**. We have a lot of flexibility since we can specify the **Frequency**, **Duty Cycle**, **Initial Value**, and **Time Delay**. The dialog box below specifies a 1 kHz clock with a 50% duty cycle:

To view the waveform with the Stimulus Editor, click the **OK** button:

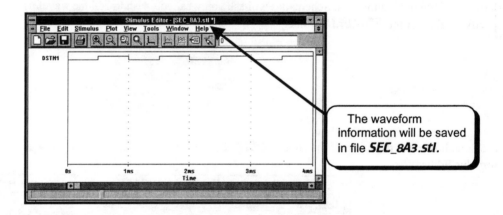

The waveform information will be saved in file **SEC_8A3.stl**. If you copy this circuit to a floppy, make sure you copy the files sec_8a3.sch and **SEC_8A3.stl**. If you do not copy the file **SEC_8A3.stl**, you will lose the waveform information.

    To apply the waveform to the DSTM1 part in the schematic and return to the schematic, select **File** and then **Exit** from the Stimulus Editor menu bar:

Click the *LEFT* mouse button on the **Yes** button to save the changes. You will return to the schematic:

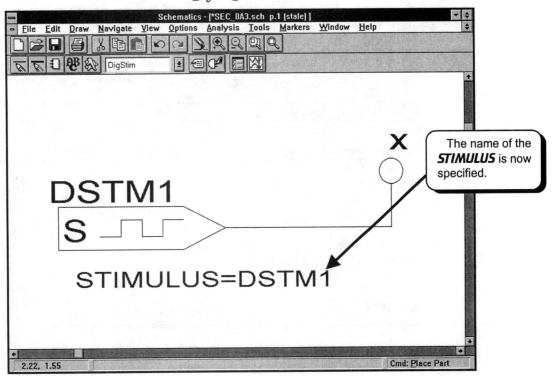

Notice that the name of the stimulus is now specified in the schematic, **STIMULUS=DSTM1**. If you run a Transient Analysis, the waveform at node **X** will have the waveform created with the Stimulus Editor.

As a second example, we will create a two-phase non-overlapping clock. Add two more DigStim parts to the circuit as shown:

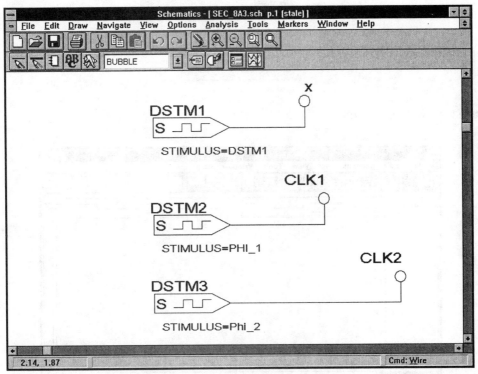

Notice that we have labeled the two **STIMULUS** signals **PHI_1** and **Phi_2**. This is done so that we do not have to specify the names when we run the Stimulus Editor. Double click the **LEFT** mouse button on the graphic for **DSTM2**, ⌐. The Stimulus Editor will run and bring up the dialog box for waveform **PHi_1**:

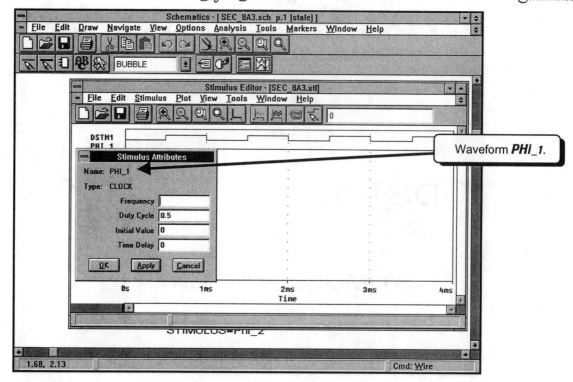

We will create a 1 kHz clock with a duty cycle of 25%. Fill in the dialog box as shown:

Click the **OK** button to view the waveform:

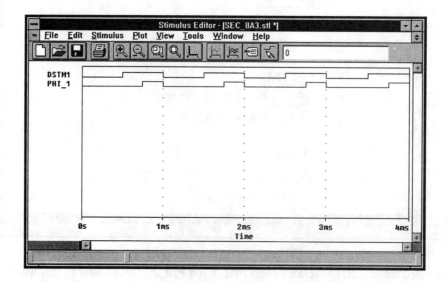

Next we need to create the waveform for Phi_2. Select **Stimulus** and then **New** from the Stimulus Editor menu bar:

We want the waveform name to be Phi_2 and the type to be a digital clock. Fill in the dialog box as shown:

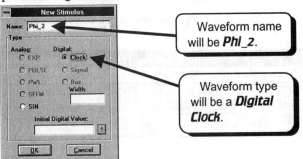

Click the **OK** button to specify the properties of the clock waveform:

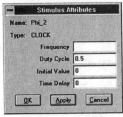

We want **Phi_2** to have the same frequency and duty cycle as PHI_1, but the pulse should be delayed by 0.25 ms. Fill in the dialog box as shown:

Click the **OK** button to view the waveforms:

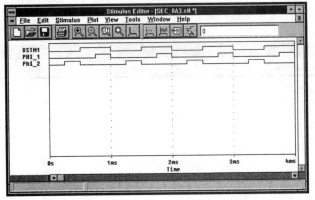

To save the waveform information and return to the schematic, select **File** and then **Exit** from the Stimulus Editor menu bar:

Click the **LEFT** mouse button on the **Yes** button to save the changes. You will return to the schematic. If you run a Transient Analysis, the waveforms at **X**, **CLK1**, and **CLK2** will be the same as those seen on the Stimulus Editor window.

8.B. Mixed Analog and Digital Simulations

The first circuit we will look at is an op-amp circuit that drives the clock of a J-K flip-flop. Wire the circuit shown below:

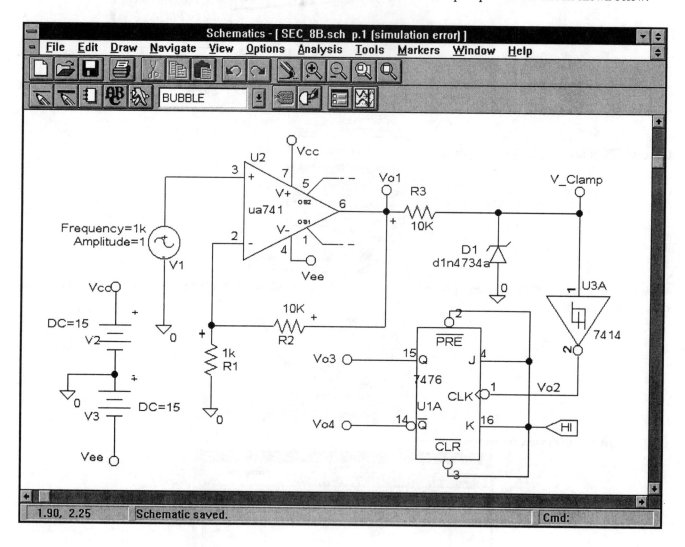

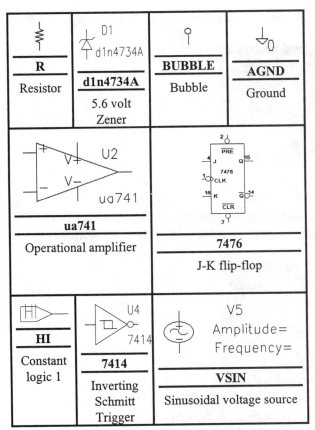

R	**d1n4734A**	**BUBBLE**	**AGND**
Resistor	5.6 volt Zener	Bubble	Ground
ua741		**7476**	
Operational amplifier		J-K flip-flop	
HI	**7414**	**VSIN**	
Constant logic 1	Inverting Schmitt Trigger	Sinusoidal voltage source	

The circuit is drawn as though we were going to wire the circuit in the lab. No special circuits are required between the analog circuitry and the digital logic gates. Note that the J and K inputs of the flip-flop are held high so that the flip-flop toggles at each negative clock edge.

The sinusoidal voltage waveform produces a 1 kHz sine wave with voltages between –1 and 1 volts. The **ua741** op-amp circuit has a gain of 11 and produces a ±11 volt sine wave of 1 kHz at node **Vo1**. This waveform goes into a Zener clipping circuit that limits the voltage to approximately +5.6 and –0.7 volts at node **V_Clamp**. This voltage is TTL compatible and can be connected to the Schmitt Trigger input. The output of the Schmitt Trigger should be a 0 to 5 volt square wave at 1 kHz. The J-K flip-flop is wired as a divide by two counter, so Q and $\overline{Q}$ should be 0 to 5 volt square waves at 500 Hz. They should also be 180 degrees out of phase.

Since the frequency of the pulse generator is 1 kHz, we will run a Transient simulation for 15 ms to allow 15 cycles. To run a digital simulation we must set up the Transient Analysis, as well as specify parameters for the digital simulation. We will first set up the Transient Analysis. Select **Analysis**, and then **Setup** from the Schematics menus:

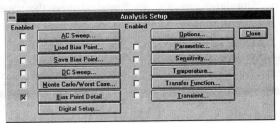

Click the *LEFT* mouse button on the **Transient** button to set up the Transient Analysis. Fill in the dialog box as shown to run the simulation for 15 ms:

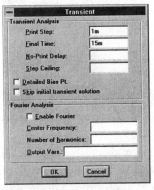

Click the **OK** button to accept the settings:

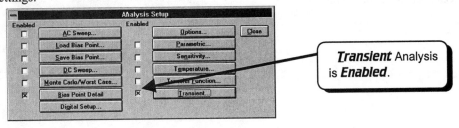

Transient Analysis is *Enabled*.

Notice that the **Transient** Analysis is **Enabled**.

We must also set up the parameters for the Digital Analysis. Click the *LEFT* mouse button on the **Digital Setup** button:

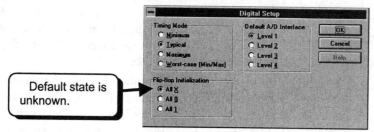

Default state is unknown.

This dialog box allows us to specify timing and initial conditions for the digital circuits. We are interested in the initial state of the flip-flops. The default initial state for all flip-flops is **All X** (unknown). We wish to set the initial state to all zeros. Click the *LEFT* mouse button on the circle ○ next to the text **All 0**. The circle should fill with a black dot ◉, indicating that the option has been specified:

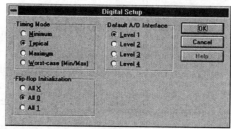

Click the **OK** button to accept the settings and then click the **Close** button to return to the schematic.

Run the simulation (**Analysis**, **Simulate**) and then run Probe. Display the traces **Vo1**, **V_Clamp**, **Vo2**, **Vo3**, and **Vo4**. To display traces, select **Trace** and then **Add** from the Probe menu bar:

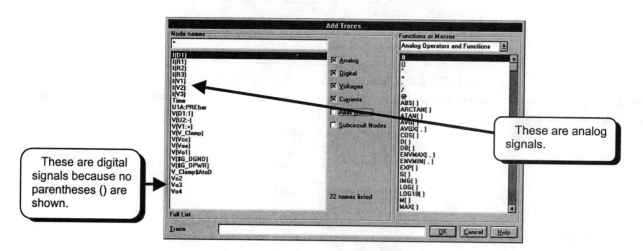

These are digital signals because no parentheses () are shown.

These are analog signals.

Some of the traces are displayed a little differently in this dialog box. At analog nodes we see that traces are displayed as **V(Vo1)** or **V(V_Clamp)**. The currents through analog components are shown as **I(D1)** or **I(V1)**. The waveforms at digital nodes are shown as **Vo3** or **Vo4**. This is how Probe allows you to distinguish between digital and analog nodes. Display the traces at all nodes of interest.

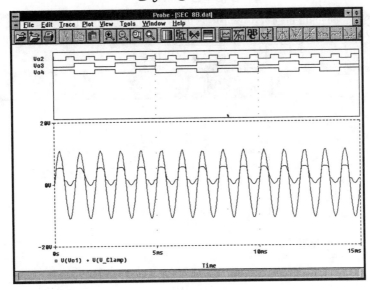

We see that the digital traces are shown on a different graph than the analog traces, but they share the same time axis. These traces show results similar to the expected results.

For a second example, we will simulate the circuit below:

R	**d1n4734A**	**BUBBLE**	**AGND**	**7414**	**HI**	**VSIN**
Resistor	5.6 volt Zener	Bubble	Ground	Inverting Schmitt Trigger	Constant logic 1	Sinusoidal voltage source

UA741	**7476**	**74161** •	**VDC**
Operational amplifier	J-K flip-flop	Binary counter	Independent DC voltage source

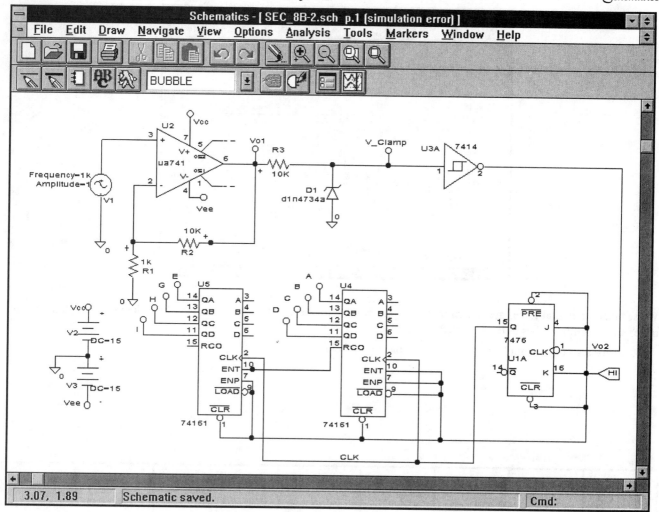

As in the previous example, the flip-flops in the 74161 must be initialized at the start of the simulation. Fill in the Digital Setup as shown on page 386 to set the initial state of all flip-flops to zero.

We will run a Transient Analysis for 130 ms. Fill in the Transient setup dialog box as shown below (**Analysis**, **Setup**, **Transient**) and run the simulation (**Analysis**, **Simulate**).

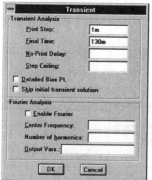

The results of the simulation are displayed with Probe. To add a trace, select **Trace** and then **Add** from the Probe menus.

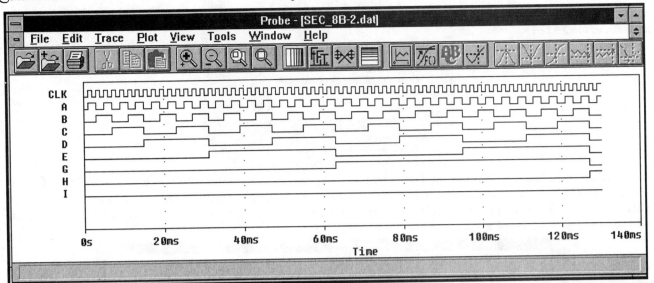

EXERCISE 8-1: Design a 60 Hz sync circuit. The input to the circuit is a 12 V amplitude, 60 Hz sine wave. The output of the circuit should be a 1 ms 5 V pulse that occurs when the sine wave crosses zero with a positive slope.

SOLUTION:

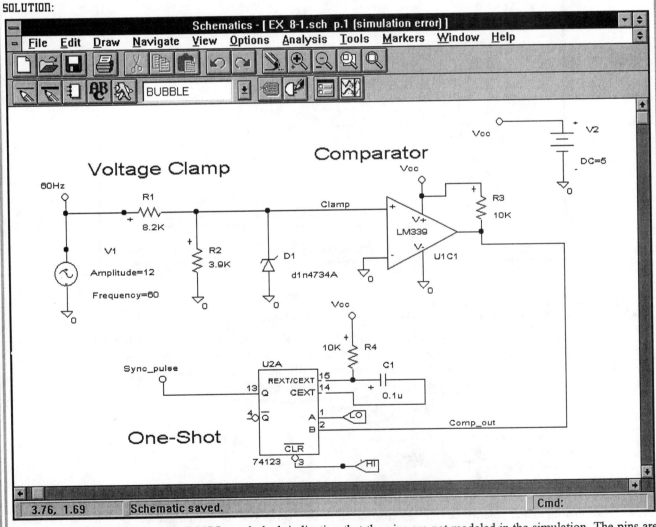

Note that pins **14** and **15** on the **74123** are dashed, indicating that the pins are not modeled in the simulation. The pins are available for connection, but the circuit connected to the pins is not used in the simulation. Thus, **C1** and **R4** do not determine the pulse timing in the simulation. They are included for completeness and could be omitted if only a simulation is required. If you were constructing a PC board, you would need to include them. The pulse width is determined by one of the attributes of the **74123**:

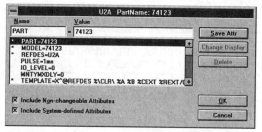

Notice the attribute **PULSE=1ms**. This attribute determines the pulse width rather than the time constant of **R4** and **C1**. The results are shown on the following screen capture. To plot traces on different plots, select **Plot** and then **Add Plot** from the Probe menu bar. Add the first trace, select **Plot** and then **Add Plot** to add a new plot; add the second trace, select **Plot** and then **Add Plot** to add a new plot; and then add the third trace.

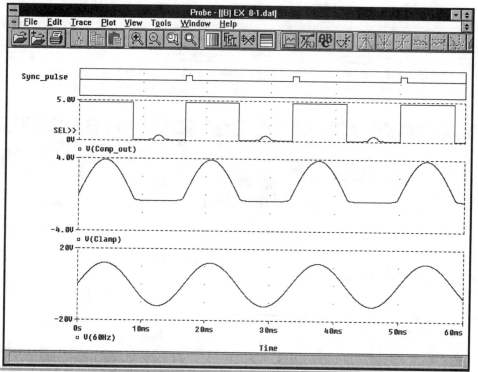

8.C. Effect of Not Initializing Flip-Flops

In the previous section we used the Digital Setup to initialize all flip-flops to the zero state. Suppose that instead of clearing the flip-flops, we specify the initial states as unknown (All X). We will be using the circuit shown on page 384. Follow the procedure for running the analysis, except fill in the Digital Setup as shown below:

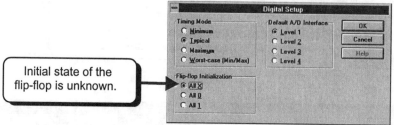

Initial state of the flip-flop is unknown.

Note that the initial state of the flip-flop is specified as unknown. Run the simulation and then display the traces **Vo1**, **V_clamp**, **Vo2**, **Vo3**, and **Vo4**. To display traces, select **Trace** and then **Add** from the Probe menu bar:

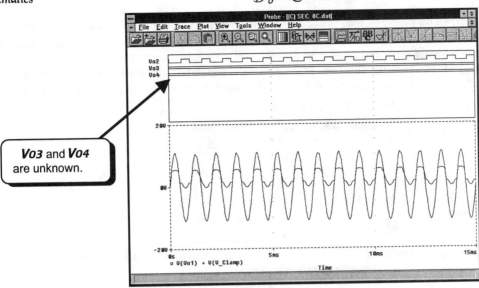

Note that the flip-flop outputs (**Vo3** and **Vo4**) appear as double lines. The double lines indicate that PSpice does not know if the output we are plotting is a logic one or zero. This is because if PSpice does not know the initial state, it cannot determine any of the following states. To run digital simulations you need to specify the initial states using the All 1 or All 0 parameter in the Digital Setup dialog box, or create a circuit that initializes flip-flops. The following section gives a circuit for initializing digital circuits.

8.C.1. Start-Up Clear Circuit

Suppose we want to initialize all flip-flops to an initial state but do not want to use the Schematics All 0 or All 1 parameter. The circuit below clears the flip-flop at the beginning of the simulation:

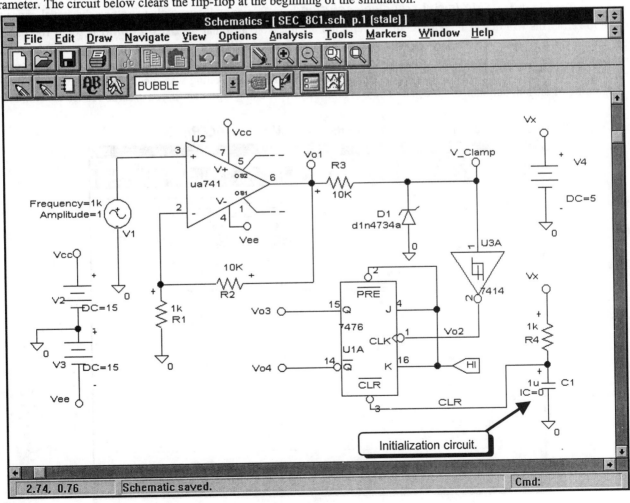

This is the same circuit as that shown on page 384 with the addition of the **R4**, **C1**, and a 5 volt DC source. The resistor and capacitor create the initialization circuit. The initial condition of the capacitor is specified as 0 volts. When the simulation runs, the capacitor starts at 0 volts (logic zero) and charges to 5 volts. While the capacitor is close to 0 volts, it provides a logic zero input to the flip-flop, which clears the flip-flop. For most of the simulation the capacitor is charged to 5 volts, providing a logic 1 to the flip-flop clear input which has no effect. See Section 6.B for specifying capacitor initial conditions.

The Digital Setup should be specified as **All X** as shown below:

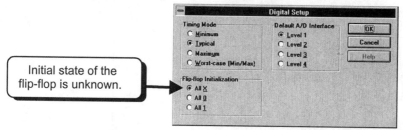

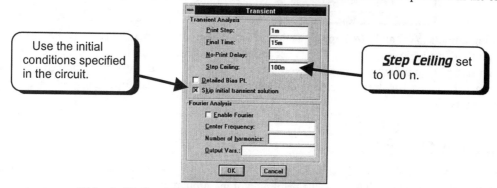

When you set up the Transient dialog box you must tell PSpice to use the initial conditions specified in the circuit:

If you forget to check the **Skip initial transient solution** box, PSpice will run without setting the initial capacitor voltage to zero. This will defeat the purpose of the capacitor circuit. Note also that the **Step Ceiling** is set to 100 ns. The Step Ceiling is the largest time step between simulation points. A **Step Ceiling** of 100 ns means that there will be at least one data point every 100 ns. Since the ending time of this simulation is 15 ms, there will be at least 150,000 simulation points (probably more). This simulation will take a long time to complete. The Step Ceiling was specified so that we can observe the beginning of the simulation in great detail. You can omit the Step Ceiling argument if you wish to reduce the simulation time.

Run the simulation and then run Probe. Add the traces **Vo3**, **Vo4**, and **V(CLR)**:

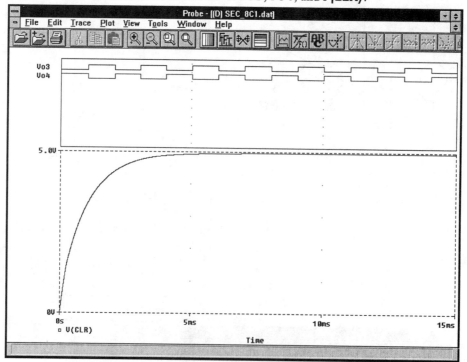

We see that the flip-flop is initialized to an initial state of zero. If we want to see that the initial state of the flip-flop is unknown, we can zoom in to the beginning of the simulation. The plot below shows the first 150 ns of the simulation:

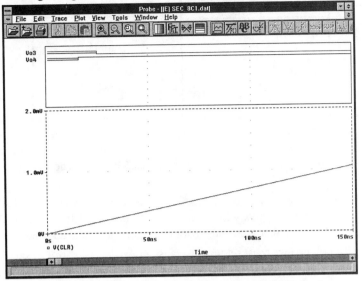

We see that the initial state of the flip-flop is unknown, but the flip-flop is quickly cleared.

8.D. Pure Digital Simulations

PSpice can be used as a logic simulator as well as a mixed analog/digital simulator. We will now simulate a switch-tail counter. Wire the circuit as shown:

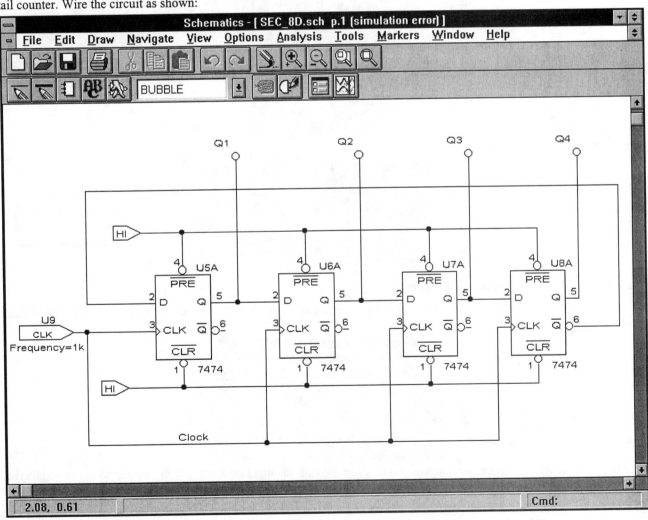

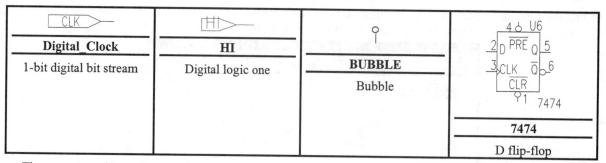

Digital_Clock	HI	BUBBLE	
1-bit digital bit stream	Digital logic one	Bubble	7474
			D flip-flop

There are two things that are different about this circuit. The first is that there are no ground or positive supply connections in the circuit. When MicroSim made the graphics for the digital parts, they included the ground and power connections in the graphics but hid them to keep schematics from getting cluttered. The power and ground connections are in the circuit but are not shown.

Instead of a bubble having been placed on the **CLOCK** wire, the wire has been labeled. To label a wire, double click the **LEFT** mouse button on the wire. A dialog box will appear, asking you for a label. Type in the appropriate label for the wire. It is sometimes better to label a wire rather than add a bubble, because a bubble adds more clutter to a circuit drawing.

When we run a digital simulation we must set up the Transient Analysis, as well as specify parameters for the digital simulation. We will first set up the Transient Analysis. Select **Analysis**, and then **Setup** from the Schematics menus, and then click the **LEFT** mouse button on the **Transient** button to set up the Transient Analysis. Fill in the dialog box as shown to run the simulation for 20 ms:

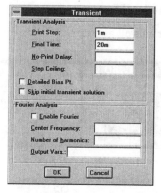

Click the **OK** button to accept the settings:

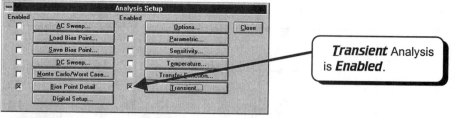

Notice that the **Transient** Analysis is **Enabled**.

We must also set up the parameters for the Digital Setup. Click the **LEFT** mouse button on the **Digital Setup** button:

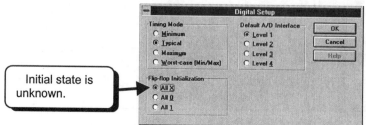

This dialog box allows us to specify timing and initial conditions for the digital circuits. We are interested in the initial state of the flip-flops. The default initial state for all flip-flops is **All X** (unknown). We wish to set the initial state to all zeros. Click the **LEFT** mouse button on the circle ○ next to the text **All 0**. The circle should fill with a black dot ◉, indicating that the option has been specified:

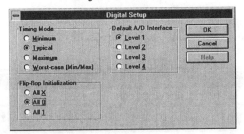

Click the **OK** button to accept the settings and then click the **Close** button to return to the schematic.

Run the simulation (**Analysis**, **Simulate**) and then run Probe. Add the traces **Q1**, **Q2**, **Q3**, and **Q4**:

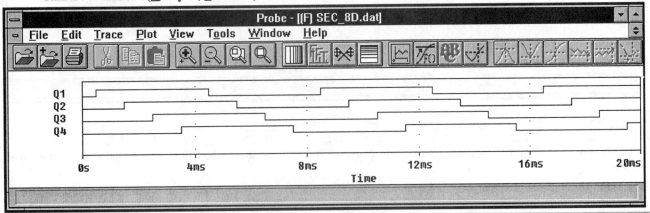

EXERCISE 8-2: Design and simulate a 4-bit ring counter. The counter should be initialized to 1000.

SOLUTION: Wire the circuit as shown. A 1 kHz clock is used. Use the capacitor startup circuit on page 391 to preset the first flip-flop to 1 and the remaining flip-flops to 0. Note that the initial condition of the capacitor is set to zero.

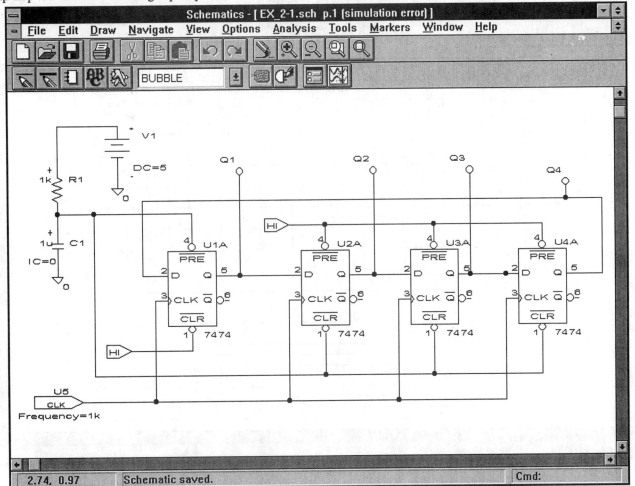

Run a Transient Analysis for several clock cycles and display the results in Probe.

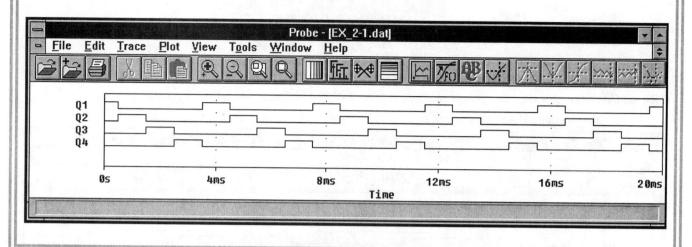

EXERCISE 8-3: Design and simulate a 4-bit decade counter. The counter should be initialized to 0000 at the start of the simulation.

SOLUTION: Wire the circuit as shown. A 1 kHz clock is used. Use the Digital Setup dialog box to set all flip-flops to an initial state of 0.

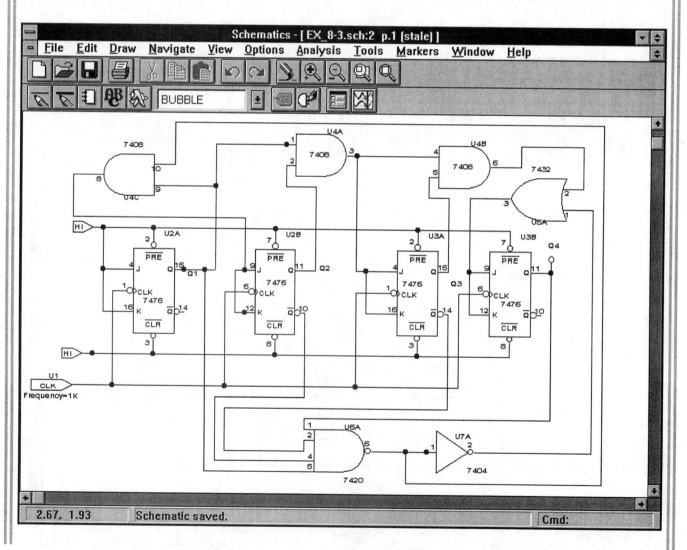

The circuit is a bit large to fit on a screen capture, but it does show the capabilities of the digital simulations. Run a Transient Analysis for several clock cycles and display the results in Probe.

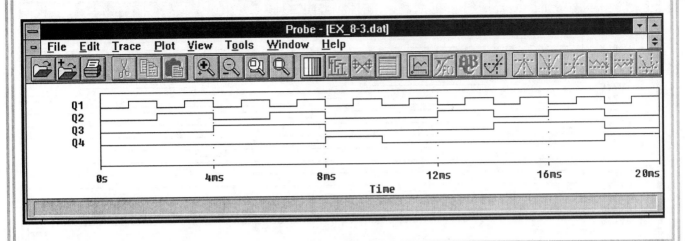

EXERCISE 8-4: Construct and simulate a circuit that uses a 555 astable multivibrator as the clock and a binary counter to drive a decoder. The counter should be initialized to 0000 at the start of the simulation.

SOLUTION: Wire the circuit as shown. Use the Digital Setup dialog box to set all flip-flops to an initial state of 0.

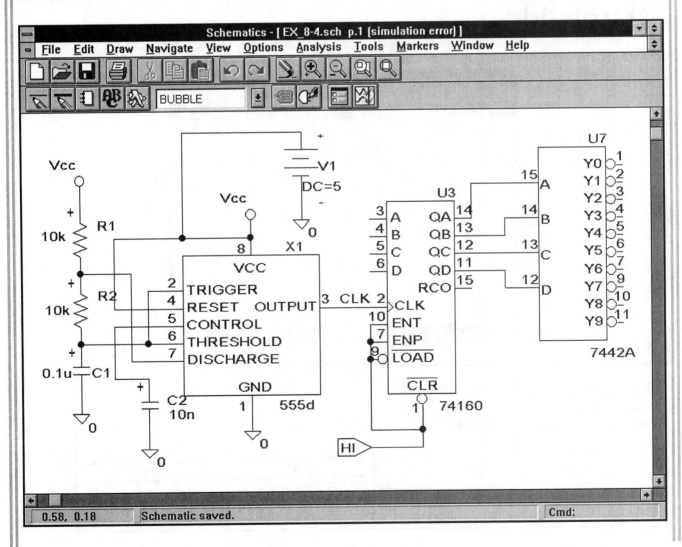

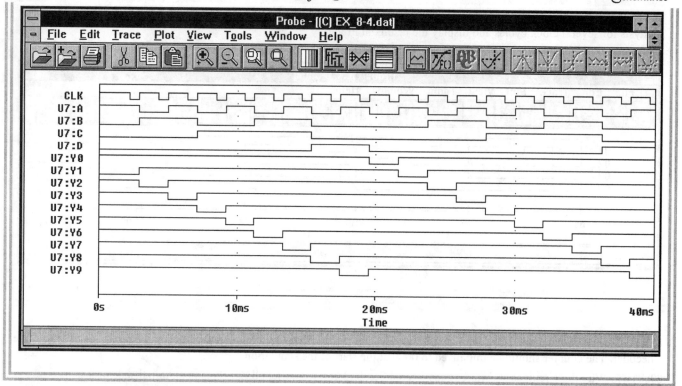

8.E. Gate Delays

PSpice includes gate delays in its digital simulations. To illustrate gate delays, we will create a two-phase, non-overlapping clock using gate delays. Wire the circuit below:

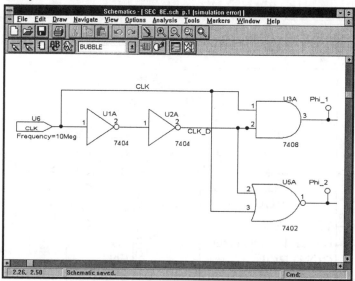

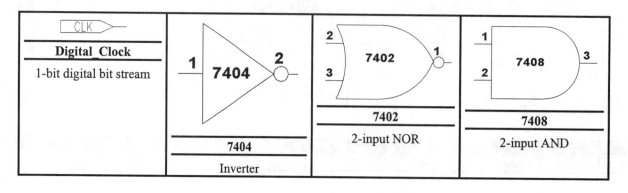

Since gate delays are on the order of nanoseconds, we will set the clock frequency to 10 MHz. We will run the simulation for 500 ns. Fill in the Transient dialog box as shown:

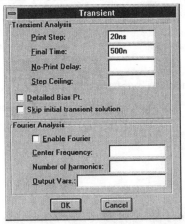

Run the simulation and then run Probe. Display the traces **CLK**, **CLK_D**, **Phi_1**, and **Phi_2**:

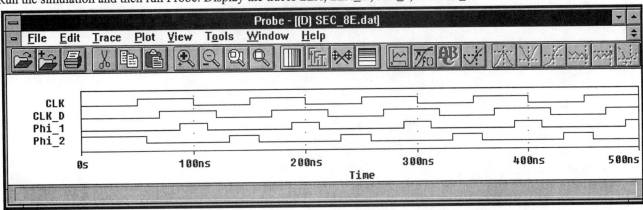

We can use the cursors to find the delay through the two inverters. Select **Tools**, **Cursor**, and then **Display** to view the cursors:

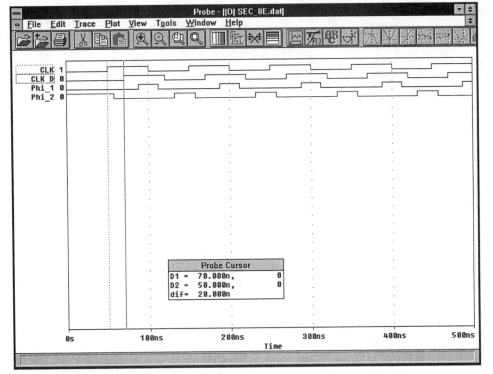

The **LEFT** mouse button controls cursor 1 and the **RIGHT** mouse button controls cursor 2. The cursor coordinates are displayed in the dialog box as **A1** and **A2** (**D1** and **D2** on my screen capture). The difference in time between the two cursors is given as **dif=20** nanoseconds. Thus, the total delay through the two inverters is 20 ns.

8.F. Summary

- PSpice can be used for pure digital simulations, pure analog simulations, or mixed analog/digital simulations.
- Always preset or clear all flip-flops before running a simulation. Use the digital setup dialog box or a start-up clear circuit to set the initial state of the flip-flops.
- PSpice simulations include gate delays.

8.G. Bibliography

[1]   MicroSim Corporation. *MicroSim PSpice A/D- Circuit Analysis Software - Reference Manual*, Version 7.1, Irvine, CA, October, 1996, p. 3-78

PART 9
Monte Carlo Analyses

The Monte Carlo analyses are used to observe how device tolerances can affect a design. There are two analyses that can be performed. The Worst Case Analysis is used to find the maximum or minimum value of a parameter given device tolerances. Device tolerances are varied to their maximum or minimum limits such that the maximum or minimum of the specified parameter is found. The Monte Carlo analysis is used to find production yield. If the Worst Case Analysis shows that not all designs will pass a specific criterion, the Monte Carlo analysis can be used to estimate what percentage will pass. The Monte Carlo analysis varies device parameters within the specified tolerance. The analysis randomly picks a value for each device that has tolerance and simulates the circuit using the random values. A specified output can be observed.

This part discusses the syntax of models with device tolerance and how to use the models in a simulation. This part does not discuss how to easily create models within Schematics. After you have understood the syntax for the models and how to run the simulations, you may wish to look at Part 7 to see how to create new models to specify the tolerance you need.

9.A. Device Models

The first thing we need to know is how to create devices with tolerance. The libraries already have a number of parts with tolerance, but these are usually not enough. PSpice allows you to create several types of distributions.

9.A.1. Uniform Distribution

The uniform distribution specifies that the part is equally likely to have a value anywhere in the specified tolerance range. The first example we will look at is the 5% resistor model included in class.lib. If you look at Appendix E on page 467, you will see the following line in the file class.lib:

.MODEL R5pcnt RES(R=1 DEV/UNIFORM 5%)

The model name is R5pcnt. It is a resistor model because the model type is RES. The nominal value of the model is R=1 and it has a 5% tolerance. The distribution is uniform. A uniform distribution means that the model parameter R is equally likely to have a value of 1.05, 1, 0.95, or any other value between 1.05 and 0.95. When you use this model, the actual value of the resistor is the value specified in the schematic times the parameter R. Thus, if the value of a 5% resistor is specified as 1k in the schematic, it may have a value anywhere between 950 and 1050 when used with the Monte Carlo analyses. An equivalent model is:

.MODEL R5pcnt RES(R=1 DEV/UNIFORM 0.05)

This model has an absolute tolerance rather than a percentage tolerance. The parameter R can have a value in the range 1±0.05.

The next model we will look at is an NPN bipolar junction transistor. The model below is a transistor with a value of β_F that may vary between 50 and 350:

.MODEL QBf NPN(Bf=200 DEV/UNIFORM 150)

The nominal value of β_F is 200. The range of β_F is 200±150. The transistor is equally likely to have a value anywhere in this range. The name of the model is QBf. The text NPN specifies the model as an NPN bipolar transistor. This model is included in class.lib. A limitation of this model is that none of the other transistor parameters have been specified. This model is almost ideal. Its limitations will become apparent when you observe that the high frequency response does not roll off, even at frequencies beyond 1 GHz.

Another transistor model that includes tolerance is the Q2N3904B model shown below:

.MODEL Q2N3904B   NPN(Is=6.734f Xti=3 Eg=1.11 Vaf=74.03
+   Bf=416.4 DEV/UNIFORM 80% Ne=1.259
+   Ise=6.734f Ikf=66.78m Xtb=1.5 Br=.7371 Nc=2 Isc=0 Ikr=0 Rc=1
+   Cjc=3.638p Mjc=.3085 Vjc=.75 Fc=.5 Cje=4.493p Mje=.2593 Vje=.75
+   Tr=239.5n Tf=301.2p Itf=.4 Vtf=4 Xtf=2 Rb=10)

This is a copy of the standard Q2N3904 model, except that tolerance has been added to the β_F parameter. This model will have the same properties as the Q2N3904 but will allow variations in β_F.

We are not limited to tolerance in one parameter. The above model has tolerance only in β_F. The model shown below specifies tolerance in β_F as well as I_S:

.MODEL Q2N3904B NPN(Is=6.734f DEV/UNIFORM 10% Xti=3 Eg=1.11

+   Vaf=74.03

+   Bf=416.4 DEV/UNIFORM 80% Ne=1.259

+   Ise=6.734f Ikf=66.78m Xtb=1.5 Br=.7371 Nc=2 Isc=0 Ikr=0 Rc=1

+   Cjc=3.638p·Mjc=.3085 Vjc=.75 Fc=.5 Cje=4.493p Mje=.2593 Vje=.75

+   Tr=239.5n Tf=301.2p Itf=.4 Vtf=4 Xtf=2 Rb=10)

The last model we will look at is a capacitor model with +80% and −20% tolerance. This model is called CAP20_80 and is included in class.lib:

.MODEL CAP20_80 CAP(C=1.3 DEV/UNIFORM 38.461538%)

Tolerances in PSpice are set up to have equal plus and minus ranges. With a +80% and −20% capacitor we have to fudge things a little. A capacitor with a +80% and −20% tolerance has a maximum value 1.8 times the nominal value, and has a minimum value 0.8 times the nominal value. This range is equivalent to a capacitor with a nominal value of (1.8 + 0.8)/2 = 1.3 and a ±38.461538% tolerance. Note that 1.3*(1+0.38461538)=1.8, and 1.3*(1−0.38461538)=0.8. Thus, a capacitor with a nominal value of 1 and +80% and −20% tolerance has the same capacitance range as a capacitor with a nominal value of 1.3 and ±38.461538% tolerance. The only difference is the nominal value.

When you use the CAP20_80 model in a schematic, the actual value of the capacitor is the value specified in the schematic times the parameter C. If the value of a +80% and −20% capacitor is specified as 470 μF in the schematic, its nominal value is 470μF*1.3=611μF. Its maximum value is 611μF*(1+0.38461538)=846μF. Its minimum value is 611μF*(1−0.38461538)=376μF.

9.A.2. Gaussian Distribution

The Gaussian tolerance distribution generates the distribution shown below. The graph shows a 1 kΩ resistor with a ±5% tolerance.

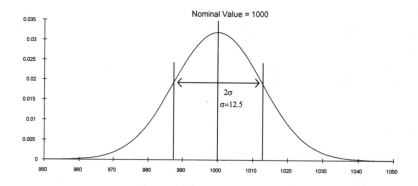

The mean value of the distribution is 1000 and the standard deviation, σ, is 12.5. A 1 kΩ, 5% resistor will have values from 950 Ω to 1050 Ω. Deviations of ±4σ achieve this spread. This distribution shows us that almost all of the resistors are within plus or minus three standard deviations from the nominal value, ±3σ=±37.5. It is very rare that we see a resistor value above 1040 Ω or below 960 Ω.

The PSpice Gaussian distribution specifies the nominal value and the standard deviation. All part distributions are limited to ±4σ since the probability of finding a resistor outside this range is extremely small. The model below describes a 5% resistor with a Gaussian distribution:

.MODEL R5gauss RES(R=1 DEV/GAUSS 1.25%)

For this model, the nominal value is 1 Ω and the standard deviation is σ=1.25%. The distribution has a maximum distribution of ±4σ=±5%.

9.B. Voltage Divider Analysis

To illustrate the basic operation of the Monte Carlo and Worst Case Analyses, we will simulate a voltage divider. Create a voltage divider using 5% resistors with Gaussian distributions as follows.

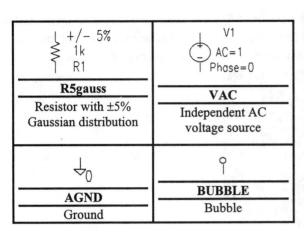

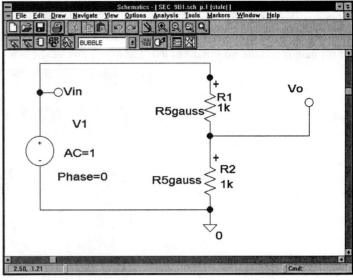

9.B.1. Voltage Divider Minimum and Maximum Voltage Gain

We will now use PSpice to find the worst case minimum and maximum voltage gain of this circuit. Since this circuit is very simple, we will do the calculations by hand and compare the results to PSpice.

The nominal voltage gain of this divider network is:

$$\frac{V_o}{V_{in}} = \frac{R_2}{(R_1 + R_2)} = \frac{1000}{1000 + 1000} = 0.5$$

The worst case maximum gain of the network is:

$$\frac{V_o}{V_{in}} = \frac{R_{2_{max}}}{\left(R_{1_{min}} + R_{2_{max}}\right)} = \frac{1050}{950 + 1050} = 0.525$$

The worst case minimum gain of the network is:

$$\frac{V_o}{V_{in}} = \frac{R_{2_{min}}}{\left(R_{1_{max}} + R_{2_{min}}\right)} = \frac{950}{1050 + 950} = 0.475$$

Now that we know what limits to expect we can set up the analysis. Since we want to find gain, we need to set up an AC Sweep. Let us find the gain of the circuit at 1 kHz. Obtain the AC Sweep dialog box by selecting **Analysis**, **Setup**, and then clicking the **AC Sweep** button. Fill in the dialog box as shown. The AC Sweep is set up to simulate one point at 1 kHz.

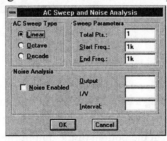

Click the **OK** button to accept the settings:

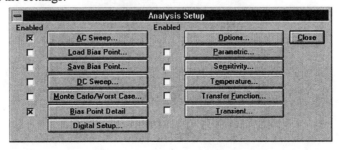

We must also set up the Worst Case Analysis. Click the **Monte Carlo/Worst Case** button. We will now set up the dialog box to find the worst case maximum gain. Since the gain is **Vo/Vin** and **Vin** is a 1 volt magnitude AC source, the magnitude of the gain is just the magnitude of **Vo**. The dialog box below is set to find the worst case maximum gain:

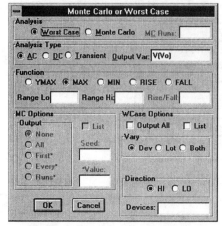

The **Analysis** is set to **Worst Case**. The **Analysis Type** is set to **AC** because we wish to find the worst case AC gain. Note that we must have the AC Sweep analysis enabled. The **Output Var**iable is **V(Vo)** because we are interested in the maximum voltage at this node. The **Function** type is **MAX** since we wish to find the worst case maximum value of **Vo**. Note that **Vary** is set to **Dev**. If you look at the model R5gauss, you will notice that it contains the text DEV. Click the **OK** button to accept the settings:

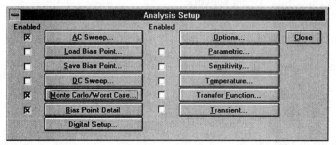

Notice that both the **AC Sweep** and the **Monte Carlo/Worst Case** Analyses are enabled. Click the **Close** button to return to the schematic.

Run PSpice by selecting **Analysis** and then **Simulate** from the Schematics menu bar. The PSpice simulation window will appear:

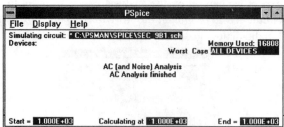

Notice that the window indicates that we are doing a **Worst Case** analysis and the analysis type is **AC Analysis**.

The results of the Worst Case Analysis are saved in the output file. Select **Analysis** and then **Examine Output** from the Schematics menu bar. The results are given at the bottom of the output file.

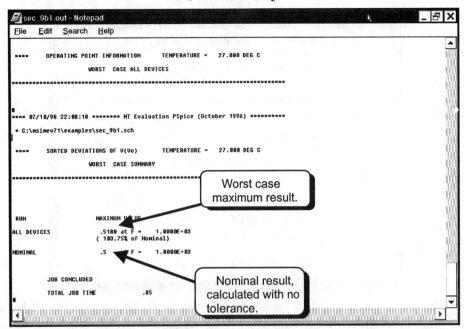

The nominal value of **Vo** is **0.5**, which is the expected result. However, the simulation says that the maximum value is **0.5188**, which is less than the expected value. Remember that for the resistor with the 5% Gaussian distribution, the standard deviation was 1.25% and the absolute limits on the distribution were ±4σ=±5%. **In the Worst Case Analysis, a device with a Gaussian distribution is varied by only ±3σ.** Had we calculated the maximum value with a 3.75% resistor variation, we would have come up with a maximum gain of 0.51875, which agrees with the PSpice result. **To obtain the worst case limits, I prefer to use the uniform distribution.**

We will now change the voltage divider circuit to use the resistor model ***R5pcnt***. Change the model reference of the resistors to ***R5Pcnt***. See Section 7.A for instructions on changing the model reference.

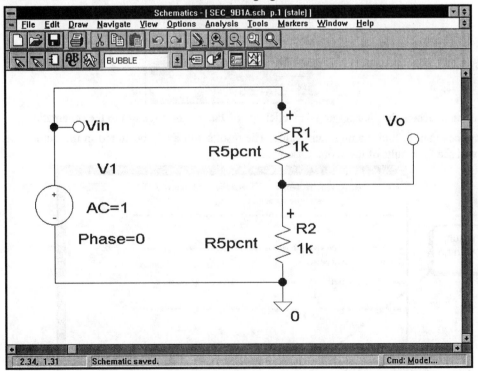

Run PSpice. The results will again be stored in the output file. At the end of the output file you will see the results of the Worst Case Analysis:

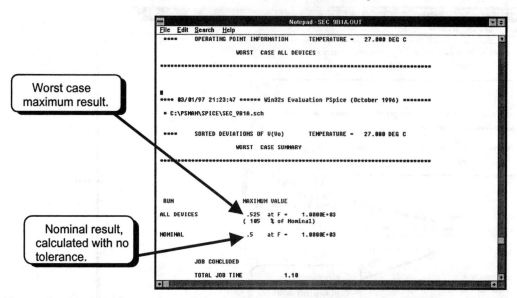

The results show that the maximum value is *.525* and the nominal value is *.5*. These are the expected results.

We would now like PSpice to find the minimum value. Obtain the Monte Carlo setup dialog box by selecting **Analysis, Setup**, and then clicking the ***Monte Carlo/Worst Case*** button. To find the worst case minimum value of ***Vo*** we need to change the ***Function*** from ***MAX*** to ***MIN***. Click the *LEFT* mouse button on the circle ○ next to the text ***MIN***. A black dot will appear in the circle ◉, as shown below.

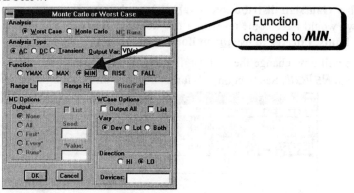

Click the ***OK*** button to accept the changes and then click the ***Close*** button to return to the schematic.

Run PSpice again to find the minimum value. The results will again be stored in the output file. At the end of the output file you will see the results of the Worst Case Analysis:

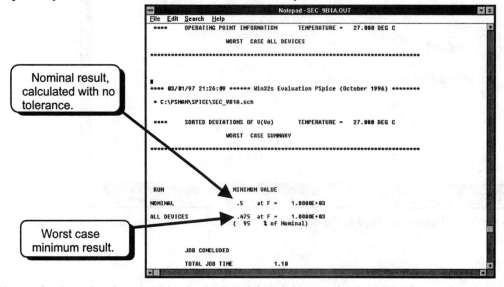

The results show that the nominal value is *.5* and the minimum value is *.475*. These values agree with our calculations.

9.B.2. Voltage Divider Monte Carlo Analysis

The Monte Carlo analysis is used to answer the question, "What percentage of my circuits will achieve or exceed my specifications?" Usually you would run a Worst Case Analysis to see if all of the circuits pass the specifications. If they all pass, there is no need to run the Monte Carlo analysis. If they do not all pass, the Monte Carlo analysis is used to estimate what percentage of the circuits will pass.

An example would be the gain of our voltage divider. We may ask the question, "A minimum gain of 0.4 is required; what percentage of the circuits will have a gain of 0.4 or higher?" From the Worst Case Analysis we know that the worst case minimum gain is 0.475, so we know that all of the circuits will achieve the specification. There is no need to run a Monte Carlo analysis to see if the circuit passes this specification since the Worst Case Analysis told us that all of the circuits will achieve the specification. However, we may ask the question, "A minimum gain of 0.49 is required; what percentage of the circuits will achieve the specification?" The Worst Case Analysis told us that some of the circuits may have a gain as low as 0.475. Not all of the circuits may achieve the specification, and we need to run a Monte Carlo analysis to answer the question.

The accuracy of the Monte Carlo analysis depends greatly on knowing the tolerance distributions of your parts. The Gaussian distribution is considered a better model of a part's distribution than the uniform distribution [1]. To make an accurate simulation you should find out the distributions of your parts. If the Gaussian or uniform distributions are not good models for your parts, you may have to make up your own distributions using the PSpice ".Distribution" statement [2]. See the PSpice Circuit Analysis manual for more information on this statement. To illustrate the effect of different tolerance distributions on the Monte Carlo analysis, we will simulate the voltage divider using resistors with Gaussian and uniform distributions.

9.B.2.a. Voltage Divider Gain Analysis with Uniform Tolerance Distribution

We will run a Monte Carlo analysis on the voltage divider circuit previously described. The circuit and parts are repeated below:

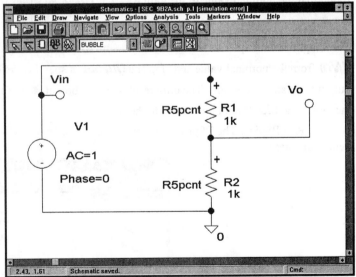

It is important to note that the resistor part used is **R5pcnt** and not R5gauss. The model **R5pcnt** has a 5% uniform distribution. We would like to find the gain of this amplifier, so we must set up an AC Sweep. Obtain the AC Sweep dialog box by selecting **Analysis**, **Setup**, and then click the **AC Sweep** button. Fill in the dialog box as shown:

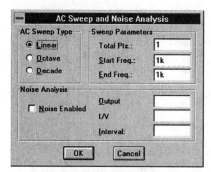

The AC Sweep is set up to simulate the circuit at a single frequency of 1 kHz.

Next, we need to set up the Monte Carlo analysis. Select **Analysis** and **Setup** from the Schematics menu bar, and then click the ***Monte Carlo/Worst Case*** button. Fill in the dialog box as shown:

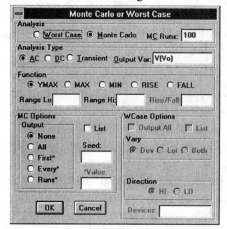

The ***Analysis*** is set up for ***Monte Carlo***. The number of runs is ***100***. The AC Sweep will run 100 times. For each run, each part that has tolerance will have a value randomly chosen within its tolerance range. The ***Analysis Type*** is ***AC*** since we are interested in gain. Note that the AC Sweep must be enabled. The ***Output Var***iable is ***V(Vo)***. The ***Function*** is ***YMAX***. This function instructs PSpice to sort the output according to the maximum difference from the nominal value. We found that the nominal value of the gain was 0.5. ***YMAX*** specifies the output function:

$$f = |V(Vo) - 0.5|$$

The output from PSpice is the function f sorted in descending order. Since we know the nominal value, we can obtain ***V(Vo)*** from the nominal value and f. The ***Output*** is specified as ***None***. This tells PSpice that we do not want intermediate output from any of the runs. The only result we will look at is the function f. Click the ***OK*** button to accept the settings and then click the ***Close*** button to return to the schematic.

Run PSpice. The PSpice simulation window shown below will appear. Note that the screen tells you which pass is being simulated.

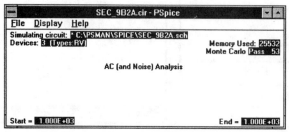

The results of the Monte Carlo analysis are stored in the output file. Examine the output file: select **Analysis** and then **Examine Output** from the Schematics menu bar. As you page down through the output file you will see a screen similar to the one shown below:

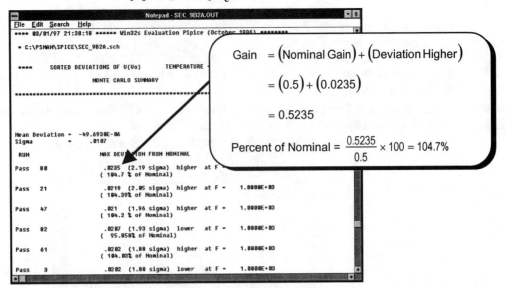

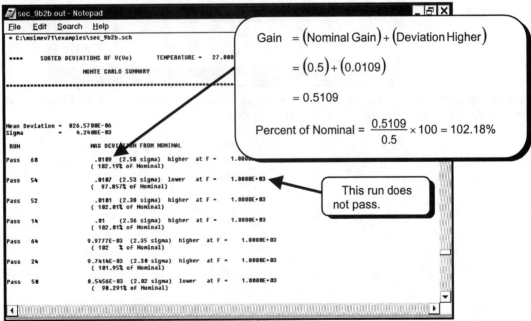

The results are given as deviations from the nominal. From previous runs we know that the nominal value is **0.5**. We wanted to know how many circuits would have a gain of 0.49 or greater. This means that all circuits with a *lower* deviation of more than 0.01 from the nominal will not pass. Circuits with a *higher* deviation have a gain greater than the nominal and pass the specification. All we need to do is count the number of *lower* deviations larger than 0.01. These will be the circuits that do not pass the specification. Since this is a random distribution, your simulation will give slightly different results. In my simulation I found that 17 runs had *lower* deviations greater than 0.01. This means that 17% of my circuits will have a gain less than 0.49. It is important to note that the more runs you do, the more accurate the simulation will be. I chose 100 runs to reduce the simulation time. To get a more accurate estimate, you may wish to perform more than 100 runs. The more runs, the more accurate your results.

9.B.2.b. Voltage Divider Gain Analysis with Gaussian Tolerance Distribution

We will now simulate the voltage divider using resistors with a Gaussian distribution. Change the model reference of the two resistors from **R5pcnt** to **R5gauss**. See Section 7.A for instructions on changing the model reference. The R5gauss resistors have a Gaussian tolerance distribution of ±5% with a standard deviation of 1.25%. The setup is the same as in the previous simulation, so all we have to do is run the simulation. Run PSpice and then examine the output file.

The output is similar to the previous simulation except that the deviations from the nominal value are smaller. Only one run did not pass the specification. The run had a lower deviation of 0.0107 which corresponds to a gain of 0.5 − 0.0107 = 0.4893. Smaller deviations should be expected since, in the Gaussian distribution, the bulk of the resistors were within plus or minus one standard deviation. From the results of the two previous simulations, we conclude that the tolerance distribution has a large impact on the Monte Carlo results.

To get more accurate results, the above simulation was run again with 1000 runs. With 1000 runs, 16 of the runs did not pass the gain specification.

EXERCISE 9-2: What percentage of the voltage divider circuits will have a gain of 0.49 or greater if 10% resistors with a Gaussian distribution are used?

SOLUTION: First, create a model with a 10% Gaussian distribution. See **EXERCISE 7-3** on page 341 to change the models of the resistors to 10%. The model for a 10% resistor with a Gaussian distribution is:

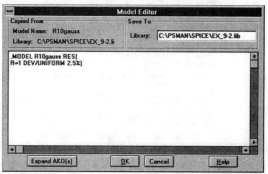

Remember that in a Gaussian distribution, the tolerance specified is the standard deviation σ, and the full distribution extends $\pm 4\sigma$. Change both resistors to R10gauss:

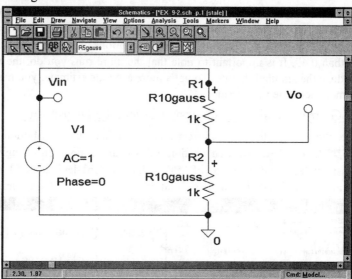

Run the simulation as shown in the previous section. In the output file, count the number of runs that have a lower deviation greater than 0.01. The results below are for 100 Monte Carlo runs:

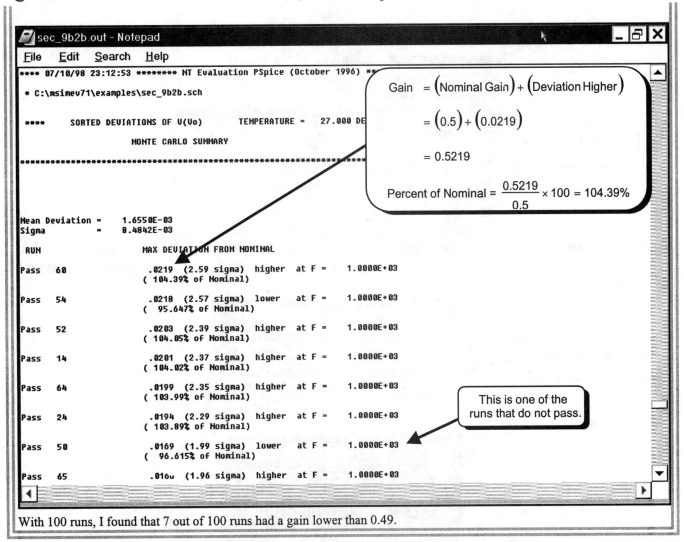

Gain $= \left(\text{Nominal Gain}\right) + \left(\text{Deviation Higher}\right)$

$= \left(0.5\right) + \left(0.0219\right)$

$= 0.5219$

Percent of Nominal $= \dfrac{0.5219}{0.5} \times 100 = 104.39\%$

```
**** 07/10/98 23:12:53 ******** NT Evaluation PSpice (October 1996) **
 * C:\msimev71\examples\sec_9b2b.sch

 ****    SORTED DEVIATIONS OF V(Vo)      TEMPERATURE =   27.000 DE
                  MONTE CARLO SUMMARY

 **************************************************************

Mean Deviation =    1.6550E-03
Sigma          =    8.4842E-03

 RUN               MAX DEVIATION FROM NOMINAL

Pass   60              .0219  (2.59 sigma)  higher  at F =   1.0000E+03
                  ( 104.39% of Nominal)

Pass   54              .0218  (2.57 sigma)  lower   at F =   1.0000E+03
                  ( 95.647% of Nominal)

Pass   52              .0203  (2.39 sigma)  higher  at F =   1.0000E+03
                  ( 104.05% of Nominal)

Pass   14              .0201  (2.37 sigma)  higher  at F =   1.0000E+03
                  ( 104.02% of Nominal)

Pass   64              .0199  (2.35 sigma)  higher  at F =   1.0000E+03
                  ( 103.99% of Nominal)

Pass   24              .0194  (2.29 sigma)  higher  at F =   1.0000E+03
                  ( 103.89% of Nominal)

Pass   50              .0169  (1.99 sigma)  lower   at F =   1.0000E+03
                  ( 96.615% of Nominal)

Pass   65              .016o  (1.96 sigma)  higher  at F =   1.0000E+03
```

This is one of the runs that do not pass.

With 100 runs, I found that 7 out of 100 runs had a gain lower than 0.49.

9.B.3. Performance Analysis — Voltage Divider Gain Spread

The performance analysis can be used in conjunction with the Monte Carlo analysis to view the distribution of a parameter as a function of device tolerances. For this example, we will display how the spread of the gain V(Vo)/V1 varies with resistor tolerances. We will use the voltage divider of the previous section and 5% resistors with a uniform distribution:

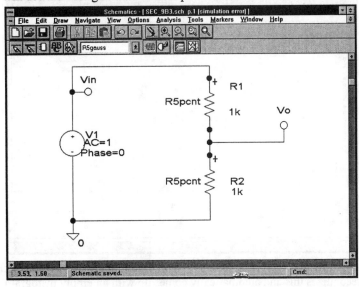

Set up the AC Sweep and Monte Carlo analysis as in the previous section. For this simulation, set the number of Monte Carlo runs to 399. This is the largest number of runs we can view with Probe.

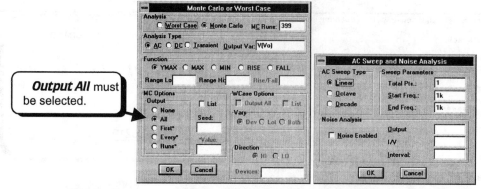

Output All must be selected.

Note that in the **MC Options** of the **Monte Carlo** setup dialog box, **Output All** is selected. If this option is not selected, you will not see any results in the performance analysis. Make sure that both the **AC Sweep** and the **Monte Carlo/Worst Case An**alysis are enabled:

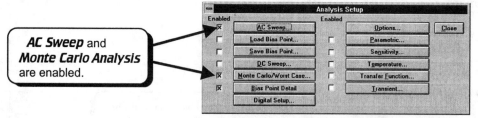

AC Sweep and **Monte Carlo Analysis** are enabled.

Simulate the circuit and then run Probe :

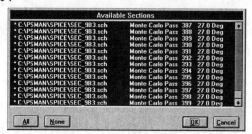

By default, Probe selects all of the runs. There are a total of 400 runs, the nominal run plus the 399 Monte Carlo runs. We would like to compose a histogram of all of the results so click the **OK** button to enter Probe and use the 400 selected traces.

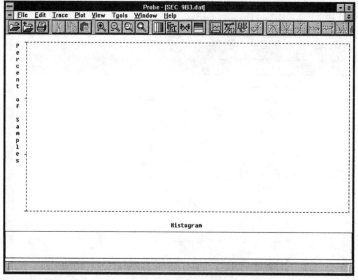

Since we ran a Monte Carlo analysis and each run contains only a single point (the gain at 1 kHz), the only graph we can create is a histogram. Probe recognizes this and automatically comes up with an empty histogram. Add the trace **V(Vo)** :

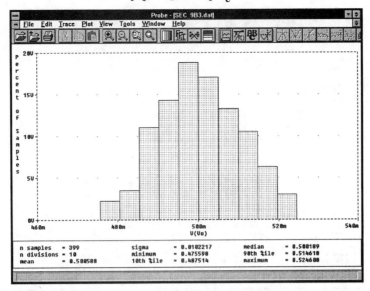

This plot shows us the distributions of gains for the specified tolerances in the circuit. The plot gives us a great deal of information: typical maximum and minimum gain to expect (not necessarily worst case maximum and minimum), mean and median gains, as well as the standard deviation (sigma).

This histogram has 10 bars (**N divisions = 10**). If you would like a histogram with more bars, select **Tools** and then **Options** from the Probe menu:

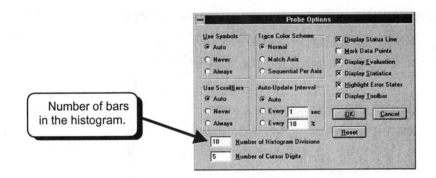

The number of bars in the histogram is set to **10**. We would like to see more divisions, so set the **Number of Histogram Divisions** to **100**:

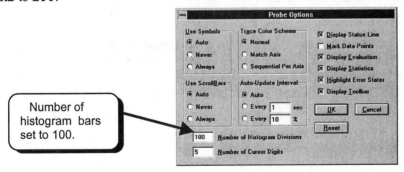

Click the **OK** button to view the graph:

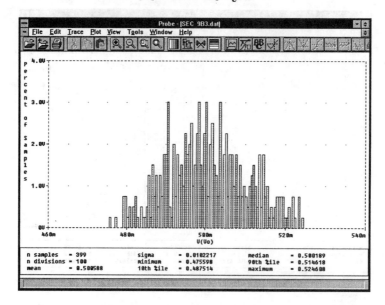

n samples	= 399	sigma	= 0.0102217	median	= 0.500189
n divisions	= 100	minimum	= 0.475598	90th %ile	= 0.514618
mean	= 0.500588	10th %ile	= 0.487514	maximum	= 0.524608

EXERCISE 9-3: Find the histogram for the gain of the voltage divider if 5% resistors with a Gaussian distribution instead of a uniform distribution are used. Compare the standard deviation, and minimum and maximum values of the histogram to the histogram using a uniform distribution.

SOLUTION: Change the model of the resistors to use the model **R5gauss**:

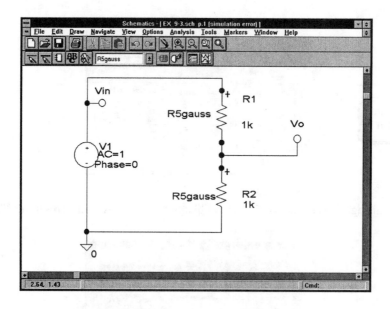

Run the simulation. When Probe runs, select all of the traces and then add the trace V(Vo):

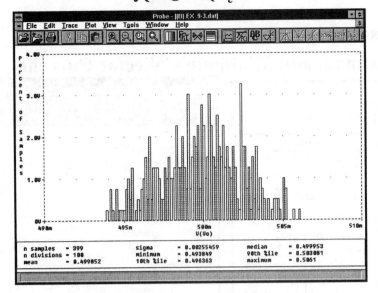

	Uniform Distribution	Gaussian Distribution
Sigma	0.01022	0.00255
Minimum	0.47559	0.4939
Maximum	0.5246	0.5061

9.B.4. Voltage Divider Summary

We can draw a few conclusions about using the Monte Carlo and Worst Case Analyses from the results of the voltage divider circuit.

1. For Worst Case Analyses, use uniform distributions. In a uniform distribution, PSpice uses the absolute limits of the distribution to find the worst case limits. In a Gaussian distribution, PSpice uses $\pm 3\sigma$ to calculate the worst case limits, even though the part can have a maximum deviation of $\pm 4\sigma$.

2. In the Monte Carlo analysis the tolerance distribution has a major effect on the results. It is best to know the distribution of your parts before you trust the results of the Monte Carlo analysis. You can use the PSpice ".Distribution" statement to define your own probability distributions. If you do not know the distribution of your parts, the Gaussian distribution is a better representation of parts than the uniform distribution.

3. When using a Gaussian distribution in a model, the specified deviation is one standard deviation, σ. The limits of the distribution are $\pm 4\sigma$. If you have $\sigma > 25\%$, the specified parameter could become negative.

9.C. BJT Bias Analysis
9.C.1. BJT Maximum and Minimum Collector Current

+/− 1% 1k R1 **R1pcnt**	+/− 5% 1k R1 **R5pcnt**
Resistor with 1% uniform tolerance distribution	Resistor with 5% uniform tolerance distribution

Q1 qbf **QBF** Small-signal NPN BJT with $350 \geq \beta_F \geq 50$	V1 AC=1 Phase=0 **VAC** AC voltage source	C **C** Capacitor
+ − **VDC** DC source	0 **AGND** Ground	**BUBBLE** Bubble

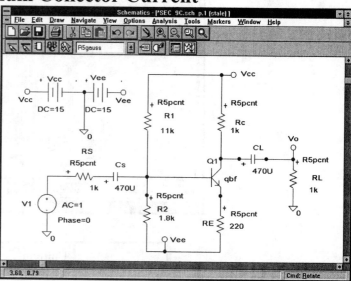

In amplifier design it is important to know how your bias will change with device tolerances. In this section we will find the minimum and maximum collector current of a BJT when we include variations in the transistor current gain, β_F, and resistor tolerances. The circuit above was previously simulated in the Transient Analysis and AC Sweep parts. We will use the same resistor values as before, but we will change the resistor models to include tolerance. The BJT is also changed to the model QBf. This model allows β_F to have a uniform distribution between 50 and 350.

Note that all resistors have ±5% tolerance. All tolerance distributions are uniform. In addition, note that all resistors in the schematic have a plus sign at one of the terminals. This plus sign indicates PSpice's designation of the positive current reference of the device. This reference becomes important if we wish to know the current through a two-terminal device in PSpice. If we wish to know the current through a resistor, Rc, for example, we would specify I(Rc). This text string does not specify a direction for the current, so we need to indicate a positive direction of current in the schematic. The current direction is specified by the plus sign. Current is positive when it enters the positive terminal of the device, as shown in Figure 9-1.

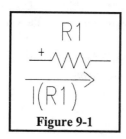

Figure 9-1

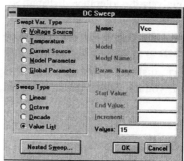

In the previous schematic, the plus sign on Rc is at the top of the resistor. This indicates that I(Rc) will be positive if it flows down. Remember, by definition, the collector current Ic is positive when it enters the collector terminal. Thus, I(Rc) = Ic. If the plus sign were at the bottom of Rc, then I(Rc) would equal –Ic.

The bias collector current is a DC value, so we must run a DC Sweep to find its minimum or maximum value. Obtain the DC Sweep dialog box by selecting **Analysis**, **Setup**, and then clicking the **DC Sweep** button. Fill in the dialog box as shown above, at the right. The dialog box is set up to simulate the circuit at Vcc=15 V. This was already specified in the circuit and appears to be redundant. It must be specified because the Worst Case Analysis requires an AC Sweep, a DC Sweep, or a Transient Analysis to be enabled. Since we are interested in the DC collector current we must enable a DC Sweep. Click the **OK** button to accept the settings.

Next we need to set up the Worst Case Analysis. Click the **Monte Carlo/Worst Case** button. Fill in the dialog box as shown:

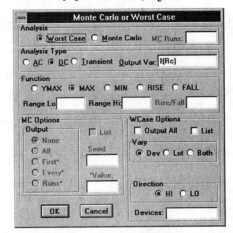

The dialog box is set up to find the worst case maximum value of *I(RC)*. Since the current through Rc is the collector current, we are asking for the Worst Case maximum value of the collector current.[*]

Run PSpice. The results are saved in the output file. Open the output file by selecting **Analysis**, and then **Examine Output**. At the bottom of the output file you will see the results of the Worst Case Analysis:

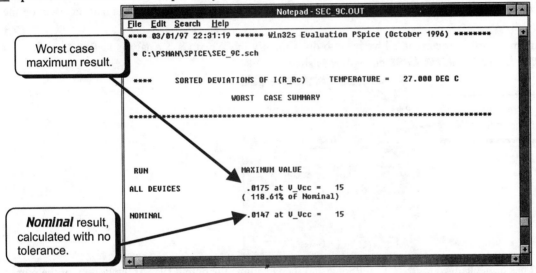

The results show us that the nominal value of the collector current is **14.7** mA and the maximum value is **17.5** mA.

Note: When you run the simulation you may come up with negative values for the current through Rc. This result is not wrong and is due to the polarity reference of the resistor. The resistor graphic has a positive reference, indicated by the plus sign. If the plus sign for Rc is at the bottom of the graphic, you will get negative values for I(Rc) because current is flowing down into the collector terminal. If you do not like negative values for the current, rotate the Rc graphic by 180 degrees, until the plus sign appears at the top of the resistor graphic. Also note that PSpice finds the numerical minimum and maximum values of the specified parameter. Negative currents for Rc switch the interpretation of minimum and maximum as well.

The output file also contains information on how each element was changed to achieve the result. If you page up in the output file, you will see the text shown below:

---

[*]We could also have asked for Ic(Q1). However, I(Rc) does not yield the same answer as Ic(Q1). I(Rc) appears to agree with the calculated results.

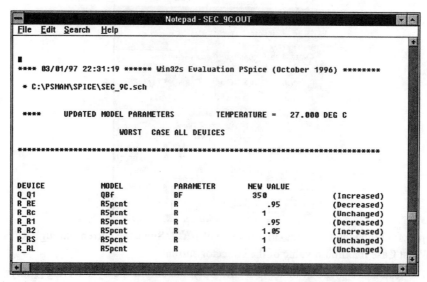

This information tells us that to achieve maximum collector current, β_f was increased to 350, **RE** was decreased by 5%, **R1** was decreased by 5%, and **R2** was increased by 5%. The other resistors had tolerance, but their values had no effect on the collector current.

To find the minimum collector current, all we have to do is modify the Worst Case Analysis to find the minimum value. Modify the **Monte Carlo or Worst Case** dialog box as shown:

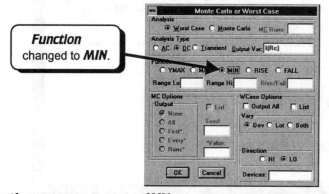

Note that the **Function** has been changed to **MIN**. Run PSpice and then examine the output file. At the bottom of the output file you will see the results:

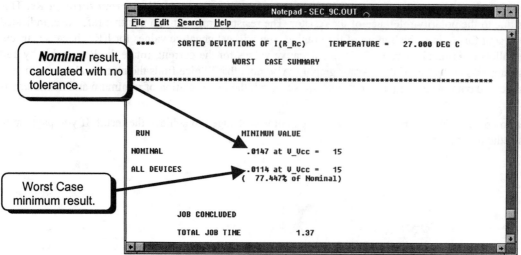

The results show that the minimum collector current is **11.4** mA.

EXERCISE 9-4: Find the maximum and minimum collector current of the circuit in this section if the base resistors have 1% tolerance rather than 5% tolerance.

SOLUTION:

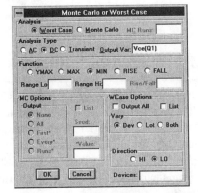

Set up the DC Sweep and the Monte Carlo analyses as in the previous sections. When we run the simulations, we find that the maximum collector current is 16.1 mA and the minimum collector current is 12.4 mA.

9.C.2. BJT Minimum V_{CE}

When biasing a BJT, we are also interested in the collector to emitter voltage, V_{CE}. The minimum or maximum value of V_{CE} can also be easily found using the Worst Case Analysis. We can use the same setup that was used to find the collector current. All we have to do is modify the **Monte Carlo or Worst Case** dialog box. Fill in the dialog box as shown below:

Note that the **Output Var**iable is **Vce(Q1)**, the collector-emitter voltage of **Q1**. The **Function** is **MIN**, so we are asking for the minimum value of V_{CE}. Run the analysis. The results are given at the end of the output file:

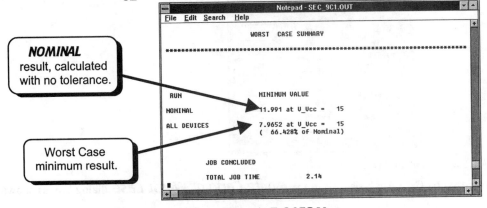

The results show that the minimum value of V_{CE} is **7.9652** V.

EXERCISE 9-5: Find the minimum value of V_{CE} of the circuit in this section if the base resistors have a 1% tolerance rather than a 5% tolerance.

SOLUTION: Use the circuit of **EXERCISE 9-4**. The result of the simulation shows that the minimum value of V_{CE} is 9.7233 V.

9.D. BJT Amplifier Minimum and Maximum Gain

We will now find the minimum and maximum gain of the amplifier of the previous section, shown on page 416. The amplifier circuit is repeated below:

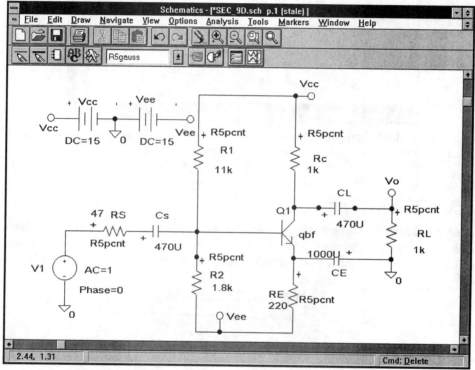

Note that Rs has been changed to 47 Ω and an emitter bypass capacitor (CE) has been added. The frequency response of this amplifier was found in Section 5.D on page 219. We found the mid-band gain to be 45.7 dB and the upper and lower 3 dB frequencies to be 66 Hz and 6.3 MHz, respectively. The amplifier above is the same as that in Section 5.D except that the BJT is described by the model **qbf** rather than Q2N3904. The model **qbf** has $350 \geq \beta_F \geq 50$ and a nominal value of $\beta_F=200$. The Q2N3904 model has no tolerance and a nominal value of $\beta_F=416$. Thus, we should expect the nominal value of the gain of the amplifier above to be slightly different from the nominal gain of the amplifier from Section 5.D.

We would like to find the minimum and maximum gain at mid-band. We will set up an AC Sweep at 1 kHz. Fill in the AC Sweep dialog box as shown:

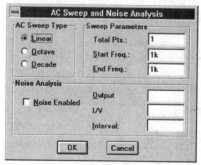

The dialog box is set to run an AC Sweep at a single frequency of 1 kHz.

Next we need to set up the Worst Case Analysis. Fill in the **Monte Carlo or Worst Case** dialog box as shown below.

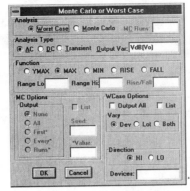

The dialog box is set up to find the Worst Case maximum value of the output variable **VdB(Vo)**. VdB is the voltage at the specified node in decibels, $VdB(V_o)=20\log_{10}(V_o)$. Since our only input is **V1** and the magnitude of **V1** is **1**,

$$VdB(V_o) = 20\log_{10}(V_o) = 20\log_{10}\left(\frac{V_o}{V_1}\right)$$

Thus, **VdB(Vo)** gives us the gain of our amplifier in decibels.

If you ran the DC Sweep of the previous section, you may want to disable the DC Sweep. The DC Sweep will not affect the results of the Worst Case Analysis, but it will fill the output file with unwanted results. Disable the DC Sweep by removing the **X** from the box next to the text **DC Sweep**, as shown below:

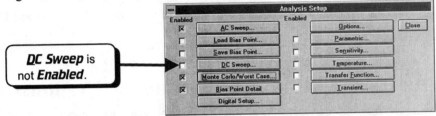

Run PSpice and then examine the output file. The results of the Worst Case Analysis are shown at the end of the file:

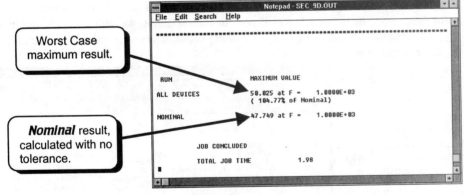

The results show a maximum gain of **50.025** dB.

To find the minimum gain, change the **Function** in the **Monte Carlo or Worst Case** dialog box from **MAX** to **MIN**:

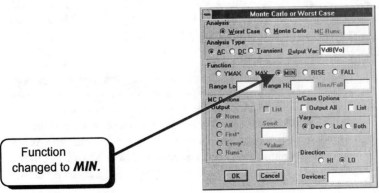

Run PSpice and then examine the output file. The results of the analysis are given at the bottom of the output file:

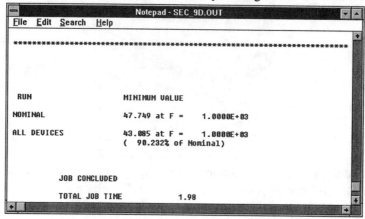

The results show a minimum gain of **43.085** dB.

EXERCISE 9-6: Find the minimum and maximum gain of the circuit in this section if the base resistors have 1% tolerance rather than 5% tolerance.

SOLUTION: The results of the simulation show that the minimum and maximum values of the gain are 43.591 dB and 49.356 dB.

9.E. Performance Analysis — Amplifier Frequency Response

When we design a circuit with tolerance, we may sometimes want to find the worst case upper or lower 3 dB frequency with component tolerances. Unfortunately, calculating a 3 dB frequency requires that we find the mid-band gain and then find the frequency where the gain is 3 dB less than the mid-band. This type of calculation cannot be specified in the Monte Carlo and Worst Case dialog box. However, we can run a Monte Carlo analysis and then determine the 3 dB frequency using the Performance Analysis capabilities available in Probe. In this example, we will illustrate finding the maximum lower 3 dB frequency (F_L), minimum upper 3 dB frequency (F_H), and maximum and minimum bandwidth (F_H-F_L) for a common-emitter amplifier. Wire the circuit below:

+80% − 20% 1U C1 **Cap20_80** Capacitor with −20%, +80% tolerance	+/− 5% 1k R1 **R5pcnt** Resistor with 1% uniform tolerance distribution
Q3 q2n3904b **Q2N3904B** Small-signal NPN BJT with $350 \geq \beta_F \geq 50$	V1 AC=1 Phase=0 **VAC** AC voltage source
VDC DC source	**AGND** Ground
BUBBLE Bubble	**Voltage Level Marker** In Schematics type **CTRL-M** to place marker

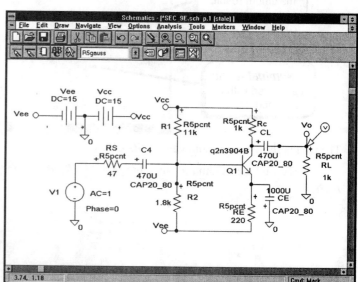

This circuit contains 5% resistors, capacitors with +80%, −20% tolerance, and a 2N3904 transistor with tolerance in β. All distributions are uniform. We will run an AC Sweep in conjunction with the Monte Carlo analysis. Fill in the AC Sweep and Monte Carlo dialog boxes as shown below:

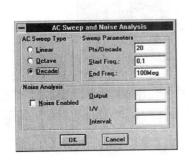

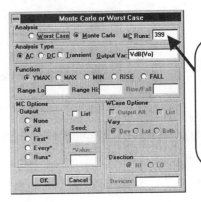

You may wish to set the number of Monte Carlo runs to 50 to reduce simulation time. 399 was chosen to get a better sampling for the distribution.

The AC Sweep dialog box is set up to sweep frequency from 0.1 Hz to 100 MHz with 20 points per decade. Note that the number of Monte Carlo runs is set to 399. This will take a large amount of simulation time but will give us more accurate results. You may wish to reduce the number of runs in order to reduce the simulation time. The maximum number of data segments allowed by Probe is 400. Each Monte Carlo run will give us a data segment. 399 runs will reach the maximum number of segments because the total number of runs is the nominal run plus the number of Monte Carlo runs.

Make sure that both the **AC Sweep** and the **Monte Carlo / Worst Case** Analysis are **Enabled**:

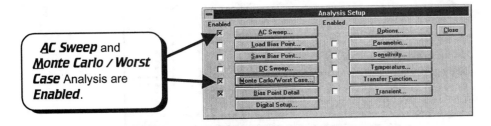

AC Sweep and **Monte Carlo / Worst Case** Analysis are **Enabled**.

Since we are running 400 simulations and the circuit is fairly large, the data file created by PSpice could be huge if we collect voltage and current data for all circuit elements. To reduce the size of the data file we will collect data only for the output voltage. This can be done by placing a marker at the output. In the circuit you may notice that the pointer of a marker, , is pointing to the output. To place a marker, select **Markers** and then **Mark Voltage/Level** from the Schematics menus, or type CTRL-M.

After the marker is placed, we must tell Probe to collect data only at the markers. Select **Analysis** and then **Probe Setup** from the Schematics menu:

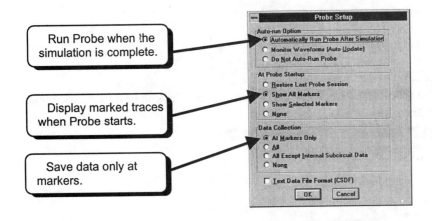

Run Probe when the simulation is complete.

Display marked traces when Probe starts.

Save data only at markers.

Note that the dialog box is set up so that data will be collected only at the markers. Also note that when Probe starts, the traces indicated by the markers will automatically be displayed. Run the simulation. When the simulation is complete, Probe will automatically run:

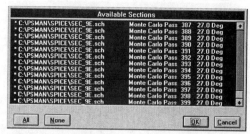

Click the **OK** button to select all of the traces. The gain plot for all traces will be displayed by Probe :

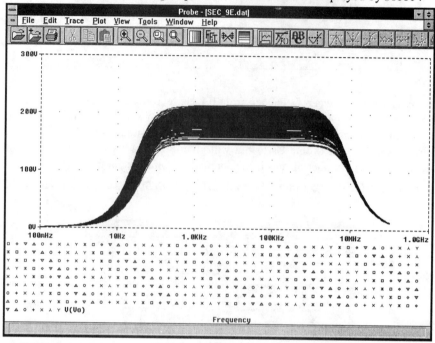

Next, we would like to display a histogram of the lower 3 dB frequency. We will first delete the displayed trace. Click the **LEFT** mouse button on the text **V(Vo)** to select the trace. When the trace is selected, the text **V(Vo)** will be highlighted in red. Press the **DELETE** key to delete the trace. (Your trace will most likely be labeled differently than **V(Vo)**.) A blank Probe screen will result.

To plot a function like bandwidth, we must use the Performance Analysis. Select **Plot** and then **X Axis Settings**:

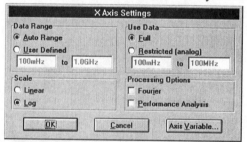

Under **Processing Options** we notice that **Performance Analysis** is not selected. Click the **LEFT** mouse button on the square □ next to **Performance Analysis**. It should fill with an x, ☒:

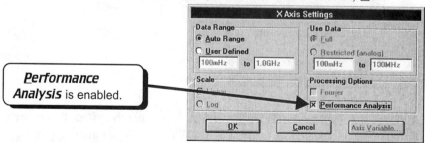

Click the **OK** button. You will return to Probe with a blank histogram displayed:

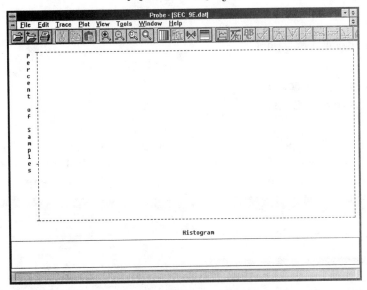

To add a trace, select **Trace** and then **Add**:

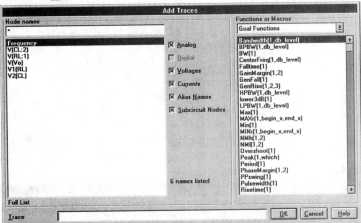

The left pane shows the normal voltage and current traces that we are familiar with. The right pane displays the goal functions. These functions are available using the Performance Analysis. The functions are defined in a file called msim.prb. If you view this file using the Windows Notepad, you will see the function below at the top of the file:

upper3dB(1) = x1

* Find the upper 3 dB frequency

{

1|sfle(max-3dB,n) !1;

}

The name of the function is **upper3dB**. The text **(1)** indicates that it has **1** input argument. **1|sfle** means search the first input forward and find a level. The level we are looking for is 3 dB less than the maximum (**max-3dB**). The **n** means find the specified level when the trace has a negative-going slope. When the point is found, the text **!1** designates its coordinates as x1 and y1. The function returns the x-coordinate of the point (**upper3dB(1) = x1**). The x-axis of a frequency trace is frequency, so this function returns the frequency of the upper 3 dB point.

       A second function is :

lower3dB(1) = x1

* Find the lower 3 dB frequency

{

1|sfle(max-3dB,p) !1;

}

> Indicates a positive-going slope.

This function is similar to the upper3dB function except that it finds the 3 dB point when the trace has a positive-going slope. This will mark the coordinates of the lower 3 dB point. A third function is:

BW(1) = x2-x1

* Find the 3 dB bandwidth of a signal

{

1|Search forward level(max-3dB,p) !1

Search forward level(max-3dB,n) !2;

}

This function finds the lower 3 dB frequency and marks the coordinates of the point x1 and y1. It then finds the upper 3 dB frequency and marks the coordinates of the point x2 and y2. The function returns the bandwidth, **x2 - x1**.

We will first plot a histogram of the lower 3 dB frequency. Enter the trace `lower3dB(V(Vo))`:

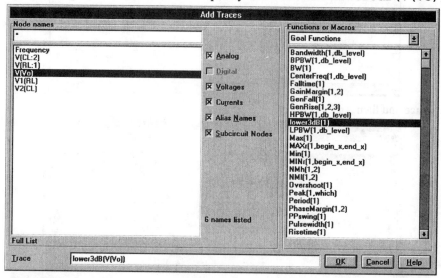

Click the **OK** button to plot the histogram:

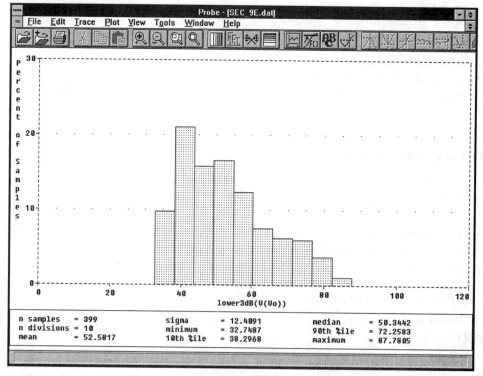

We see that the maximum lower 3 dB frequency is at **87.7805** Hz. To view the upper 3 dB frequencies, add the trace `upper3dB(V(Vo))`:

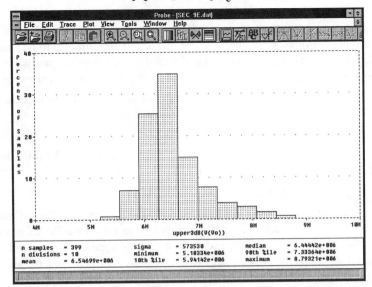

We see that the minimum upper 3 dB frequency is **5.18334** MHz. To view the bandwidth of the amplifier, add the trace `BW(V(Vo))`:

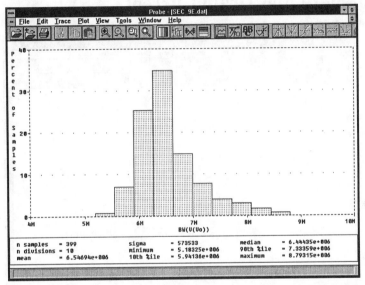

Although the BW function is not too helpful for a circuit of this type, it is very useful for bandpass filters.

9.F. jFET Minimum and Maximum Drain Current

The jFET provided in the libraries is the 2N5951. This jFET has the characteristics:

$$7 \ \text{mA} \leq I_{DSS} \leq 13 \ \text{mA}$$

$$-5 \ \text{V} \leq V_P \leq -2 \ \text{V}$$

The equation that governs the jFET's operation in the saturation region is:

$$I_D = I_{DSS} \left(1 - \frac{V_{GS}}{V_P}\right)^2$$

I_D is maximum when $I_{DSS} = 13$ mA and $V_P = -5$ V. I_D is minimum when $I_{DSS} = 7$ mA and $V_P = -2$ V. We shall let $I_{DSS(max)} = 13$ mA, $I_{DSS(min)} = 7$ mA, $V_{P(max)} = -5$ V, and $V_{P(min)} = -2$ V.

The equation used by PSpice to describe the jFET is:

$$I_D = \beta \left(V_{GS} - V_{TO}\right)^2$$

where $V_{TO} = V_P$ and $\beta = I_{DSS}/(V_P)^2$. When we are running a Worst Case Analysis on a jFET circuit, we would like to let both I_{DSS} and V_P vary at the same time. However, because of the non-linear relationship of β to I_{DSS} and V_P, a spread in the values of I_{DSS} and V_P does not correspond to an equivalent spread in β and V_{TO}. To model the worst case limits of the jFET we will have to make two models, one that corresponds to $I_{DSS(max)}$ and $V_{P(max)}$, and a second that corresponds to $I_{DSS(min)}$ and $V_{P(min)}$.

In the library class.lib there are two jFET models called jMAX and jMIN. These models correspond to the minimum and maximum limits of the model J2n5951. The model jMAX is shown below:

> .model jMAX NJF(Vto=–5 Beta=.52m)

In the jMAX model $V_{TO} = V_{P(max)} = -5$ V, and $\beta = I_{DSS(max)}/(V_{P(max)})^2 = 0.00052$. The model jMIN is:

> .model jMIN NJF(Vto=–2 Beta=1.75m)

In the jMIN model $V_{TO} = V_{P(min)} = -2$ V, and $\beta = I_{DSS(min)}/(V_{P(min)})^2 = 0.00175$.

To find the minimum and maximum of a quantity, you will have to run the Worst Case simulation four times. To find the maximum of the quantity, you will have to run the simulation once with the jMAX model and once with the jMIN model. To find the minimum of the quantity, you will also have to run the simulation once with the jMAX model and once with the jMIN model. If you happen to know which of the two models will give you the maximum or minimum of the quantity you are looking for, you may be able to reduce the number of simulations.

9.F.1. jFET Minimum Bias Drain Current

We would like to find the minimum drain current of the circuit shown below:

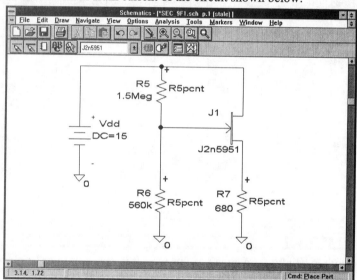

In order to include the tolerance of the jFET we must use the models jMAX and jMIN. Replace the J2n5951 jFET with the jMAX part:

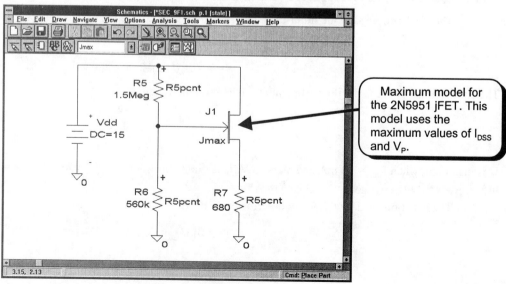

We must now set up a DC Sweep since the bias drain current is a DC quantity. Fill in the DC Sweep dialog box as shown:

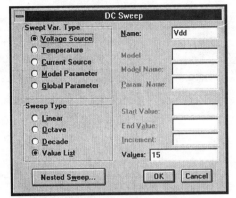

The dialog box is set up to run a DC Sweep with Vdd equal to 15 V. Setting Vdd to 15 V is redundant, but a DC Sweep is required for the Worst Case Analysis and Vdd is the only DC source in the circuit.

Next, we must set up the Worst Case Analysis. Fill in the **Monte Carlo or Worst Case** dialog box as shown:

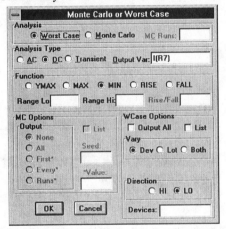

The dialog box is set up to find the worst case minimum value of the quantity **I(R7)**. This is the current through resistor R7 as well as the drain current of device **J1**.*

Run PSpice and then examine the output file. The results are stored at the bottom of the output file. The screen below tells us that the minimum drain current with the model jMAX is **7.1235** mA.

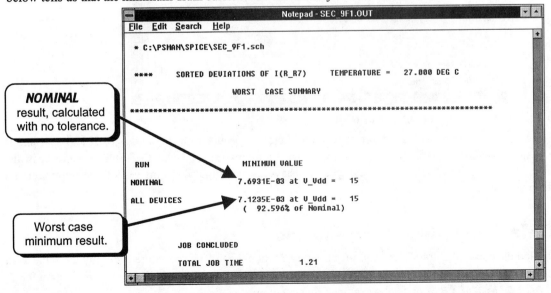

---

*We could also have asked for ID(J1). However, I(R7) does not yield the same answer as ID(J1). I(R7) appears to agree with the calculated results.

We must now run the simulation again with the model jMIN. Change the jFET model from jMAX to jMIN as shown in the circuit below:

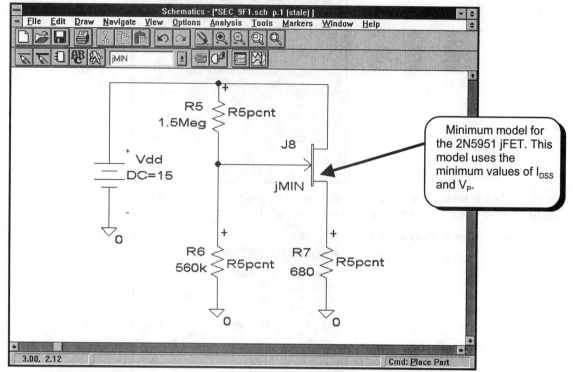

Run PSpice and then examine the output file:

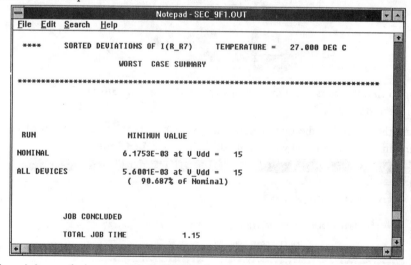

The results show that the minimum drain current is **5.6001** mA.

From these two runs we determined that the minimum drain current occurs with the model jMIN and its value is **5.6001** mA. There is no general rule to determine if the maximum or minimum model of the jFET will give you the maximum or minimum of the output variable. In general, you will have to make multiple runs to find the minimum and multiple runs to find the maximum.

9.F.2. jFET Maximum Bias Drain Current

We will now find the maximum drain current of the circuit in Section 9.F.1. The procedure is exactly the same as in Section 9.F.1, except that we must change the worst case *Function* to *MAX*. Follow the procedure given in Section 9.F.1, but use the *Monte Carlo or Worst Case* dialog box settings shown below:

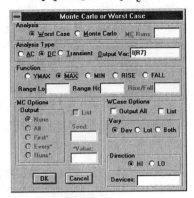

The only difference between this dialog box and the dialog box in Section 9.F.1 is that the **Function** is set to **MAX**. If you run the simulation with the jMAX and jMIN models as shown in Section 9.F.1, you should get the results shown below:

Maximum Drain Current	
Model	Drain Current (mA)
jMAX	8.3277
jMIN	6.8221

EXERCISE 9-7: In the circuit below, the jFET has the following parameters: 6 mA ≤ IDSS ≤ 14 mA, and –6 V ≤ VP ≤ –1 V. Find the minimum and maximum drain current.

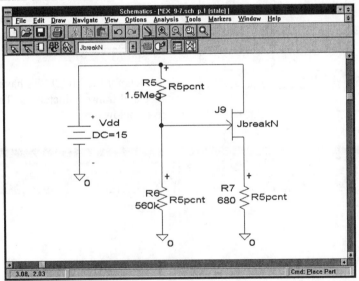

SOLUTION: Create two new models. Let model Jmx use IDSS = 14 mA and VP = –6 V, and let Jmn use IDSS = 6 mA and VP = –1 V. The models are shown below. See Part 7 for instructions on creating models.

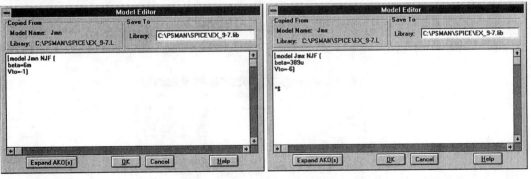

Use these two models and run worst case minimum and maximum simulations with each model in the circuit above. The results show that the minimum drain current is 5.3789 mA and the maximum drain current is 8.7350 mA.

9.G. Performance Analysis — Inverter Switching Speed

When designing digital circuits we are usually concerned with the rise and fall times of the design, given device tolerances. The example given here is for a CMOS inverter, but the procedure used can be applied to any switching circuit with device tolerances. Wire the circuit below:

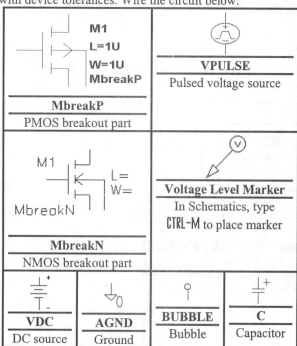

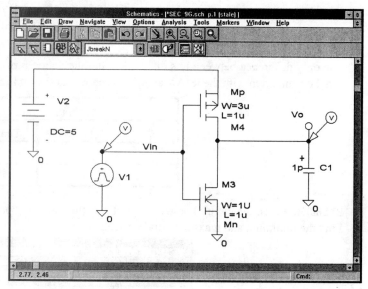

The models for the MOSFETs have tolerances in their threshold voltages and transconductances. The models for the MOSFETs are given below:

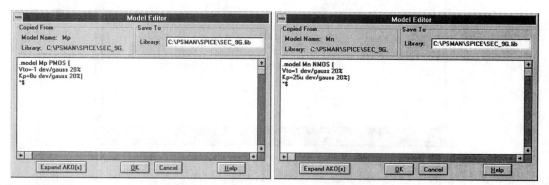

See Part 7 for instructions on creating models.

We would like to see how the rise and fall times vary with random device tolerances. We must set up the Transient Analysis to view waveforms versus time, and the Monte Carlo analysis to allow for device variations. First we will look at the input pulsed waveform. The attributes for **V1** are:

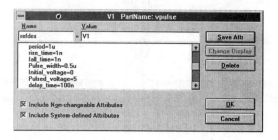

The attributes specify a 0 to 5 V pulse with a 500 ns pulse width and 1 µs period. A delay time of 100 ns is specified so that the pulse does not start until 100 ns after the beginning of the simulation. The rise and fall times are 1 ns. We would like to set up a Transient Analysis to simulate one cycle of the input:

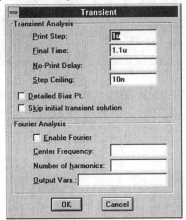

Next, we need to set up the Monte Carlo analysis. We will run the simulation 399 times to get a good sampling. The dialog box below selects the Monte Carlo analysis and specifies 399 Transient Analysis runs:

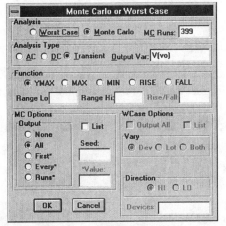

      Transient Analyses create a large amount of data. Even though the circuit is fairly small, running the analysis 399 times will generate a large amount of data. To reduce the size of the data file, markers are added to the circuit and the **Probe Setup** is set to collect data at markers only. To specify the Probe Setup, select **Analysis** and then **Pro<u>b</u>e Setup** from the Schematics menus. Fill in the options as shown:

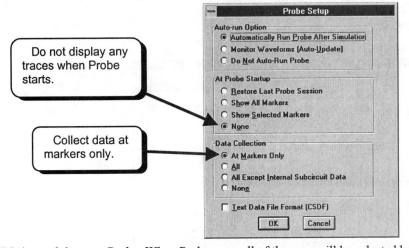

Run PSpice and then run Probe. When Probe runs, all of the runs will be selected by default:

Click the **OK** button to enter Probe :

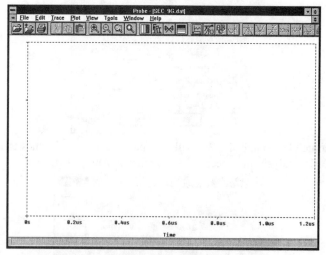

Since we have so many traces, we will instruct Probe not to use symbols to mark the traces. Select **Tools** and then **Options** from the Probe menus. Fill in the dialog box as shown:

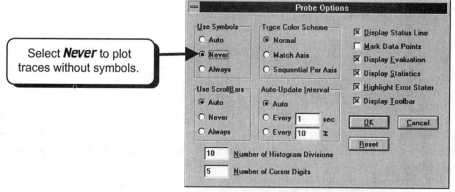

Click the **OK** button and add the trace **V(Vin)** :

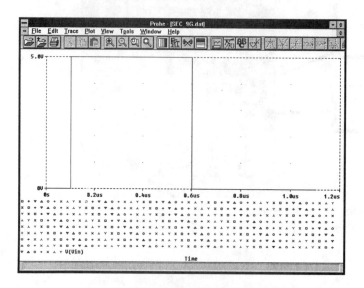

This is our input trace. We see that it has no variation with device tolerances, as should be expected. We would like to display the output on a different window. Select **Window** and then **New** to open a new window, and then add the trace V(Vo):

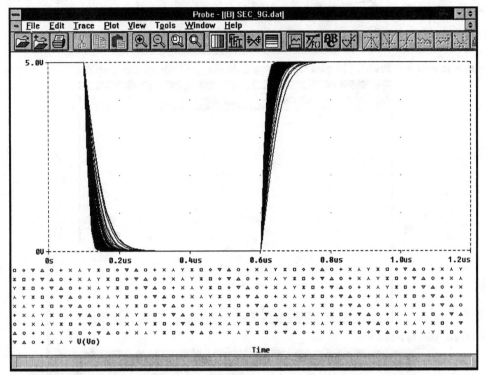

This is how the output pulse looks with time.

We would like to find out what the minimum and maximum rise and fall times are from this data. We can view the results as a histogram. First, add a new window by selecting **Window** and then **New**:

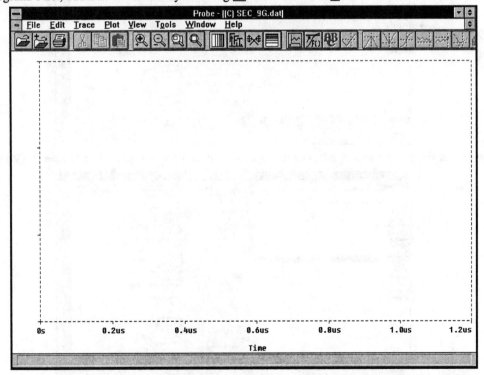

We must use the Performance Analysis to plot the information in which we are interested. Select **Plot** and then **X Axis Settings** to obtain the *X Axis Settings* dialog box, and then specify *Performance Analysis* as follows:

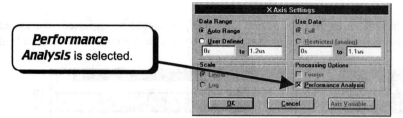

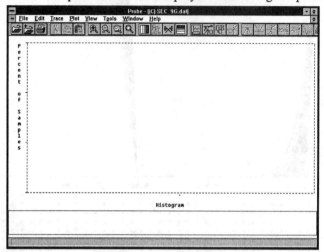

Click the **OK** button to return to Probe. The plot window will display a blank histogram plot:

Select **Trace** and then **Add** to add a trace:

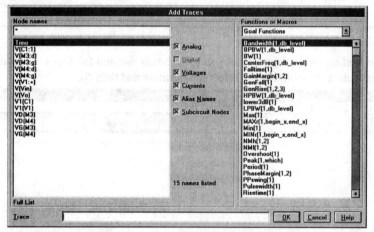

The screen shows all of the Performance Analysis traces available to us. Enter the trace **Risetime(V(Vo))**:

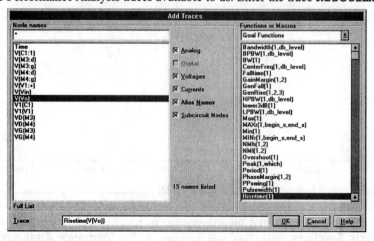

Click the **OK** button to plot the histogram:

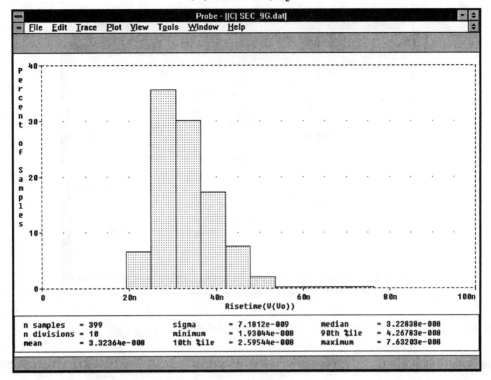

We see that the minimum and maximum rise times are 19.3 ns and 76.3 ns. To find the fall times, add the trace **Falltime(V(Vo))**:

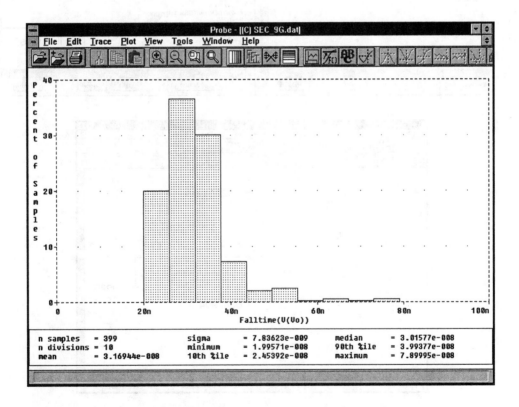

The minimum and maximum fall times are approximately 20 ns and 79 ns, respectively.

EXERCISE 9-8: Find the minimum and maximum rise and fall times for the BJT inverter studied in Section 6.K. Let the resistors have 20% Gaussian distributions and let β_f have a uniform distribution from 50 to 350. Start with the circuit from Section 6.K, but use the 5% resistor model and the 2n3904 BJT model (R5pcnt and Q2n3904):

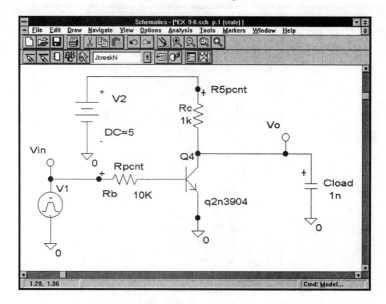

SOLUTION: Change the resistor and BJT models to the ones shown below:

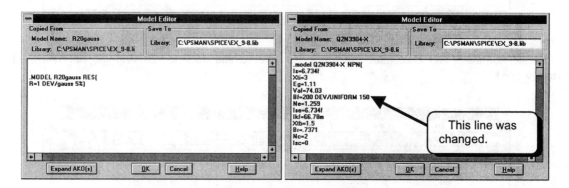

The results of the simulation yield the histograms below:

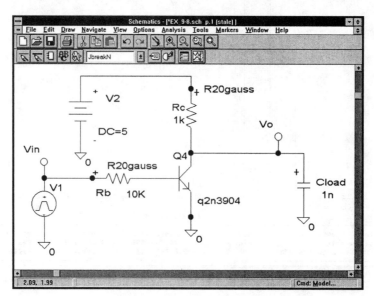

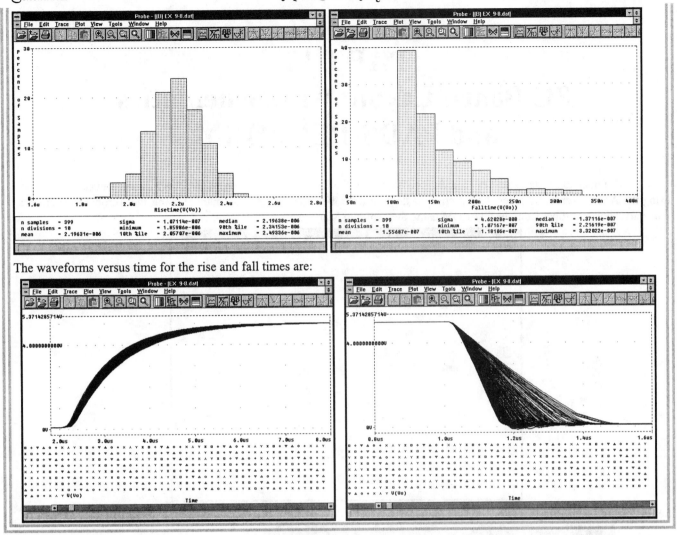

The waveforms versus time for the rise and fall times are:

9.H. Summary

- The Monte Carlo analyses (Monte Carlo and Worst Case) are used to determine how tolerance in component values will affect circuit performance.
- The Worst Case Analysis determines the absolute maximum or minimum value of a parameter for given component tolerances.
- If the Worst Case Analysis determines that not all circuits will pass a specified performance parameter, the Monte Carlo analysis may be used to estimate what percentage of the circuits will pass.
- The uniform distribution is the easiest to use for the Worst Case Analysis since this distribution specifies the upper and lower limits of a component's tolerance.
- The Gaussian distribution is the best to use with the Monte Carlo analysis since it more closely matches the actual distribution of components.
- For accurate results in a Monte Carlo simulation, the actual distribution of the components should be used rather than the Gaussian or uniform distributions. Use the PSpice ".Distribution" command to specify a nonstandard distribution.
- The Performance Analysis can be used together with the Monte Carlo Analysis to display a histogram. The histogram will display properties such as minimum and maximum values of an output versus random variations. The Performance Analysis can find the minimum and maximum values of quantities not available with the Worst Case Analysis, such as bandwidth and rise time.

9.I. Bibliography

[1]   P.R. Gray and R.G. Meyer. *Analysis and Design of Analog Integrated Circuits*, 2nd ed. New York: Wiley, 1984, p. 225.

[2]   MicroSim Corporation. *MicroSim PSpice A/D- Circuit Analysis Software - Reference Manual*, Version 7.1, Irvine, CA, October, 1996, p. 1-10.

PART 10
PC Board Layout with Schematics and PADS-PERFORM

The manual for designing PC boards is contained on the CD-ROM that accompanies this manual, and can be viewed using the Adobe Acrobat Reader. This section of the text describes how to view the manual using the Acrobat Reader. We will start Adobe Acrobat from the CD-ROM using the Program Manager:

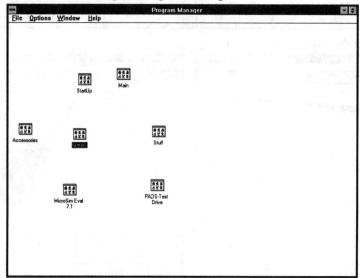

Select **File** and then **Run** from the Program Manager menus:

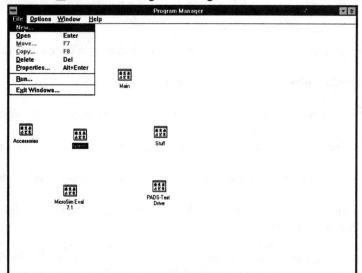

The Adobe Acrobat reader is located on the CD-ROM in directory Adobe. The name of the file is acroread.exe. Assuming that your CD-ROM is labeled d:, enter the text d:\adobe\acroread.exe. My CD-ROM is labeled as F: drive.

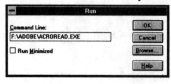

Click the **OK** button to run Adobe Acrobat Reader:

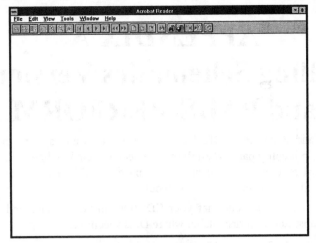

We now need to load the PC board layout file. Select **File** and the **Open** from the Acrobat Reader Menus:

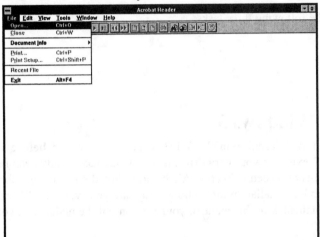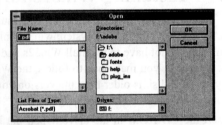

The name of the file we wish to load is named pcb_lay.pdf and is located in the root directory of the CD-ROM:

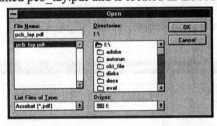

Click the **OK** button to display the file:

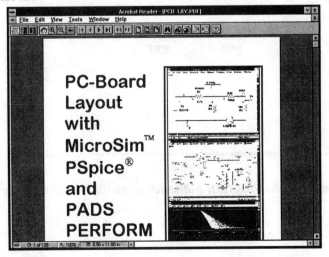

You can now view and print the manual.

APPENDIX A
Installing Schematics Version 7.1
and PADS-PERFORM

Schematics version 7.1 and PADS-PERFORM can be installed on any system that runs Windows 3.1. A color monitor is not required but will help identify parts when they have been selected. Schematics requires Win32s to be installed on your system. PADS-PERFORM requires 7 MB of hard disk space, 16 MB of RAM, and a math coprocessor. A color monitor is required to distinguish different levels of a board layout.

In the installation instructions we assume that your CD-ROM drive is designated as drive D:. If your CD-ROM drive has a different drive label, substitute your drive label where D: is specified.

A.1.   Installing Schematics

To install Schematics properly you must do the following three tasks:

1. Install Win32s.
2. Install the Schematics software.
3. Install the PSpice libraries for this text.

These tasks are outlined in sections A.1.a, A.1.b, and A.1.c.

A.1.a.   Installing Win32s for Windows 3.x

To install PSpice with Schematics you must first install Win32s. Win32s must be installed **before** you install Schematics. If the file "c:\windows\system\win32s.ini" exists on your hard drive, then Win32s has already been installed on your system. You do not have to reinstall Win32s and may proceed to Section A.1.b and follow the instructions for installing Schematics. However, you may wish to follow the Win32s installation instructions to update your version of Win32s. If the version on your hard disk is older than the version contained with this manual, your version will be updated. If your version is newer, your system will remain unchanged.

Insert the CD-ROM in the drive. Open the Program Manager by double clicking the *LEFT* mouse button on the

Program Manager icon, 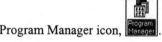. The **Program Manager** window will open as shown:

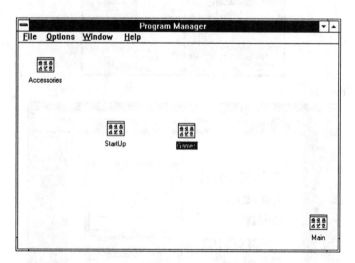

Click the *LEFT* mouse button on **File** in the **Program Manager** menu bar. The **File** pull-down menu will appear:

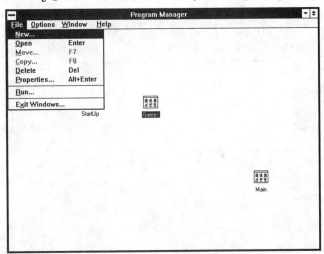

Click the *LEFT* mouse button on the **Run** menu selection. The **Run** dialog box will appear:

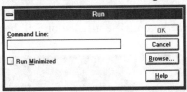

Enter the text **d:\ole32s\130\disk1\setup.exe** in the text field below the text **Command Line**:

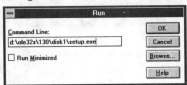

Click the **OK** button to start the setup program. A dialog box will appear, notifying you that the Win32s Setup is being initialized.

After a few moments the dialog box below will appear:

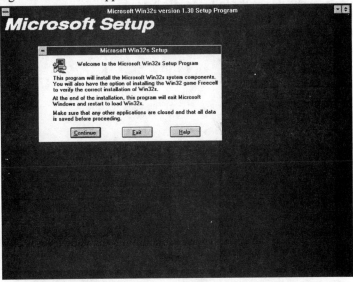

Click the *LEFT* mouse button on the **Continue** button:

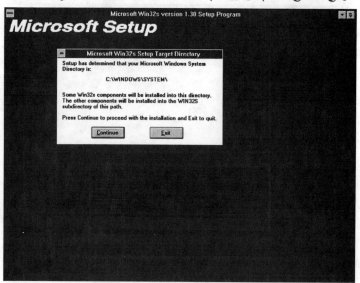

Click the *LEFT* mouse button on the **Continue** button. The installation will begin. A dialog box will appear, showing you the progress of the installation:

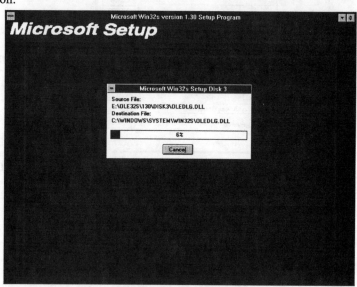

When the installation is complete, you should see the following dialog box:

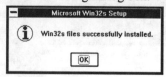

Click the *OK* button.

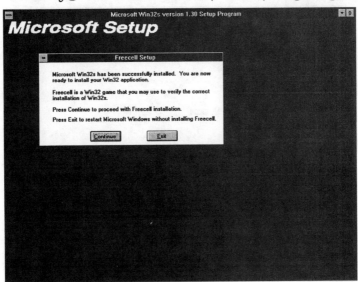

The setup program is asking us if we wish to install a game called *Freecell*. This is a game that uses Win32s. Click the *Exit* button to bypass installing the game and continue with the installation process for Win32s.

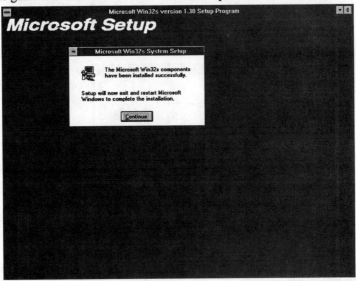

Click the *Continue* button to exit the installation program and restart Windows. **After Windows has been restarted, you may continue with the installation below.**

A.1.b.   Installing the Schematics Software

We will start with the Program Manager:

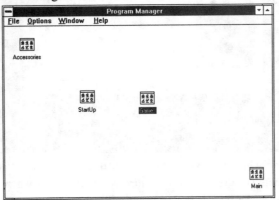

Click the *LEFT* mouse button on <u>F</u>ile in the *Program Manager* menu bar. The <u>F</u>ile pull-down menu will appear:

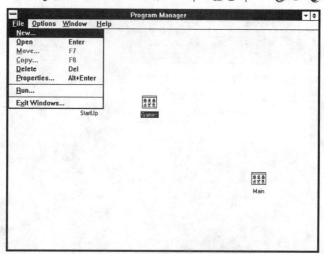

Click the *LEFT* mouse button on the **Run** menu selection. The *Run* dialog box will appear:

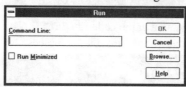

Enter the text **d:setup** in the text field below the text *Command Line*:

Click the *OK* button to run the setup program.

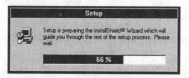

After a few seconds, the dialog box will disappear and the *Setup* dialog box will appear:

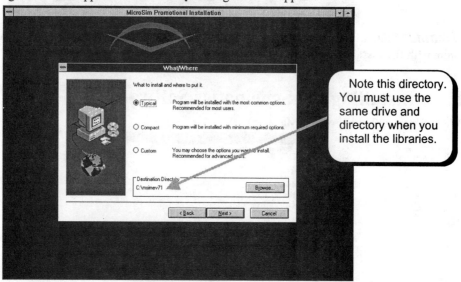

This dialog box allows you to change where you would like to install the Schematics software. If you specify a directory other than *C:\msimev71*, note the name of the directory because you will need it when you install the PSpice libraries. To install PSpice, click the *Next>* button. The installation program will ask you to specify the name of the group where you want to place the icons:

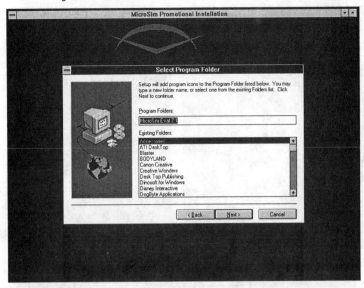

The default group name is **MicroSim Eval 7.1** and is highlighted. Click the *LEFT* mouse button on the **Next>** button to accept the name.

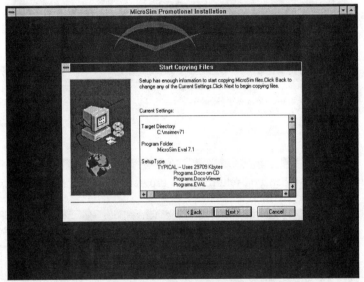

This dialog box tells us that the typical installation will use **29709 Kbytes** of disk space. Click the **Next>** button.

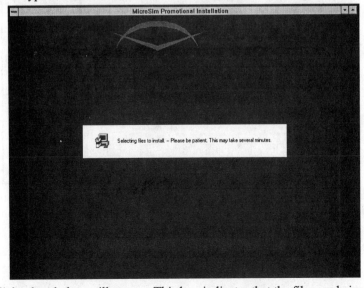

After a few moments, the dialog box below will appear. This box indicates that the files are being copied.

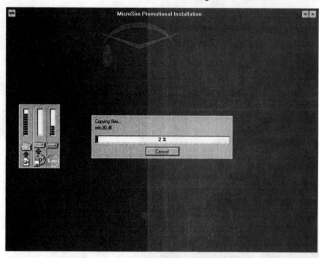
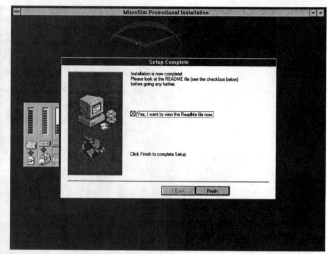

Click the **Finish** button. The **README.WRI** file will be displayed:

You can look at the file if you like. When you return to the Program manager, the MicroSim Eval 7.1 window will be displayed:

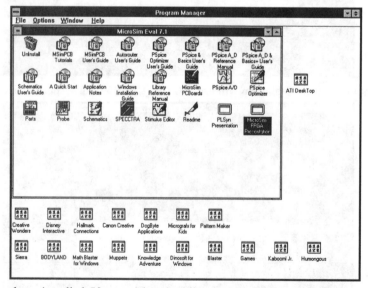

The Schematics software has been installed. If you wish to install only Schematics, you must continue the installation process by following the instructions in Section A.1.c. If you wish to install PADS-PERFORM, then you must continue with the next section.

A.1.c.   Installing the PSpice Libraries

A second installation program is provided to install the PSpice libraries. Open the Program Manager by double

clicking the *LEFT* mouse button on the Program Manager icon, 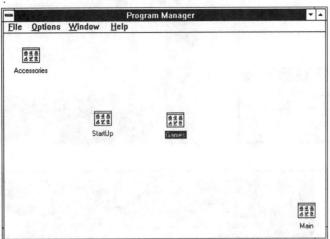. The **Program Manager** window will open as shown:

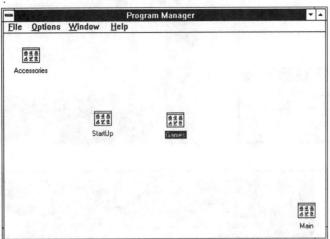

Click the *LEFT* mouse button on **File** in the **Program Manager** menu bar. The **File** pull-down menu will appear:

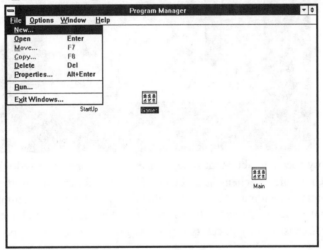

Click the *LEFT* mouse button on the **Run** menu selection. The **Run** dialog box will appear:

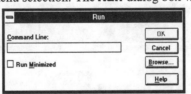

Enter the text **D:\BOOK\PSPICE\SETUP.EXE** in the text field below the text **Command Line**:

Click the **OK** button to run the library installation program:

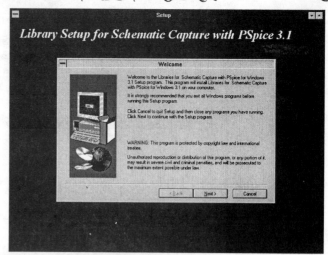

Click the **Next>** button:

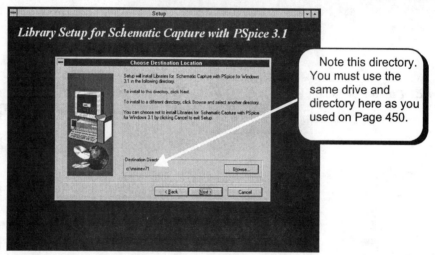

Note this directory. You must use the same drive and directory here as you used on Page 450.

Note that the drive and directory specified above must be the same as the one you used when installing Schematics. If you selected all of the defaults when installing Schematics, you can click the **Next>** button. If you specified a different drive or directory when installing Schematics, you must specify the same directory here. For example, if you installed Schematics in directory F:\msimev71, you must specify directory F:\msimev71 in the dialog box above. If you installed Schematics in directory d:\junk\msimev71, you must specify directory d:\junk\msimev71 in the dialog box above. Click the **Next>** button to continue.

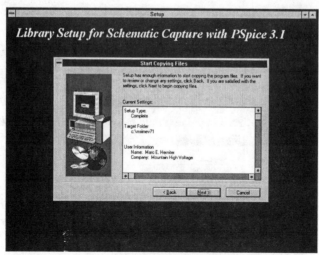

Click the **Next>** button to continue. The installation will begin:

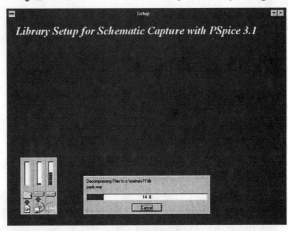

After a few moments, the library files will be copied to your computer.

You should restart your computer, so click the **Finish** button.

A.1.d.   Circuit Files Used in the Text

The library installation program copies all of the circuit files used as examples in the book to your hard drive. The files are placed in a directory named "book_ckt" in the directory you specified for installation.

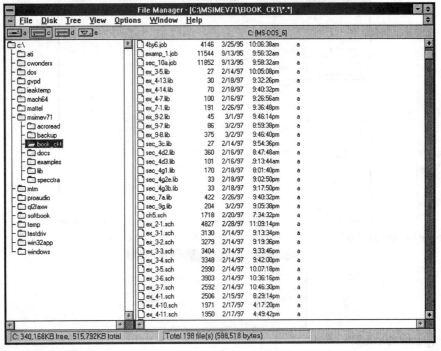

You may run these files and modify them.

A.2.   Installing PADS-PERFORM

To install PADS properly you must do the following four tasks:

1. Rename file c:\windows\system\vbrun300.dll if it exists.
2. Install the PADS software.
3. Install the PADS libraries for this text.
4. Run Schematics and modify the PC board layout setup.

These tasks are outlined in sections A.2.a, A.2.b, and A.2.c..

A.2.a.   Installing PADS-PERFORM Software

Before installing PADS-PERFORM, you must rename file c:\windows\system\vbrun300.dll to c:\windows\system\vbrun300.old. Use the File Manager to make this change. The installation program will not run properly if you do not make this change. If the file c:\windows\system\vbrun300.dll does not exist on your system, then you may continue with the installation given below.

Insert the CD-ROM in the drive. Open the Program Manager by double clicking the **LEFT** mouse button on the

Program Manager icon, [icon]. The **Program Manager** window will open:

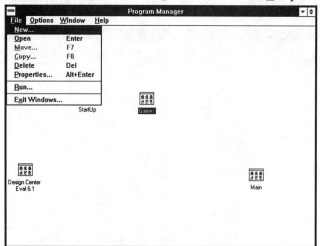

Click the **LEFT** mouse button on **File** in the **Program Manager** menu bar. The **File** pull-down menu will appear:

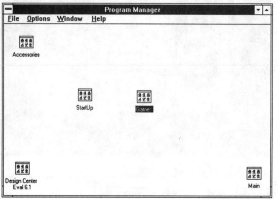

Click the **LEFT** mouse button on the **Run** menu selection. The **Run** dialog box will appear:

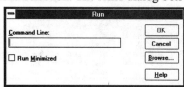

Type in the text **d:\pads\setup**:

Click the **OK** button to start the installation.

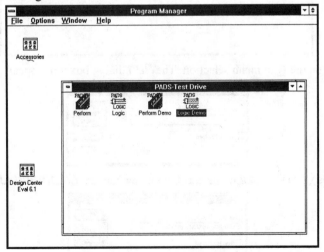

This dialog box allows you to change where you would like to install the PADS files. If you specify a directory other than **C:\TESTDRIV**, note the name of the directory because you will need it when you install the PADS libraries. Click the **Continue** button to begin the installation. A dialog box will appear that indicates the progress of the installation:

After a few minutes the installation will be complete:

Click the **OK** button. You will return to the Program Manager. You will notice that a new group called **PADS-Test Drive** has been created in the Program Manager window:

Installation of PADS-PERFORM is now complete. You must continue with the next section.

A.2.b.   Installing the PADS Libraries

A second installation program is provided to install the PADS libraries. Open the Program Manager by double

clicking the *LEFT* mouse button on the Program Manager icon,   . The **Program Manager** window will open as shown:

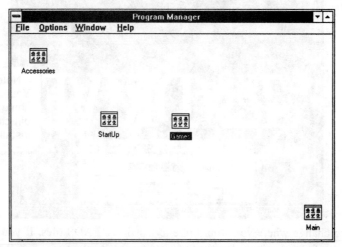

Click the *LEFT* mouse button on **File** in the **Program Manager** menu bar. The **File** pull-down menu will appear:

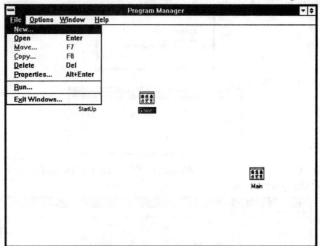

Click the *LEFT* mouse button on the **Run** menu selection. The **Run** dialog box will appear:

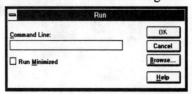

Enter the text **D:\BOOK\PADS\SETUP.EXE** in the text field below the text **Command Line**:

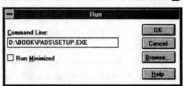

Click the **OK** button to run the library installation program:

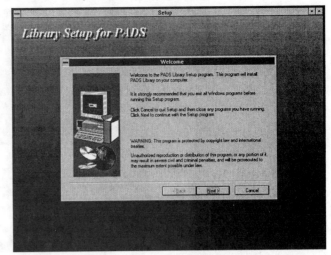

Click the **Next>** button:

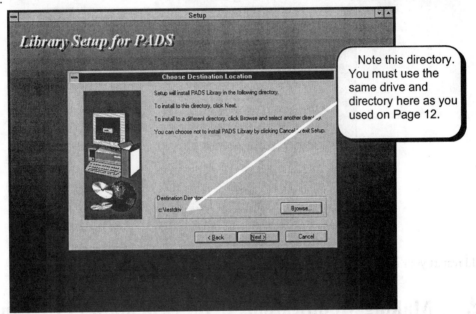

Note that the drive and directory specified above must be the same as the one you used when installing PADS. If you selected all of the defaults when installing PADS, you can click the **Next>** button. If you specified a different drive or directory when installing PADS, you must specify the same directory here. For example, if you installed PADS in directory F:\testdriv, you must specify directory F:\testdriv in the dialog box above. If you installed Schematics in directory d:\junk\pads, you must specify directory d:\junk\pads in the dialog box above. Click the **Next>** button to continue.

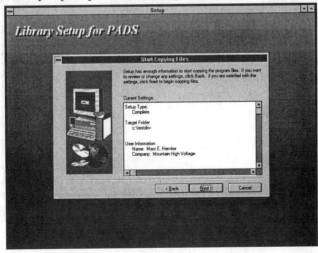

Click the **Next>** button to continue. The installation will begin:

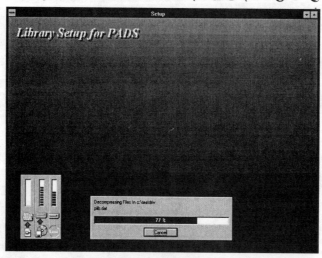

After a few moments, the library files will be copied to your computer.

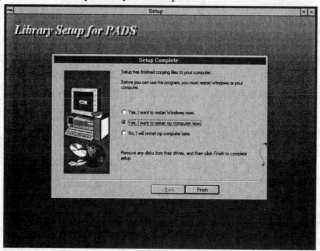

You should restart your computer, so click the **Finish** button.

A.2.c.   Making Modifications to the Schematics Program

The last thing we need to do is modify Schematics so that it will start up PADS when you invoke the layout editor. Run Schematics:

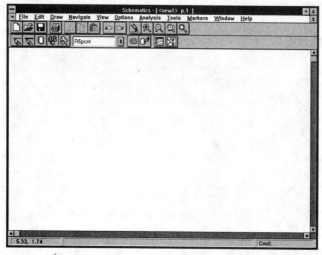

Select **Tools** from the Schematics menu bar:

Select **Configure Layout Editor**:

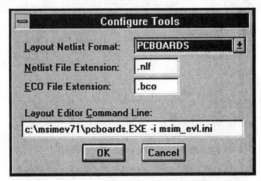

The default layout editor is MicroSim **PCBOARDS**. We need to change the editor to PADS. Click on the down arrow as shown below. A list of available editors will be displayed:

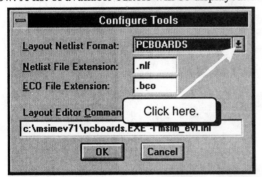 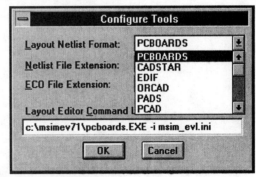

Click the *LEFT* mouse button on the text **PADS** to select the editor:

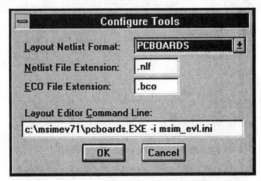

Next, we must change the **Layout Editor Command Line**. Change the line to `C:\TESTDRIV\PPERFTDW.EXE /x=C:\TESTDRIV\`

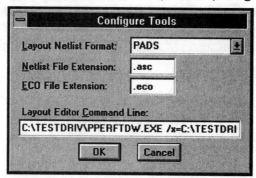

If you installed the PADS software in a directory other than c:\testdriv, you will need to change the directory in the **Layout Editor Command Line** above. For example, if you installed PADS in directory f:\pads\testdriv, the line would be F:\PADS\TESTDRIV\PPERFTDW.EXE /X=F:\PADS\TESTDRIV\. Click the OK button to accept the changes:

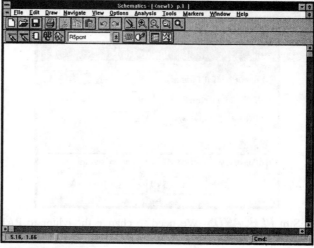

The installation of the PADS software is now complete.

APPENDIX B
Scale Multipliers for PSpice and Schematics

SYMBOL [1]	SCALE	NAME
F	10^{-15}	femto-
P	10^{-12}	pico-
N	10^{-9}	nano-
U	10^{-6}	micro-
MIL	25.4×10^{-6}	
M	10^{-3}	milli-
K	10^{+3}	kilo-
MEG	10^{+6}	mega-
G	10^{+9}	giga-
T	10^{+12}	tera-
C		Clock cycle

Bibliography

[1]   MicroSim Corporation. *The Design Center - Circuit Analysis - Reference Manual*, Version 6.1. Irvine, CA, July, 1994, p. 3-1.

APPENDIX C
Functions Available with Probe

Function [1]	Meaning	Comments
ABS(x)	$\lvert x \rvert$	
SGN(x)	+1 (if x>0), 0 (if x=0) −1 (if x<0)	
SQRT(x)	$x^{1/2}$	
EXP(x)	e^x	
LOG(x)	ln(x)	log base e
LOG10(x)	log(x)	log base 10
M(x)	magnitude of x	
P(x)	phase of x	result in degrees
R(x)	real part of x	
IMG(x)	imaginary part of x	
G(x)	group delay of x	result in seconds
PWR(x,y)	$\lvert x \rvert^y$	
SIN(x)	sin(x)	x in radians
COS(x)	cos(x)	x in radians
TAN(x)	tan(x)	x in radians
ATAN(x)	$\tan^{-1}(x)$	result in radians
ARCTAN(x)	$\tan^{-1}(x)$	result in radians
d(x)	derivative of x with respect to the x-axis variable	
s(x)	integral of x over the range of the x-axis variable	
AVG(x)	running average of x over the range of the x-axis variable	
AVGX(x,d)	running average of x (from x-d to x) over the range of the x-axis variable	
RMS(x)	running RMS average of x over the range of the x-axis variable	
DB(x)	magnitude of x in decibels	
MIN(x)	minimum of the real part of x	
MAX(x)	maximum of the real part of x	

Bibliography

[1]   MicroSim Corporation. *The Design Center - Circuit Analysis - Reference Manual*, Version 6.1. Irvine, CA, July, 1994, pp. 8-17–8-18.

APPENDIX D
Schematic Errors

This appendix contains a brief discussion of the common errors encountered in drawing schematics. There are two types of errors you may encounter when you use schematic capture with PSpice. The first type are drawing errors, which are detected by the Schematics program. When you create a netlist, any drawing errors will be detected by Schematics and the circuit will not be simulated. Thus, all drawing errors must be corrected before the circuit can be simulated.

The second type are run-time errors, which the PSpice program detects. Once you have a circuit free of drawing errors, the PSpice program will run. Errors generated with PSpice are identified in the output file. The PSpice program will exit and you will be returned to the Schematics program. To see the run-time errors you must look at the output file, select **Analysis**, and then **Examine Output** from the Schematics menu bar.

This appendix will discuss only drawing errors. We will start with the circuit below. This circuit has some obvious errors and a few you might never spot.

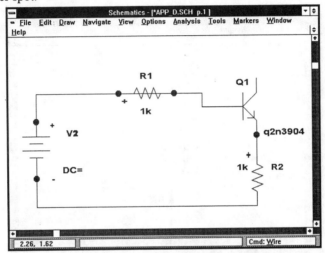

To check for errors, click the *LEFT* mouse button on **Analysis**. The **Analysis** pull-down menu will appear:

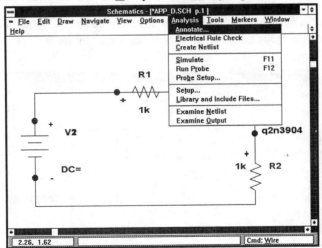

To check for drawing errors we could either do an electrical rule check or create a netlist. Creating a netlist automatically performs an electrical rule check, so we shall attempt to create a netlist. Click the *LEFT* mouse button on **Create Netlist**. The *Error* dialog box will appear.

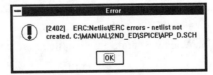

This box informs us that there are errors. To see the errors, click the *LEFT* mouse button on the **OK** button. The following *Error List* dialog box will appear:

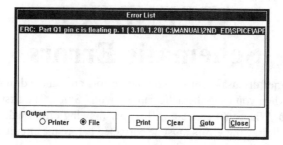

The dialog box tells us that **pin c** of part **Q1** is **floating**. Floating means that **pin c** is not connected to anything. This is a fairly obvious error and it is easily spotted. However, if the error is not easily seen, click the **LEFT** mouse button on the **Goto** button. When you press the **Goto** button, the dialog box disappears and the mouse pointer points at the error in the circuit. The screen below shows the mouse pointer pointing to pin c of **Q1**.

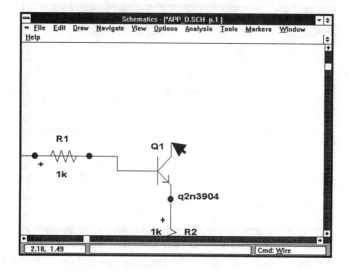

The **Goto** button can be very useful when two circuit elements appear to be connected but are not. To fix the problem I will make a connection as shown below. Although this may not result in a useful circuit, it does fix the problem.

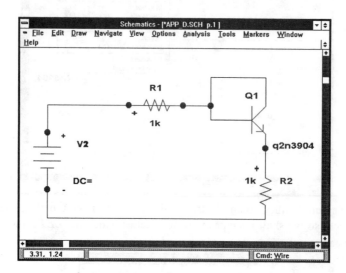

We shall now attempt to create a netlist again. When we create a netlist, there are once again errors, as shown below:

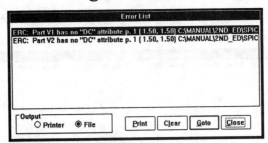

This error message says that the **DC** attribute is undefined. For some reason the message is displayed twice. Also note that the error box specifies **Part V1** and also specifies **Part V2**. Where is **V1** on the schematic anyway? When we click the **Goto** button, the mouse points to the graphic of the DC voltage source, ⊣⊦.

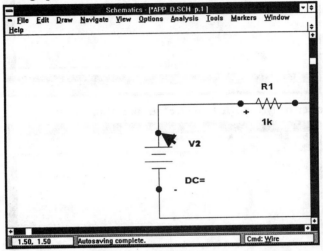

The DC attribute of this source must be specified before the circuit can be simulated. I will edit the attributes of the DC source and change the attribute to **DC=12**, as shown below:

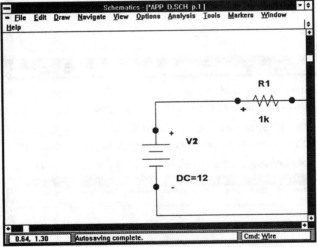

We now check the circuit again by creating a netlist. Once again there are errors:

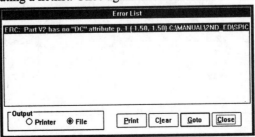

This is the same error as before, but it now appears only once. When we click the **Goto** button, the mouse indicates that the DC source is again the culprit. **It turns out that there are two voltage sources placed directly on top of each other, but only one can be seen.** I will now delete the DC voltage source. This is done by clicking the **LEFT** mouse button on the DC source graphic, ⊣⊩⊢. The graphic will turn red, indicating that it has been selected. When the graphic is selected press the **DELETE** key. At first the DC source seems to disappear:

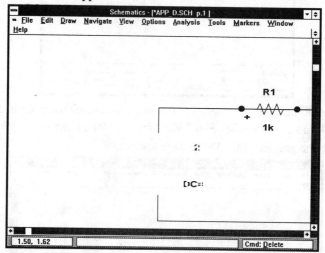

However, when we redraw the screen by typing **CTRL-L**, the second source appears:

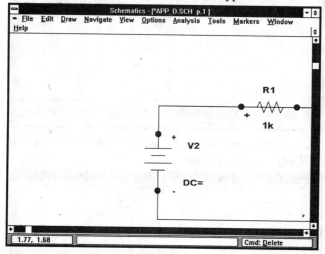

Notice that there is still a DC voltage source and that it is labeled **V2**. Also note that the DC attribute of this source is undefined. This undefined attribute was the cause of the error. You can now define it as we did above.

This error was caused by two identical parts being placed directly on top of each other. It was impossible to see the bottom source because the two graphics were the same. This is a very hard problem to spot.

After we delete the duplicate DC source, the circuit is free of drawing errors as far as Schematics is concerned:

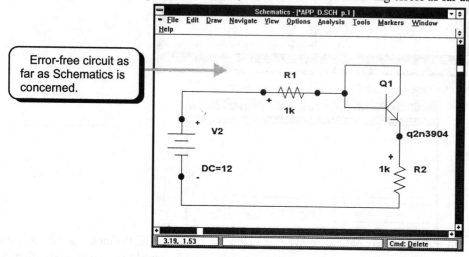

Error-free circuit as far as Schematics is concerned.

No more errors will be indicated when we create a netlist. However, there are still two errors in the drawing. The first is that the circuit is not grounded. The entire circuit is floating. This error would have been caught when we ran the simulation. You must add a part called AGND to your circuit:

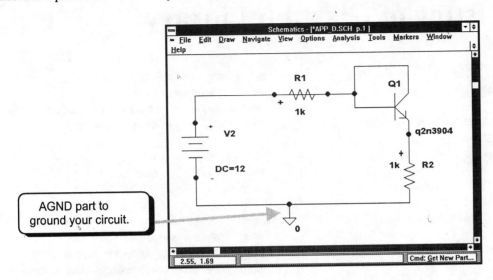

AGND part to
ground your circuit.

The second error is a little more subtle. If we look closely at the resistor, we note that a wire is drawn through it. If we zoom in around *R1*, the error becomes more obvious:

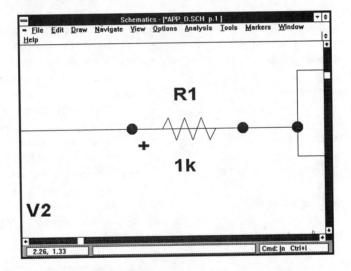

The wire through *R1* is shorting out the resistor. This wire was probably not intentional since *R1* will have no effect. This error would not be detected by either Schematics or PSpice. This is a valid connection as far as the simulator is concerned. The simulation would run but the results would be incorrect. The wire through *R1* should be deleted. To delete the wire, click the *LEFT* mouse button on the wire segment going through *R1*. When the segment has been selected, it will be highlighted in red. Press the DELETE key to delete the wire.

APPENDIX E
Listing of Class.lib Library

```
*
.model D1N4734A D(Is=1.085f Rs=.7945 Ikf=0 N=1 Xti=3 Eg=1.11
Cjo=157p M=.2966
+        Vj=.75 Fc=.5 Isr=2.811n Nr=2 Bv=5.6 Ibv=.37157 Nbv=.64726
+        Ibvl=1m Nbvl=6.5761 Tbv1=267.86u)
*        Motorola    pid=1N4734    case=DO-41
*        89-9-19 gjg
*        Vz = 5.6 @ 45mA, Zz = 40 @ 1mA, Zz = 4.5 @ 5mA, Zz = 1.9 @
20mA
.model D1N4148  D(Is=0.1p Rs=16 CJO=2p Tt=12n Bv=100 Ibv=0.1p)
*        85-??-??    Original library
.model D1N5401  D(Is=11.5f Rs=8.254m Ikf=3.87 N=1 Xti=3 Eg=1.11
Cjo=130.4p
+        M=.3758 Vj=.75 Fc=.5 Isr=79.29u Nr=2)
*        Motorola    pid=1N5400    case=267-01
*        88-09-12 rmn
.model D1N914  D(Is=0.1p Rs=16 CJO=2p Tt=12n Bv=100 Ibv=0.1p)
*        85-??-??    Original library
.model D1N4007  D(Is=0.1p Rs=4 CJO=2p Tt=3n Bv=1000 Ibv=0.1p)
*        85-??-??    Original library
.model D1N4001  D(Is=0.1p Rs=4 CJO=2p Tt=3n Bv=50 Ibv=0.1p)
*        85-??-??    Original library
.model D1N4004  D(Is=0.1p Rs=4 CJO=2p Tt=3n Bv=400 Ibv=0.1p)
*        85-??-??    Original library
*-------------------------------------------------------------
.MODEL Q2N2222    NPN(IS=3.108E-15 XTI=3 EG=1.11 VAF=131.5
+     BF=200 DEV/UNIFORM 150 NE=1.541
+     ISE=190.7E-15 IKF=1.296 XTB=1.5 BR=6.18 NC=2 ISC=0 IKR=0
RC=1
+     CJC=14.57E-12 VJC=.75 MJC=.3333 FC=.5 CJE=26.08E-12
VJE=.75
+     MJE=.3333 TR=51.35E-9 TF=451E-12 ITF=.1 VTF=10 XTF=2)
*.MODEL Q2N2907 PNP(IS=9.913E-15 XTI=3 EG=1.11 VAF=90.7
BF=197.8 NE=2.264
.MODEL Q2N2907  PNP(IS=9.913E-15 XTI=3 EG=1.11 VAF=90.7
+        BF=200
+        ISE=6.191E-12 IKF=.7322 XTB=1.5 BR=3.369 NC=2 ISC=0
IKR=0 RC=1
+        CJC=14.57E-12 VJC=.75 MJC=.3333 FC=.5 CJE=20.16E-12
VJE=.75
+        MJE=.3333 TR=29.17E-9 TF=405.7E-12 ITF=.4 VTF=10 XTF=2)
*        88-09-09 bam    creation
     pwt    change Rb
.model Q2N3904  NPN(Is=6.734f Xti=3 Eg=1.11 Vaf=74.03
+        Bf=416.4 Ne=1.259
+        Ise=6.734f Ikf=66.78m Xtb=1.5 Br=.7371 Nc=2 Isc=0 Ikr=0 Rc=1
+        Cjc=3.638p Mjc=.3085 Vjc=.75 Fc=.5 Cje=4.493p Mje=.2593
Vje=.75
+        Tr=239.5n Tf=301.2p Itf=.4 Vtf=4 Xtf=2 Rb=10)
*        National    pid=23    case=TO92
*        88-09-08 bam    creation
```

```
*
.model Q2N3906  PNP(Is=1.41f Xti=3 Eg=1.11 Vaf=18.7
+        Bf=180.7 Ne=1.5 Ise=0
+        Ikf=80m Xtb=1.5 Br=4.977 Nc=2 Isc=0 Ikr=0 Rc=2.5 Cjc=9.728p
+        Mjc=.5776 Vjc=.75 Fc=.5 Cje=8.063p Mje=.3677 Vje=.75
Tr=33.42n
+        Tf=179.3p Itf=.4 Vtf=4 Xtf=6 Rb=10)
*        National    pid=66    case=TO92
*        88-09-09 bam    creation
*
.model Q2N3904B NPN(Is=6.734f Xti=3 Eg=1.11 Vaf=74.03
+        Bf=416.4 DEV/UNIFORM 80% Ne=1.259
+        Ise=6.734f Ikf=66.78m Xtb=1.5 Br=.7371 Nc=2 Isc=0 Ikr=0 Rc=1
+        Cjc=3.638p Mjc=.3085 Vjc=.75 Fc=.5 Cje=4.493p Mje=.2593
Vje=.75
+        Tr=239.5n Tf=301.2p Itf=.4 Vtf=4 Xtf=2 Rb=10)
*        National    pid=23    case=TO92
*        88-09-08 bam    creation
*
.model Q2N3906B PNP(Is=1.41f Xti=3 Eg=1.11 Vaf=18.7
+        Bf=180.7 DEV/UNIFORM 73% Ne=1.5 Ise=0
+        Ikf=80m Xtb=1.5 Br=4.977 Nc=2 Isc=0 Ikr=0 Rc=2.5 Cjc=9.728p
+        Mjc=.5776 Vjc=.75 Fc=.5 Cje=8.063p Mje=.3677 Vje=.75
Tr=33.42n
+        Tf=179.3p Itf=.4 Vtf=4 Xtf=6 Rb=10)
*        National    pid=66    case=TO92
*        88-09-09 bam    creation
*
.model MPF102   NJF( Betatce=-.5 Rd=1 Rs=1 Lambda=2m
+     Vto=-3.41 Beta=1.04m
+        Vtotc=-2.5m Is=33.57f Isr=322.4f N=1 Nr=2 Xti=3 Alpha=311.7
+        Vk=243.6 Cgd=1.6p M=.3622 Pb=1 Fc=.5 Cgs=2.414p
Kf=11.73E-18
+        Af=1)
.model J2N5951  NJF( Betatce=-.5 Rd=1 Rs=1 Lambda=7.25m
+     Vto=-3.427 Beta=736.9u
+        Vtotc=-2.5m Is=33.57f Isr=322.4f N=1 Nr=2 Xti=3 Alpha=311.7
+        Vk=243.6 Cgd=1.6p M=.3622 Pb=1 Fc=.5 Cgs=2.414p
Kf=5.642E-18
+        Af=1)
*        National    pid=50    case=TO92
*        88-08-02 rmn    BVmin=30
*
*
* Create a transistor with Beta variations of 50 to 350.
.model QBf NPN( Bf=200 DEV/UNIFORM 150)

* QLM3046 model created using Parts version 4.04 on 01/01/80 at 02:25
*
.model QLM3046   NPN(Is=2p Xti=3 Eg=1.11 Vaf=16.64 Bf=128.3
Ise=17.62p Ne=1.667
+        Ikf=31.29m Nk=.4878 Xtb=1.5 Br=1 Isc=2p Nc=2 Ikr=0 Rc=0
```

```
+       Cjc=991.8f Mjc=.3333 Vjc=.75 Fc=.5 Cje=1.026p Mje=.3333
Vje=.75
+       Tr=10n Tf=274.7p Itf=.4434 Xtf=31.73 Vtf=10)


.subckt diff_pair c1 c2 b1 b2 e
Q1 c1 b1 e QLM3046
Q2 C2 B2 e QLM3046
.ends
*
*
*
*_____
*_____
.SUBCKT Ideal_OPAMP    v_plus v_minus v_out
*
* This is an ideal OpAmp model with the output voltage
* limited to +/- 15V.
*
Rin       v_plus v_minus 10MEG
Eamp  v_out  0    value={LIMIT(1MEG*v(v_plus,v_minus),-15,15)}
R0        v_out  0     10MEG
.ENDS
*
*
*_____
.SUBCKT Ideal_OPAMP_2  v_plus v_minus v_out
*
* This is an ideal OpAmp model with the output voltage
* limited to +/- 15V.
*
Rin       v_plus v_minus 10MEG
Eamp  v_out  0    TABLE {1MEG*(v(v_plus,v_minus)))}=(-11.25,-11.25)
(11.25,11.25)
R0        v_out  0     10MEG
.ENDS
*
*
*_____
* The following models simulate the max and min transistors for
* a 2n5951 N-jFET
.model Jmax NJF(Vto=-5 Beta=.52m)
.model Jmin NJF(Vto=-2 Beta=1.75m)
*
* The following models simulate resistors with tolerances
*
* The two models below are 1% and 5% resistors with uniform
* probability distributions
.MODEL R1pcnt RES(R=1 DEV/UNIFORM 1%)
.MODEL R5pcnt RES(R=1 DEV/UNIFORM 5%)
* The models below are 1% and 5% resistors with Gaussian
* probability distributions
.MODEL R1gauss RES(R=1 DEV/gauss 0.25%)
.MODEL R5gauss RES(R=1 DEV/gauss 1.25%)
* THe following lines simulate a Capacitor with -10% and +50 %
* tolerance. Note that to achieve this asymmetric tolerance the
* nominal value is not the actual value.
.model CAP10_50 CAP(C=1.2 DEV/UNIFORM 25%)
* THe following lines define a -20 + 80% tolerance Capacitor.
```

```
.model CAP20_80 CAP(C=1.3 DEV/UNIFORM 38.461538%)
*
* UA741 operational amplifier "macromodel" subcircuit
* created using Parts release 4.01 on 07/05/89 at 09:09
* (REV N/A)
* connections:   non-inverting input
*           | inverting input
*           | | positive power supply
*           | | | negative power supply
*           | | | | output
*           | | | | |
.subckt UA741   1 2 3 4 5
*
  c1   11 12 4.664E-12
  c2   6  7 20.00E-12
  dc   5 53 dx
  de   54 5 dx
  dlp  90 91 dx
  dln  92 90 dx
  dp   4  3 dx
  egnd 99 0 poly(2) (3,0) (4,0) 0 .5 .5
  fb   7 99 poly(5) vb vc ve vlp vln 0 10.61E6 -10E6 10E6 10E6 -10E6
  ga   6  0 11 12 137.7E-6
  gcm  0  6 10 99 2.574E-9
  iee  10 4 dc 10.16E-6
  hlim 90 0 vlim 1K
  q1   11 2 13 qx
  q2   12 1 14 qx
  r2   6  9 100.0E3
  rc1  3 11 7.957E3
  rc2  3 12 7.957E3
  re1  13 10 2.740E3
  re2  14 10 2.740E3
  ree  10 99 19.69E6
  ro1  8  5 150
  ro2  7 99 150
  rp   3  4 18.11E3
  vb   9  0 dc 0
  vc   3 53 dc 2.600
  ve   54 4 dc 2.600
  vlim 7  8 dc 0
  vlp  91 0 dc 25
  vln  0 92 dc 25
.model dx D(Is=800.0E-18)
.model qx NPN(Is=800.0E-18 Bf=62.50)
.ends
* LF411C operational amplifier "macromodel" subcircuit
* created using Parts release 4.01 on 06/27/89 at 08:19
* (REV N/A)
* connections:   non-inverting input
*           | inverting input
*           | | positive power supply
*           | | | negative power supply
*           | | | | output
*           | | | | |
.subckt LF411C   1 2 3 4 5
*
  c1   11 12 3.498E-12
```

```
c2   6  7 15.00E-12
dc   5 53 dx
de  54  5 dx
dlp 90 91 dx
dln 92 90 dx
dp   4  3 dx
egnd 99  0 poly(2) (3,0) (4,0) 0 .5 .5
fb   7 99 poly(5) vb vc ve vlp vln 0 28.29E6 -30E6 30E6 30E6 -30E6
ga   6  0 11 12 282.8E-6
gcm  0  6 10 99 1.590E-9
iss  3 10 dc 195.0E-6
hlim 90  0 vlim 1K
j1  11  2 10 jx
j2  12  1 10 jx
r2   6  9 100.0E3
rd1  4 11 3.536E3
rd2  4 12 3.536E3
ro1  8  5 50
ro2  7 99 25
rp   3  4 15.00E3
rss 10 99 1.026E6
vb   9  0 dc 0
vc   3 53 dc 2.200
ve  54  4 dc 2.200
vlim 7  8 dc 0
vlp 91  0 dc 30
vln  0 92 dc 30
.model dx D(Is=800.0E-18)
.model jx PJF(Is=12.50E-12 Beta=250.1E-6 Vto=-1)
.ends
* LM301A operational amplifier "macromodel" subcircuit
* created using Parts release 4.01 on 09/01/89 at 13:14
* (REV N/A)
* connections:  non-inverting input
*               | inverting input
*               | | positive power supply
*               | | | negative power supply
*               | | | | output
*               | | | | | compensation
*               | | | | | | /\
.subckt LM301A   1 2 3 4 5 6 7
*
c1  11 12 7.977E-12
dc   5 53 dx
de  54  5 dx
dlp 90 91 dx
dln 92 90 dx
dp   4  3 dx
egnd 99  0 poly(2) (3,0) (4,0) 0 .5 .5
fb   7 99 poly(5) vb vc ve vlp vln 0 42.44E6 -40E6 40E6 40E6 -40E6
ga   6  0 11 12 188.5E-6
gcm  0  6 10 99 3.352E-9
iee 10  4 dc 15.14E-6
hlim 90  0 vlim 1K
q1  11  2 13 qx
q2  12  1 14 qx
r2   6  9 100.0E3
rc1  3 11 5.305E3
```

```
rc2  3 12 5.305E3
re1 13 10 1.839E3
re2 14 10 1.839E3
ree 10 99 13.21E6
ro1  8  5 50
ro2  7 99 25
rp   3  4 16.81E3
vb   9  0 dc 0
vc   3 53 dc 2.600
ve  54  4 dc 2.600
vlim 7  8 dc 0
vlp 91  0 dc 25
vln  0 92 dc 25
.model dx D(Is=800.0E-18)
.model qx NPN(Is=800.0E-18 Bf=107.1)
.ends
* LM324 operational amplifier "macromodel" subcircuit
* created using Parts release 4.01 on 09/08/89 at 10:54
* (REV N/A)
* connections:  non-inverting input
*               | inverting input
*               | | positive power supply
*               | | | negative power supply
*               | | | | output
*               | | | | |
.subckt LM324   1 2 3 4 5
*
c1  11 12 5.544E-12
c2   6  7 20.00E-12
dc   5 53 dx
de  54  5 dx
dlp 90 91 dx
dln 92 90 dx
dp   4  3 dx
egnd 99  0 poly(2) (3,0) (4,0) 0 .5 .5
fb   7 99 poly(5) vb vc ve vlp vln 0 15.91E6 -20E6 20E6 20E6 -20E6
ga   6  0 11 12 125.7E-6
gcm  0  6 10 99 7.067E-9
iee  3 10 dc 10.04E-6
hlim 90  0 vlim 1K
q1  11  2 13 qx
q2  12  1 14 qx
r2   6  9 100.0E3
rc1  4 11 7.957E3
rc2  4 12 7.957E3
re1 13 10 2.773E3
re2 14 10 2.773E3
ree 10 99 19.92E6
ro1  8  5 50
ro2  7 99 50
rp   3  4 30.31E3
vb   9  0 dc 0
vc   3 53 dc 2.100
ve  54  4 dc .6
vlim 7  8 dc 0
vlp 91  0 dc 40
vln  0 92 dc 40
.model dx D(Is=800.0E-18)
```

.model qx PNP(Is=800.0E-18 Bf=250)

.ends

*——————————————————————LM7915C

.SUBCKT LM7915C Input Output Ground

  x1 Input Output Ground x_LM79XX PARAMS:

+   Av_feedback=555, R1_Value=13845,

+   Rbg_Tc1=-9.50E-7, Rbg_Tc2=-6.53E-7,

+   Rout_Value=0.01, Rreg_Value=11.3k

.ENDS

*——————————————————————LM7918C

.SUBCKT LM7815C Input Output Ground

  x1 Input Output Ground x_LM78XX PARAMS:

+   Av_feedback=550, R1_Value=3060

.ENDS

*——————————————————————LM7805C

.SUBCKT LM7805C Input Output Ground

  x1 Input Output Ground x_LM78XX PARAMS:

+   Av_feedback=1665, R1_Value=1020

.ENDS

*

*

*

*

*** Voltage regulators (positive)

.SUBCKT x_LM78XX Input Output Ground PARAMS:

+   Av_feedback=1665, R1_Value=1020

*

* SERIES 3-TERMINAL POSITIVE REGULATOR

*

* Note: This regulator is based on the LM78XX series of

*   regulators (also the LM140 and LM340). The model

*   will cause some current to flow to Node 0 which

*   is not part of the actual voltage regulator circuit.

*

* Band-gap voltage source:

*

*   The source is off when Vin<3V and fully on when Vin>3.7V.

*   Line regulation and ripple rejection) are set with

*   Rreg= 0.5 * dVin/dVbg. The temperature dependence of this

*   circuit is a quadratic fit to the following points:

*

T	Vbg(T)/Vbg(nom)
0	.999
37.5	1
125	.990

*

*   The temperature coefficient of Rbg is set to 2 * the band gap

*   temperature coefficient. Tnom is assumed to be 27 deg. C and

*   Vnom is 3.7V

*

Vbg 100 0 DC 7.4V

Sbg (100,101) (Input,Ground) Sbg1

Rbg 101 0 1 TC=1.612E-5,-2.255E-6

Ebg (102,0) (Input,Ground) 1

Rreg 102 101 7k

.MODEL Sbg1 VSWITCH (Ron=1 Roff=1MEG Von=3.7 Voff=3)

*

* Feedback stage

*

*   Diodes D1,D2 limit the excursion of the amplifier

*   outputs to being near the rails. Rfb, Cfb Set the

*   corner frequency for roll-off of ripple rejection.

*

*   The opamp gain is given by: Av = (Fores/Freg) * (Vout/Vbg)

*   where Fores = output impedance corner frequency

*   with CI=0 (typical value about 1MHz)

*   Freg = corner frequency in ripple rejection

*   (typical value about 600 Hz)

*   Vout = regulator output voltage (5,12,15V)

*   Vbg = bandgap voltage (3.7V)

*

*   Note: Av is constant for all output voltages, but the

*   feedback factor changes. If Av=2250, then the

*   Av*Feedback factor is as given below:

*

Vout	Av*Feedback factor
5	1665
12	694
15	550

*

Rfb 9 8 1MEG

Cfb 8 Ground 265PF

* Eopamp 105 0 VALUE={2250*v(101,0)+Av_feedback*v(Ground,8)}

Vgainf 200 0 {Av_feedback}

Rgainf 200 0 1

Eopamp 105 0 POLY(3) (101,0) (Ground,8) (200,0) 0 2250 0 0 0 0 0 1

Ro 105 106 1k

D1 106 108 Dlim

D2 107 106 Dlim

.MODEL Dlim D (Vj=0.7)

VI1 102 108 DC 1

VI2 107 0 DC 1

*

* Quiescent current modelling

*

*   Quiescent current is set by Gq, which draws a current

*   proportional to the voltage drop across the regulator and

*   R1 (temperature coefficient .1%/deg C). R1 must change

*   with output voltage as follows: R1 = R1(5v) * Vout/5v.

*

Gq (Input,Ground) (Input,9) 2.0E-5

R1 9 Ground {R1_Value} TC=0.001

*

* Output Stage

*

*   Rout is used to set both the low frequency output impedance

*   and the load regulation.

*

Q1 Input 5 6 Npn1

Q2 Input 6 7 Npn1 10

.MODEL Npn1 NPN (Bf=50 Is=1E-14)

```
* Efb Input 4 VALUE={v(Input,Ground)+v(0,106)}
Efb Input 4 POLY(2) (Input,Ground) (0,106) 0 1 1
Rb 4 5 1k TC=0.003
Re 6 7 2k
Rsc 7 9 0.275 TC=1.136E-3,-7.806E-6
Rout 9 Output 0.008
*
* Current Limit
*
Rbcl 7 55 290
Qcl 5 55 9 Npn1
Rcldz 56 55 10k
Dz1 56 Input Dz
.MODEL Dz D (Is=0.05p Rs=3 Bv=7.11 Ibv=0.05u)
.ENDS
*
*--------------------------------------------------------LM7815C
*** Voltage regulators (negative)

.SUBCKT x_LM79XX Input Output Ground PARAMS:
+      Av_feedback=1660, R1_Value=4615,
+      Rbg_Tc1=8.13E-5, Rbg_Tc2=0.0,
+      Rout_Value=0.01, Rreg_Value=1.2k
*
* SERIES 3-TERMINAL NEGATIVE REGULATOR
*
* Note: This regulator is based on the LM79XX series of
*       regulators (also the LM120 and LM320).  The
*       LM79XX regulators are unstable and will
*       oscillate unless a 1 uFarad solid tantalum
*       capacitor is placed on the output with an ESR
*       betweed .5 and 1.5.  This model is stable without
*       a capacitor on the output.  When performing
*       simulations a 1 uFarad capacitor should still be
*       placed on the output.  However, it it not necessary
*       to include a resistor in series with this capacitor
*       to model the ESR of the capacitor.  See the
*       comments and circuit description of the x_LM78XX
*       regulator for more information on this model.
*
* Band-gap voltage source:
*
Vbg 100 0 DC -7.4V
Sbg (100,101) (Ground,Input) Sbg1
Rbg 101 0 Rbg1 1
.MODEL Rbg1 RES (Tc1={Rbg_Tc1},Tc2={Rbg_Tc2})
Ebg (102,0) (Input,Ground) 1
Rreg 102 101 {Rreg_Value}
.MODEL Sbg1 VSWITCH (Ron=1 Roff=1MEG Von=3.7 Voff=3)
*
* Feedback stage
*
Rfb 9 8 1MEG
Cfb 8 Ground 265PF
* Eopamp 105 0 VALUE={2250*v(101,0)+Av_feedback*v(Ground,8)}
Vgainf 200 0 {Av_feedback}
Rgainf 200 0 1
Eopamp 105 0 POLY(3) (101,0) (Ground,8) (200,0) 0 2250 0 0 0 0 0 0 1
```

```
Ro 105 106 1k
D1 108 106 Dlim
D2 106 107 Dlim
.MODEL Dlim D (Vj=0.7)
VI1 108 102 DC 1
VI2 0 107 DC 1
*
* Quiescent current modelling
*
Gq (Ground,Input) (9,Input) 9.0E-7
R1 9 Ground {R1_Value} TC=0.001
FI (Ground,0) Vmon 3.0E-4
*
* Output Stage
*
Q1 9 5 6 Npn1
Q2 9 6 7 Npn1 10
.MODEL Npn1 NPN (Bf=50 Is=1E-14)
* Efb 4 Ground VALUE={v(Input,Ground)+v(0,106)}
Efb 4 Ground POLY(2) (Input,Ground) (0,106) 0 1 1
Rb 4 5 1k TC=0.003
Re 6 7 2k
Rsc 7 Input 0.13 TC=1.136E-3,-7.806E-6
Rout 9 Imon {Rout_Value}
Vmon Imon Output DC 0.0
*
* Current Limit
*
Qcl1 54 52 53 Npn1
Qcl3 Input 54 5 Pnp1
.MODEL Pnp1 PNP (Bf=250 Is=1E-14)
Rcl3 5 54 1.8k
Qcl2 52 52 51 Npn1
Veset 53 Input DC 0.3v
Ibias Input 52 DC 300u
Rcl1 50 51 20k
Rcl2 51 7 115
Dz1 50 9 Dz
.MODEL Dz D (Is=0.05p Rs=3 Bv=7.11 Ibv=0.05u)
.ENDS
*

* Digital Circuits for EGR482
* TTL Circuit Simulation
*
.subckt ttl_gate in out VCC
Q1   4  2  in   Q2n2222
Q2   6  4  5    Q2n2222
Q3   out 5  0   Q2n2222
Q4   7  6  8    Q2n2222
Q5   8  8  out  Q2n2222
Rb   Vcc 2  4k
Re   5  0  1k
Rc4  Vcc 6  1.4k
Rc2  Vcc 7  100
.ends
*
```

```
* DTL Circuit
.subckt dtl_gate in out Vcc
Q1      2     1      3      Q2n2222
Q2      out   4      0      Q2n2222
Q3      3     3      4      Q2n2222
Q4      1     1      IN     Q2n2222
R1      Vcc   2      1.6K
R2      2     1      2.15K
Rb      4     0      5k
Rc      Vcc   out    2K
.ends
*
* RTL Circuit
*
.subckt rtl_gate in out Vcc
Rb      in    1      10K
Rc      Vcc   out    1K
Q1      out   1      0      Q2n2222
.ends
*
* Power Mosfets
*
.model IRFD1Z3 NMOS(Level=3 Gamma=0 Delta=0 Eta=0 Theta=0
Kappa=0 Vmax=0 Xj=0
+       Tox=100n Uo=600 Phi=.6 Rs=2.063 Kp=21.17u W=.2 L=2u
Vto=3.936
+       Rd=.6512 Rds=266.7K Cbd=84.39p Pb=.8 Mj=.5 Fc=.5
Cgso=932.5p
+       Cgdo=116.3p Rg=26.48 Is=1.135p N=1 Tt=470n)
*       Int'l Rectifier pid=IRFC1Z0   case=4 Pin DIP
*       88-08-26 bam   creation
.model IRFD9113 PMOS(Level=3 Gamma=0 Delta=0 Eta=0 Theta=0
Kappa=0 Vmax=0 Xj=0
+       Tox=100n Uo=300 Phi=.6 Rs=.5286 Kp=10.29u W=.25 L=2u
Vto=-3.909
+       Rd=.54 Rds=266.7K Cbd=309.5p Pb=.8 Mj=.5 Fc=.5
Cgso=3.761n
+       Cgdo=498.1p Rg=3.99 Is=8.282f N=4 Tt=8800n)
*       Int'l Rectifier pid=IRFC9110  case=4 Pin DIP
*       88-08-26 bam   creation
.model IRF252 NMOS(Level=3 Gamma=0 Delta=0 Eta=0 Theta=0
Kappa=0 Vmax=0 Xj=0
+       Tox=100n Uo=600 Phi=.6 Rs=2.081m Kp=20.86u W=1 L=2u
Vto=3.794
+       Rd=67.06m Rds=888.9K Cbd=3.481n Pb=.8 Mj=.5 Fc=.5
Cgso=1.585n
+       Cgdo=442.1p Rg=5.549 Is=168.3p N=1 Tt=340n)
*       Int'l Rectifier pid=IRFC250   case=TO3
*       88-08-25 bam   creation
.model MJE3055 NPN(Is=974.4f Xti=3 Eg=1.11 Vaf=50
+       Bf=99.49 DEV/UNIFORM 80% Ne=1.941
+       Ise=902.5p Ikf=4.029 Xtb=1.5 Br=2.949 Nc=2 Isc=0 Ikr=0 Rc=.1
+       Cjc=276p Vjc=.75 Mjc=.3333 Fc=.5 Cje=569.1p Vje=.75
Mje=.3333
+       Tr=971.7n Tf=39.11n Itf=20 Vtf=10 Xtf=2 Rb=.1)
*       Texas Inst.   pid=2N3055   case=TO3
*       Original Library
*       02 Jan 91   pwt   change Rb
```

```
.model Q2N3055  NPN(Is=974.4f Xti=3 Eg=1.11 Vaf=50
+       Bf=99.49 DEV/UNIFORM 80% Ne=1.941
+       Ise=902.5p Ikf=4.029 Xtb=1.5 Br=2.949 Nc=2 Isc=0 Ikr=0 Rc=.1
+       Cjc=276p Vjc=.75 Mjc=.3333 Fc=.5 Cje=569.1p Vje=.75
Mje=.3333
+       Tr=971.7n Tf=39.11n Itf=20 Vtf=10 Xtf=2 Rb=.1)
*       Texas Inst.   pid=2N3055   case=TO3
*       Original Library
*       02 Jan 91   pwt   change Rb
.model MJE2955  PNP(Is=66.19f Xti=3 Eg=1.11 Vaf=100
+       Bf=137.6 DEV/UNIFORM 85% Ise=862.2f
+       Ne=1.481 Ikf=1.642 Nk=.5695 Xtb=2 Br=5.88 Isc=273.5f
Nc=1.24
+       Ikr=3.555 Rc=79.39m Cjc=870.4p Mjc=.6481 Vjc=.75 Fc=.5
+       Cje=390.1p Mje=.4343 Vje=.75 Tr=235.4n Tf=23.21n Itf=71.33
+       Xtf=5.982 Vtf=10 Rb=.1)
*       National Semiconductor
*       Transistor Databook, 1982, process 5A, pg 9-30
*       30 Nov 90   pwt   creation
.model TIP31   ako:NSC_4F    NPN()  ; case TO-220

.model TIP31A  ako:NSC_4F    NPN()  ; case TO-220

.model TIP31B  ako:NSC_4F    NPN()  ; case TO-220

.model TIP31C  ako:NSC_4F    NPN()  ; case TO-220

.model TIP32   ako:NSC_5F    PNP()  ; case TO-220

.model TIP32A  ako:NSC_5F    PNP()  ; case TO-220

.model TIP32B  ako:NSC_5F    PNP()  ; case TO-220

.model NSC_5F  PNP(Is=51.23f Xti=3 Eg=1.11 Vaf=100 Bf=434.1
Ise=51.23f Ne=1.22
+       Ikf=.3883 Nk=.5544 Xtb=2.2 Br=55.47 Isc=51.23f Nc=1.205
+       Ikr=10.87 Rc=.3443 Cjc=136.9p Mjc=.3155 Vjc=.75 Fc=.5
+       Cje=179.9p Mje=.4294 Vje=.75 Tr=20.25n Tf=13.05n Itf=6.85
+       Xtf=1.573 Vtf=10 Rb=.1)
*       National Semiconductor
*       Transistor Databook, 1982, process 5F, pg 9-36
*       30 Nov 90   pwt   creation

.model NSC_4F  NPN(Is=2.447p Xti=3 Eg=1.11 Vaf=100 Bf=208.2
Ise=70.69p
+       Ne=1.565 Ikf=.9743 Nk=.6134 Xtb=1.5 Br=12.59 Isc=11.68n
+       Nc=1.835 Ikr=3.86 Rc=.4685 Cjc=142p Mjc=.4353 Vjc=.75 Fc=.5
+       Cje=188.5p Mje=.4878 Vje=.75 Tr=194.2n Tf=19.85n Itf=164.1
+       Xtf=5.945 Vtf=10 Rb=.1)
*       National Semiconductor
*       Transistor Databook, 1982, process 4F, pg 9-13
*       30 Nov 90   pwt   creation
* TL064 operational amplifier "macromodel" subcircuit
* created using Parts release 4.01 on 06/28/89 at 10:42
* (REV N/A)
* connections:  non-inverting input
*             | inverting input
```

```
*        | | positive power supply
*        | | | negative power supply
*        | | | | output
*        | | | | |
.subckt TL064/TI 1 2 3 4 5
*

  c1   11 12 3.498E-12
  c2    6  7 15.00E-12
  dc    5 53 dx
  de   54  5 dx
  dlp  90 91 dx
  dln  92 90 dx
  dp    4  3 dx
  egnd 99  0 poly(2) (3,0) (4,0) 0 .5 .5
  fb    7 99 poly(5) vb vc ve vlp vln 0 318.3E3 -300E3 300E3 300E3 -300E3
  ga    6  0 11 12 94.26E-6
  gcm   0  6 10 99 1.607E-9
  iss   3 10 dc 52.50E-6
  hlim 90  0 vlim 1K
  j1   11  2 10 jx
  j2   12  1 10 jx
  r2    6  9 100.0E3
  rd1   4 11 10.61E3
  rd2   4 12 10.61E3
  ro1   8  5 200
  ro2   7 99 200
  rp    3  4 150.0E3
  rss  10 99 3.810E6
  vb    9  0 dc 0
  vc    3 53 dc 2.200
  ve   54  4 dc 2.200
  vlim  7  8 dc 0
  vlp  91  0 dc 15
  vln   0 92 dc 15
.model dx D(Is=800.0E-18)
.model jx PJF(Is=15.00E-12 Beta=100.5E-6 Vto=-1)
.ends
.subckt LM339    1 2 3 4 5
*

  x_lm339 1 2 3 4 5 LM139
*

* the LM339 is identical to the LM139, but has a more limited temp. range
*

.ends
* connections:  non-inverting input
*            | inverting input
*            | | positive power supply
*            | | | negative power supply
*            | | | | open collector output
*            | | | | |
.subckt LM139    1 2 3 4 5
*

  f1    9  3 v1 1
  iee   3  7 dc 100.0E-6
  vi1  21  1 dc .75
  vi2  22  2 dc .75
  q1    9 21  7 qin
  q2    8 22  7 qin
```

```
  q3    9  8  4 qmo
  q4    8  8  4 qmi
.model qin PNP(Is=800.0E-18 Bf=2.000E3)
.model qmi NPN(Is=800.0E-18 Bf=1002)
.model qmo NPN(Is=800.0E-18 Bf=1000 Cjc=1E-15 Tr=475.4E-9)
  e1   10  4  9  4 1
  v1   10 11 dc 0
  q5    5 11  4 qoc
.model qoc NPN(Is=800.0E-18 Bf=20.69E3 Cjc=1E-15 Tf=3.540E-9
Tr=472.8E-9)
  dp    4  3 dx
  rp    3  4 37.50E3
.model dx D(Is=800.0E-18 Rs=1)
*

.ends

.subckt N/O_switch 1 2 PARAMS: to=0
s1 1 2 3 0 sx
V_s1 3 0 pulse(0 2 {to} 1n 1n 999 9999)
.model sx Vswitch( Ron=.001 Roff=1g Von=1 Voff=0.5)
.ends

.subckt N/C_switch 1 2 PARAMS: to=0
s1 1 2 3 0 sx
V_s1 3 0 pulse(2 0 {to} 1n 1n 999 9999)
.model sx Vswitch( Ron=.001 Roff=1g Von=1 Voff=0.5)
.ends
*

* Ideal Transformer with V2 = aV1.
*

.subckt Ideal_XFMR_1/a 1 2 3 4 PARAMS: a=1
Rs1 1 a 1U
Rs2 3 b 1U
Rp1 1 2 1G
Rp2 3 4 1T
L1 a 2 1
L2 b 4 {a*a}
K1 L1 L2 .99999999
R1 2 4 1T
.ends
*
*

* Ideal Transformer Sprcifying Vout and Vin
*

.subckt Ideal_XFMR_Vo/Vin 1 2 3 4 PARAMS: Vin=115 Vo=12
Rs1 1 a 1U
Rs2 3 b 1U
Rp1 1 2 1G
Rp2 3 4 1T
L1 a 2 1
L2 b 4 {(Vo*Vo)/(Vin*Vin)}
K1 L1 L2 .99999999
R1 2 4 1T
.ends
*

*Switched Models
.model Dswitch  D(Is=0.1p Rs=1 CJO=2p Tt=12n Bv=1000 Ibv=0.1p)
.model Sx vswitch (Voff=0.5 Von=1.5)
```

```
* Laplace Transform Block Library
*
* Vo=K/(s+a)
*
.subckt Xform1 1 2 PARAMS: K=1, a=1
E1 1 0 LAPLACE {V(2)} = { K /(s+ a )}
Ro 1 0 1k
Rin 2 0 1T
.ends
*
* Vo=K(s+a)/s
*
.subckt Xform2 1 2 PARAMS: K=1, a=1
E1 1 0 LAPLACE {V(2)} = { K *(s+ a )}
Ro 1 0 1k
Rin 2 0 1T
.ends
*
.subckt Xform3 1 2 PARAMS: K=1,
a0=1,a1=0,a2=0,a3=0,a4=0,b0=1,b1=0,b2=0,b3=0,b4=0
E1 1 0 LAPLACE {V(2)} =
+ {((K
*(a4*s*s*s*s+a3*s*s*s+a2*s*s+a1*s+a0)/(b4*s*s*s*s+b3*s*s*s+b2*s*s+b1*s
+b0))}
Ro 1 0 1k
Rin 2 0 1T
.ends
*
*
* Two input Summer
.subckt two_sum 1 2 3
E1 1 0 POLY(2) 2 0  3 0 0 1 1
Ro 1 0 1K
R1 2 0 1T
R2 3 0 1T
.ends
*
* Two input Difference amp
.subckt two_diff 1 2 3
E1 1 0 POLY(2) 2 0  3 0 0 1 -1
Ro 1 0 1K
R1 2 0 1T
R2 3 0 1T
.ends
*
*
* Gain block
*
.subckt gain_block 1 2 PARAMS: K=1
Eo 1 0 VALUE={v(2,0)*K}
Ro 1 0 1K
Ri 2 0 1T
.ends
*
* Schematics Subcircuit *
```

```
.SUBCKT lm555 Vcc Threshold Control Trigger Ground Output Discharge
R_R1    Vcc Control 5K
R_R2    Control $N_0001 5K
R_R3    $N_0001 Ground 5K
X_U1A   R $N_0002 Output $G_DPWR $G_DGND 7402 PARAMS:
+ IO_LEVEL=0 MNTYMXDLY=0
X_U2A   Output S $N_0002 $G_DPWR $G_DGND 7402 PARAMS:
+ IO_LEVEL=0 MNTYMXDLY=0
R_R4    $N_0002 $N_0003 10K
Q_Q1    Discharge $N_0003 Ground Q2N3904
X_U19   Threshold Control  R 555_comp
X_U20   $N_0001 Trigger  S 555_comp
.ENDS   lm555


*-----------------------------------------------------------------
.SUBCKT 555_COMP      v_plus v_minus v_out
*
* This is an ideal OpAmp model with the output voltage
* limited to 0 to +5 Volts+/- 15V.
*
Rin      v_plus v_minus 10MEG
Eamp   v_out  0       TABLE {100000*(v(v_plus,v_minus))}=(0,0) (5,5)
R0       v_out  0     10MEG
.ENDS
*
*
* OP-AMP_Breakout operational amplifier "macromodel" subcircuit
* This is a copy of the 741 op-amp model.
* (REV N/A)
* connections:   non-inverting input
*          | inverting input
*          | | positive power supply
*          | | | negative power supply
*          | | | | output
*          | | | | |
.subckt OP-AMP_Breakout   1 2 3 4 5
*
  c1   11 12 4.664E-12
  c2   6  7 20.00E-12
  dc   5 53 dx
  de   54 5 dx
  dlp  90 91 dx
  dln  92 90 dx
  dp   4  3 dx
  egnd 99  0 poly(2) (3,0) (4,0) 0 .5 .5
  fb   7 99 poly(5) vb vc ve vlp vln 0 10.61E6 -10E6 10E6 10E6 -10E6
  ga   6  0 11 12 137.7E-6
  gcm  0  6 10 99 2.574E-9
  iee  10 4 dc 10.16E-6
  hlim 90  0 vlim 1K
  q1   11 2 13 qx
  q2   12 1 14 qx
  r2   6  9 100.0E3
  rc1  3 11 7.957E3
  rc2  3 12 7.957E3
```

```
re1  13 10 2.740E3
re2  14 10 2.740E3
ree  10 99 19.69E6
ro1   8  5 150
ro2   7 99 150
rp    3  4 18.11E3
vb    9  0 dc 0
vc    3 53 dc 2.600
ve   54  4 dc 2.600
vlim  7  8 dc 0
vlp  91  0 dc 25
vln   0 92 dc 25
.model dx D(Is=800.0E-18)
.model qx NPN(Is=800.0E-18 Bf=62.50)
.ends

.SUBCKT MbreakN_Sub 1 2 3
M1 1 2 3 3 MbreakN
.ends

.SUBCKT MbreakP_Sub 1 2 3
M1 1 2 3 3 MbreakP
.ends *        85-??-??    Original library
```

INDEX

R

S

T

U

V